STUDENT SOLUTIONS MANUAL

ELEMENTARY LINEAR ALGEBRA

SIXTH EDITION

HOWARD ANTON
Drexel University

Prepared by
Elizabeth M. Grobe
Charles A. Grobe, Jr.
Bowdoin College

JOHN WILEY & SONS
New York · Chichester · Brisbane · Toronto · Singapore

ISBN 0-471-52741-6

Printed in the United States of America

10 9 8 7 6 5 4 3 2 1

PREFACE

This manual contains detailed solutions to substantially more than half of the exercises in the Sixth Edition of Howard Anton's *Elementary Linear Algebra*.

Routine Exercises: Since the textbook contains answers to virtually all of the routine exercises, we include solutions to only a representative sample.

Less Routine Exercises: We give solutions to well over half of the less routine exercises, including a solution to at least one of every type of exercise in each problem set. Where there is more than one exercise of a given kind, we have omitted the solution to at least one problem so that it can be assigned for credit.

Solutions to most of the exercises which are new to the Sixth Edition have been included, and most of the solutions which also appeared in the *Solutions Manual* to earlier editions have been retained.

This book was typed under the direction of Risa Batterman of Techsetters, Inc. We very much appreciate the excellent job that she has done.

Elizabeth M. Grobe
Charles A. Grobe, Jr.

CONTENTS

CHAPTER THREE

CHAPTER FOUR

CHAPTER FIVE

CHAPTER SIX

CHAPTER TEN

EXERCISE SET 1.1

1. **(b)** Not linear because of the term x_1x_3.

   (d) Not linear because of the term x_1^{-2}.

   (e) Not linear because of the term $x_1^{3/5}$.

3. **(a)** Let $x = t$. Then $y = \frac{7t - 3}{5}$.

   Alternatively, let $y = s$. Then $x = \frac{5s + 3}{7}$.

   (c) One possibility is to let $x_1 = r$, $x_2 = s$, and $x_4 = t$. Then

$$x_3 = \frac{-8r + 2s + 6t - 1}{5}$$

6. **(b)** Substituting the given expressions for x and y into the equation $x = 5 + 2y$ yields

$$t = 5 + 2\left[\frac{1}{2}t - \frac{5}{2}\right] = 5 + t - 5 = t$$

   Since this equation is valid for all values of t, the proposed solution is, indeed, the general one.

   Alternatively, we let $x = t$ and solve for y, obtaining $y = \frac{1}{2}t - \frac{5}{2}$.

7. Since each of the three given points must satisfy the equation of the curve, we have the system of equations

$$\begin{aligned} ax_1^2 + bx_1 + c &= y_1 \\ ax_2^2 + bx_2 + c &= y_2 \\ ax_3^2 + bx_3 + c &= y_3 \end{aligned}$$

If we consider this to be a system of equations in the three unknowns a, b, and c, the augmented matrix is clearly the one given in the exercise.

8. If $x - y = 3$, then $2x - 2y = 6$. Therefore, the equations are consistent if and only if $k = 6$; that is, there are no solutions if $k \neq 6$. If $k = 6$, then the equations represent the same line, in which case, there are infinitely many solutions. Since this covers all of the possibilities, there is never a unique solution.

9. **(a)** If the system of equations fails to have a solution, then there are three possibilities: The three lines intersect in (i) 3 distinct points, (ii) 2 distinct points, or (iii) not at all.

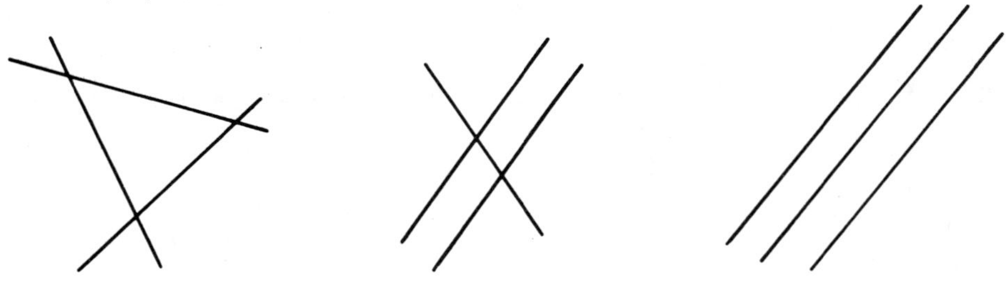

(i) (ii) (iii)

(b) If the system of equations has exactly one solution, then all of the lines must pass through a common point. Moreover, at least two of the lines must be distinct.

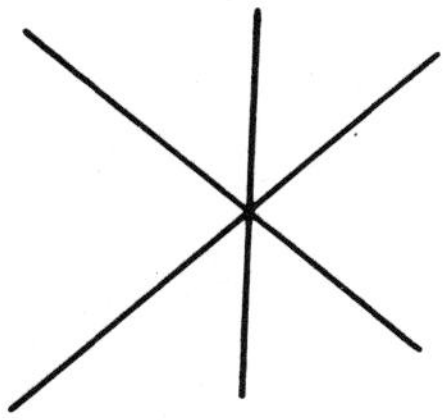

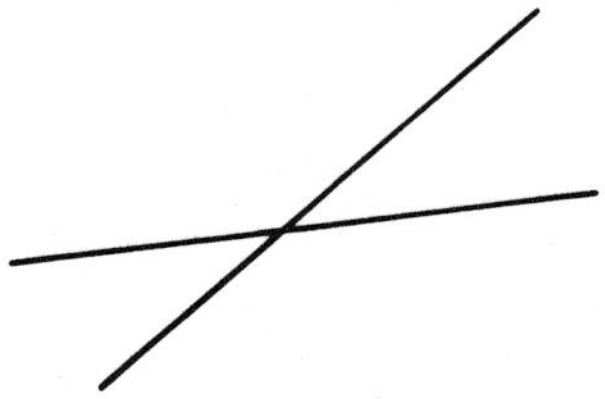

(c) If the system has infinitely many solutions, then the three lines coincide; that is, the three equations represent the same line.

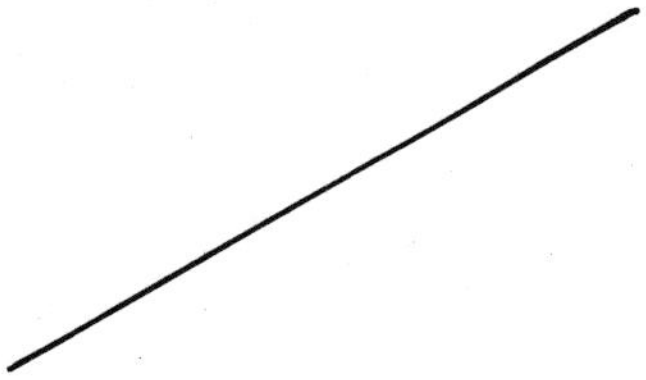

12. If the system is consistent, then we can, for instance, subtract the first two equations from the last. This yields $c - a - b = 0$ or $c = a + b$. Unless this equation holds, the system cannot have a solution, and hence cannot be consistent.

13. The solutions of $x_1 + kx_2 = c$ are $x_1 = c - kt$, $x_2 = t$ where t is any real number. If these satisfy $x_1 + \ell x_2 = d$, then $c - kt + \ell t = d$, or $(\ell - k)t = d - c$ for all real numbers t. In particular, if $t = 0$, then $d = c$, and if $t = 1$, then $\ell = k$.

EXERCISE SET 1.2

1. **(e)** Not in reduced row-echelon form because Property 2 is not satisfied.

   (f) Not in reduced row-echelon form because Property 3 is not satisfied.

   (g) Not in reduced row-echelon form because Property 4 is not satisfied.

2. **(c)** Not in row-echelon form because Property 1 is not satisfied.

   (d) In row-echelon form.

   (e) Not in row-echelon form because Property 1 is not satisfied.

4. **(b)** Let $x_4 = t$. Then we can read off the solution

$$x_1 = 8 + 7t$$
$$x_2 = 2 - 3t$$
$$x_3 = -5 - t$$
$$x_4 = t$$

   (c) Let $x_5 = t$. Then $x_4 = 8 - 5t$ and $x_3 = 7 - 4t$. Let $x_2 = s$. Then $x_1 = -2 - 3t + 6s$.

5. **(a)** The solution is

$$x_3 = 5$$
$$x_2 = 2 - 2x_3 = -8$$
$$x_1 = 7 - 4x_3 + 3x_2 = -37$$

(b) Let $x_4 = t$. Then $x_3 = 2 - t$. Therefore

$$x_2 = 3 + 9t - 4x_3 = 3 + 9t - 4(2 - t) = -5 + 13t$$
$$x_1 = 6 + 5t - 8x_3 = 6 + 5t - 8(2 - t) = -10 + 13t$$

6. **(a)** The augmented matrix is

$$\begin{bmatrix} 1 & 1 & 2 & 8 \\ -1 & -2 & 3 & 1 \\ 3 & -7 & 4 & 10 \end{bmatrix}$$

Replace Row 2 by Row 1 plus Row 2 and replace Row 3 by Row 3 plus -3 times Row 1.

$$\begin{bmatrix} 1 & 1 & 2 & 8 \\ 0 & -1 & 5 & 9 \\ 0 & -10 & -2 & -14 \end{bmatrix}$$

Multiply Row 2 by -1.

$$\begin{bmatrix} 1 & 1 & 2 & 8 \\ 0 & 1 & -5 & -9 \\ 0 & -10 & -2 & -14 \end{bmatrix}$$

Note that we resist the temptation to multiply Row 3 by -1/2. While this is sensible, it is not part of the formal elimination procedure. Replace Row 3 by Row 3 plus 10 times Row 2.

$$\begin{bmatrix} 1 & 1 & 2 & 8 \\ 0 & 1 & -5 & -9 \\ 0 & 0 & -52 & -104 \end{bmatrix}$$

Multiply Row 3 by -1/52.

$$\begin{bmatrix} 1 & 1 & 2 & 8 \\ 0 & 1 & -5 & -9 \\ 0 & 0 & 1 & 2 \end{bmatrix}$$

Now multiply Row 3 by 5 and add the result to Row 2.

$$\begin{bmatrix} 1 & 1 & 2 & 8 \\ 0 & 1 & 0 & 1 \\ 0 & 0 & 1 & 2 \end{bmatrix}$$

Next multiply Row 3 by -2 and add the result to Row 1.

$$\begin{bmatrix} 1 & 1 & 0 & 4 \\ 0 & 1 & 0 & 1 \\ 0 & 0 & 1 & 2 \end{bmatrix}$$

Finally, multiply Row 2 by -1 and add to Row 1.

$$\begin{bmatrix} 1 & 0 & 0 & 3 \\ 0 & 1 & 0 & 1 \\ 0 & 0 & 1 & 2 \end{bmatrix}$$

Thus the solution is $x_1 = 3$, $x_2 = 1$, $x_3 = 2$.

(c) The augmented matrix is

$$\begin{bmatrix} 1 & -1 & 2 & -1 & -1 \\ 2 & 1 & -2 & -2 & -2 \\ -1 & 2 & -4 & 1 & 1 \\ 3 & 0 & 0 & -3 & -3 \end{bmatrix}$$

Replace Row 2 by Row 2 plus −2 times Row 1, replace Row 3 by Row 3 plus Row 1, and replace Row 4 by Row 4 plus −3 times Row 1.

$$\begin{bmatrix} 1 & -1 & 2 & -1 & -1 \\ 0 & 3 & -6 & 0 & 0 \\ 0 & 1 & -2 & 0 & 0 \\ 0 & 3 & -6 & 0 & 0 \end{bmatrix}$$

Multiply Row 2 by 1/3 and then add −1 times the new Row 2 to Row 3 and add −3 times the new Row 2 to Row 4.

$$\begin{bmatrix} 1 & -1 & 2 & -1 & -1 \\ 0 & 1 & -2 & 0 & 0 \\ 0 & 0 & 0 & 0 & 0 \\ 0 & 0 & 0 & 0 & 0 \end{bmatrix}$$

Finally, replace Row 1 by Row 1 plus Row 2.

$$\begin{bmatrix} 1 & 0 & 0 & -1 & -1 \\ 0 & 1 & -2 & 0 & 0 \\ 0 & 0 & 0 & 0 & 0 \\ 0 & 0 & 0 & 0 & 0 \end{bmatrix}$$

Thus if we let $z = s$ and $w = t$, then we have $x = -1 + w = -1 + t$ and $y = 2z = 2s$. The solution is therefore

$$\begin{aligned} x &= -1 + t \\ y &= 2s \\ z &= s \\ w &= t \end{aligned}$$

7. **(a)** In Problem 6.(a), we reduced the augmented matrix to the following row-echelon matrix:

$$\begin{bmatrix} 1 & 1 & 2 & 8 \\ 0 & 1 & -5 & -9 \\ 0 & 0 & 1 & 2 \end{bmatrix}$$

By Row 3, $x_3 = 2$. Thus by Row 2, $x_2 = 5x_3 - 9 = 1$. Finally, Row 1 implies that $x_1 = -x_2 - 2x_3 + 8 = 3$. Hence the solution is

$$x_1 = 3$$
$$x_2 = 1$$
$$x_3 = 2$$

(c) According to the solution to Problem 6.(c), one row-echelon form of the augmented matrix is

$$\begin{bmatrix} 1 & -1 & 2 & -1 & -1 \\ 0 & 1 & -2 & 0 & 0 \\ 0 & 0 & 0 & 0 & 0 \\ 0 & 0 & 0 & 0 & 0 \end{bmatrix}$$

Row 2 implies that $y = 2z$. Thus if we let $z = s$, we have $y = 2s$. Row 1 implies that $x = -1 + y - 2z + w$. Thus if we let $w = t$, then $x = -1 + 2s - 2s + t$ or $x = -1 + t$. Hence the solution is

$$x = -1 + t$$
$$y = 2s$$
$$z = s$$
$$w = t$$

8. **(a)** The augmented matrix is

$$\begin{bmatrix} 2 & -3 & -2 \\ 2 & 1 & 1 \\ 3 & 2 & 1 \end{bmatrix}$$

Divide Row 1 by 2.

$$\begin{bmatrix} 1 & -3/2 & -1 \\ 2 & 1 & 1 \\ 3 & 2 & 1 \end{bmatrix}$$

Replace Row 2 by Row 2 plus -2 times Row 1 and replace Row 3 by Row 3 plus -3 times Row 1.

$$\begin{bmatrix} 1 & -3/2 & -1 \\ 0 & 4 & 3 \\ 0 & 13/2 & 4 \end{bmatrix}$$

Divide Row 2 by 4. Then multiply the new Row 2 by -13/2 and add the result to Row 3.

$$\begin{bmatrix} 1 & -3/2 & -1 \\ 0 & 1 & 3/4 \\ 0 & 0 & -7/8 \end{bmatrix}$$

Thus, if we multiply Row 3 by -8/7, we obtain the following row-echelon matrix:

$$\begin{bmatrix} 1 & -3/2 & -1 \\ 0 & 1 & 3/4 \\ 0 & 0 & 1 \end{bmatrix}$$

Finally, if we complete the reduction to the reduced row-echelon form, we obtain

$$\begin{bmatrix} 1 & 0 & 0 \\ 0 & 1 & 0 \\ 0 & 0 & 1 \end{bmatrix}$$

The system of equations represented by this matrix is inconsistent because Row 3 implies that $0x_1 + 0x_2 = 0 = 1$.

(c) The augmented matrix is

$$\begin{bmatrix} 4 & -8 & 12 \\ 3 & -6 & 9 \\ -2 & 4 & -6 \end{bmatrix}$$

Divide Row 1 by 4.

$$\begin{bmatrix} 1 & -2 & 3 \\ 3 & -6 & 9 \\ -2 & 4 & -6 \end{bmatrix}$$

Replace Row 2 by Row 2 plus -3 times Row 1 and replace Row 3 by Row 3 plus 2 times Row 1.

$$\begin{bmatrix} 1 & -2 & 3 \\ 0 & 0 & 0 \\ 0 & 0 & 0 \end{bmatrix}$$

If we let $x_2 = t$, then $x_1 = 3 + 2x_2 = 3 + 2t$. Hence, the solution is

$$x_1 = 3 + 2t$$
$$x_2 = t$$

9. **(a)** In Problem 8.(a), we reduced the augmented matrix of this system to row-echelon form, obtaining the matrix

$$\begin{bmatrix} 1 & -3/2 & -1 \\ 0 & 1 & 3/4 \\ 0 & 0 & 1 \end{bmatrix}$$

Row 3 again yields the equation 0 = 1 and hence the system is inconsistent.

(c) In Problem 8.(c), we found that one row-echelon form of the augmented matrix is

$$\begin{bmatrix} 1 & -2 & 3 \\ 0 & 0 & 0 \\ 0 & 0 & 0 \end{bmatrix}$$

Again if we let $x_2 = t$, then $x_1 = 3 + 2x_2 = 3 + 2t$.

10. **(a)** The augmented matrix of the system is

$$\begin{bmatrix} 5 & -2 & 6 & 0 \\ -2 & 1 & 3 & 1 \end{bmatrix}$$

Divide Row 1 by 5.

$$\begin{bmatrix} 1 & -2/5 & 6/5 & 0 \\ -2 & 1 & 3 & 1 \end{bmatrix}$$

Replace Row 2 by Row 2 plus 2 times Row 1.

$$\begin{bmatrix} 1 & -2/5 & 6/5 & 0 \\ 0 & 1/5 & 27/5 & 1 \end{bmatrix}$$

Multiply Row 2 by 5.

$$\begin{bmatrix} 1 & -2/5 & 6/5 & 0 \\ 0 & 1 & 27 & 5 \end{bmatrix}$$

Replace Row 1 by Row 1 plus 2/5 times Row 2.

$$\begin{bmatrix} 1 & 0 & 12 & 2 \\ 0 & 1 & 27 & 5 \end{bmatrix}$$

Thus, if we let $x_3 = t$, then $x_2 = 5 - 27x_3 = 5 - 27t$ and $x_1 = 2 - 12x_3 = 2 - 12t$. Therefore, the solution is

$$x_1 = 2 - 12t$$

$$x_2 = 5 - 27t$$

$$x_3 = t$$

10. **(c)** The augmented matrix of the system is

$$\begin{bmatrix} 0 & 0 & 1 & 2 & -1 & 4 \\ 0 & 0 & 0 & 1 & -1 & 3 \\ 0 & 0 & 1 & 3 & -2 & 7 \\ 2 & 4 & 1 & 7 & 0 & 7 \end{bmatrix}$$

Interchange Rows 1 and 4 and multiply the new Row 1 by 1/2.

$$\begin{bmatrix} 1 & 2 & 1/2 & 7/2 & 0 & 7/2 \\ 0 & 0 & 0 & 1 & -1 & 3 \\ 0 & 0 & 1 & 3 & -2 & 7 \\ 0 & 0 & 1 & 2 & -1 & 4 \end{bmatrix}$$

Interchange Rows 2 and 4. Then replace Row 3 by Row 3 plus -1 times the new Row 2.

$$\begin{bmatrix} 1 & 2 & 1/2 & 7/2 & 0 & 7/2 \\ 0 & 0 & 1 & 2 & -1 & 4 \\ 0 & 0 & 0 & 1 & -1 & 3 \\ 0 & 0 & 0 & 1 & -1 & 3 \end{bmatrix}$$

Replace Row 4 by Row 4 plus -1 times Row 3.

$$\begin{bmatrix} 1 & 2 & 1/2 & 7/2 & 0 & 7/2 \\ 0 & 0 & 1 & 2 & -1 & 4 \\ 0 & 0 & 0 & 1 & -1 & 3 \\ 0 & 0 & 0 & 0 & 0 & 0 \end{bmatrix}$$

Replace Row 1 by Row 1 plus -1/2 times Row 2.

$$\begin{bmatrix} 1 & 2 & 0 & 5/2 & 1/2 & 3/2 \\ 0 & 0 & 1 & 2 & -1 & 4 \\ 0 & 0 & 0 & 1 & -1 & 3 \\ 0 & 0 & 0 & 0 & 0 & 0 \end{bmatrix}$$

Replace Row 2 by Row 2 plus -2 times Row 3 and replace Row 1 by Row 1 plus -5/2 times Row 3.

$$\begin{bmatrix} 1 & 2 & 0 & 0 & 3 & -6 \\ 0 & 0 & 1 & 0 & 1 & -2 \\ 0 & 0 & 0 & 1 & -1 & 3 \\ 0 & 0 & 0 & 0 & 0 & 0 \end{bmatrix}$$

Thus, if we let $y = t$ and $v = s$, we have

$$\begin{aligned} u &= -6 - 2s - 3t \\ v &= s \\ w &= -2 - t \\ x &= 3 + t \\ y &= t \end{aligned}$$

11. **(a)** From Problem 10.(a), a row-echelon form of the augmented matrix is

$$\begin{bmatrix} 1 & -2/5 & -6/5 & 0 \\ 0 & 1 & 3 & 0 \end{bmatrix}$$

If we let $x_3 = t$, then Row 2 implies that $x_2 = -3t$. Row 1 then implies that $x_1 = (6/5)x_3 + (2/5)x_2 = 0$. Hence the solution is

$$\begin{aligned} x_1 &= 0 \\ x_2 &= -3t \\ x_3 &= t \end{aligned}$$

(c) From Problem 10.(c), a row-echelon form of the augmented matrix is

$$\begin{bmatrix} 1 & 2 & 1/2 & 7/2 & 0 & 7/2 \\ 0 & 0 & 1 & 2 & -1 & 4 \\ 0 & 0 & 0 & 1 & -1 & 3 \\ 0 & 0 & 0 & 0 & 0 & 0 \end{bmatrix}$$

If we let $y = t$, then Row 3 implies that $x = 3 + t$. Row 2 then implies that $w = 4 - 2x + t = -2 - t$. Now let $v = s$. By Row 1, $u = 7/2 - 2s - (1/2)w - (7/2)x = -6 - 2s - 3t$. Thus we have the same solution which we obtained in Problem 10(c).

12. The augmented matrix of this system is

$$\begin{bmatrix} 2 & -1 & 3 & 4 & 9 \\ 1 & 0 & -2 & 7 & 11 \\ 3 & -3 & 1 & 5 & 8 \\ 2 & 1 & 4 & 4 & 10 \end{bmatrix}$$

Its reduced row-echelon form is

$$\begin{bmatrix} 1 & 0 & 0 & 0 & -1 \\ 0 & 1 & 0 & 0 & 0 \\ 0 & 0 & 1 & 0 & 1 \\ 0 & 0 & 0 & 1 & 2 \end{bmatrix}$$

Hence the solution is

$$I_1 = -1$$
$$I_2 = 0$$
$$I_3 = 1$$
$$I_4 = 2$$

13. **(a)** The augmented matrix of the system is

$$\begin{bmatrix} 2 & 1 & a \\ 3 & 6 & b \end{bmatrix}$$

Divide Row 1 by 2 and Row 2 by 3.

$$\begin{bmatrix} 1 & \frac{1}{2} & \frac{a}{2} \\ 1 & 2 & \frac{b}{3} \end{bmatrix}$$

Subtract Row 1 from Row 2.

$$\begin{bmatrix} 1 & \frac{1}{2} & \frac{a}{2} \\ 0 & \frac{3}{2} & -\frac{a}{2} + \frac{b}{3} \end{bmatrix}$$

Multiply Row 2 by 2/3.

$$\begin{bmatrix} 1 & \frac{1}{2} & \frac{a}{2} \\ 0 & 1 & -\frac{a}{3} + \frac{2b}{9} \end{bmatrix}$$

Thus

$$y = -\frac{a}{3} + \frac{2b}{9}$$

Row 1 then implies that

$$x = \frac{a}{2} - \frac{y}{2}$$

or

$$x = \frac{a}{2} - \frac{1}{2}\left[-\frac{a}{3} + \frac{2b}{9}\right] = \frac{2a}{3} - \frac{b}{9}$$

14. The Gauss-Jordan process will reduce this system to the equations

$$x + 2y - 3z = 4$$

$$y - 2z = 10/7$$

$$(a^2 - 16)z = a - 4$$

If $a = 4$, then the last equation becomes $0 = 0$, and hence there will be infinitely many solutions – for instance, $z = t$, $y = 2t + \frac{10}{7}$, $x = -2\left[2t + \frac{10}{7}\right] + 3t + 4$. If $a = -4$, then the last equation becomes $0 = -8$, and so the system will have no solutions. Any other value of a will yield a unique solution for z and hence also for y and x.

16. One possibility is

$$\begin{bmatrix} 1 & 3 \\ 2 & 7 \end{bmatrix} \rightarrow \begin{bmatrix} 1 & 3 \\ 0 & 1 \end{bmatrix}$$

Another possibility is

$$\begin{bmatrix} 1 & 3 \\ 2 & 7 \end{bmatrix} \rightarrow \begin{bmatrix} 2 & 7 \\ 1 & 3 \end{bmatrix} \rightarrow \begin{bmatrix} 1 & 7/2 \\ 1 & 3 \end{bmatrix} \rightarrow \begin{bmatrix} 1 & 7/2 \\ 0 & 1 \end{bmatrix}$$

17. Let $x_1 = \sin \alpha$, $x_2 = \cos \beta$, and $x_3 = \tan \gamma$. If we solve the given system of equations for x_1, x_2, and x_3, we find that $x_1 = 1$, $x_2 = -1$, and $x_3 = 0$. Thus $\alpha = \frac{\pi}{2}$, $\beta = \pi$, and $\gamma = 0$.

18. We treat this system as linear in the variables x^2, y^2, and z^2. The augmented matrix is

$$\begin{bmatrix} 1 & 1 & 1 & 6 \\ 1 & -1 & 2 & 2 \\ 2 & 1 & -1 & 3 \end{bmatrix}$$

which reduces to

$$\begin{bmatrix} 1 & 0 & 0 & 1 \\ 0 & 1 & 0 & 3 \\ 0 & 0 & 1 & 2 \end{bmatrix}$$

Thus we have

$$x^2 = 1 \Rightarrow x = \pm 1$$
$$y^2 = 3 \Rightarrow y = \pm \sqrt{3}$$
$$z^2 = 2 \Rightarrow z = \pm \sqrt{2}$$

19. Using the given points, we obtain the equations

$$\begin{aligned} d &= 10 \\ a + b + c + d &= 7 \\ 27a + 9b + 3c + d &= -11 \\ 64a + 16b + 4c + d &= -14 \end{aligned}$$

If we solve this system, we find that $a = 1$, $b = -6$, $c = 2$, and $d = 10$.

20. There are eight possibilities, one of which is

$$\begin{bmatrix} 1 & 0 & r \\ 0 & 1 & s \\ 0 & 0 & 0 \end{bmatrix}$$

where r and s can be any real numbers.

21. If $a \neq 0$, then the reduction can be accomplished as follows:

$$\begin{bmatrix} a & b \\ c & d \end{bmatrix} \rightarrow \begin{bmatrix} 1 & \frac{b}{a} \\ c & d \end{bmatrix} \rightarrow \begin{bmatrix} 1 & \frac{b}{a} \\ 0 & \frac{ad - bc}{a} \end{bmatrix} \rightarrow \begin{bmatrix} 1 & \frac{b}{a} \\ 0 & 1 \end{bmatrix} \rightarrow \begin{bmatrix} 1 & 0 \\ 0 & 1 \end{bmatrix}$$

If $a = 0$, then $b \neq 0$ and $c \neq 0$, so the reduction can be carried out as follows:

$$\begin{bmatrix} 0 & b \\ c & d \end{bmatrix} \rightarrow \begin{bmatrix} c & d \\ 0 & b \end{bmatrix} \rightarrow \begin{bmatrix} 1 & \frac{d}{c} \\ 0 & b \end{bmatrix} \rightarrow \begin{bmatrix} 1 & \frac{d}{c} \\ 0 & 1 \end{bmatrix} \rightarrow \begin{bmatrix} 1 & 0 \\ 0 & 1 \end{bmatrix}$$

Where did you use the fact that $ad - bc \neq 0$? (This proof uses it twice.)

EXERCISE SET 1.3

1. **(a)** There are more unknowns than equations. By Theorem 1.3.1, the system has infinitely many solutions.

   (d) If we let $x_2 = t$, then $x_1 = \frac{2}{3}t$. Thus the system has infinitely many solutions.

2. **(b)** The augmented matrix of the homogenous system is

$$\begin{bmatrix} 3 & 1 & 1 & 1 & 0 \\ 5 & -1 & 1 & -1 & 0 \end{bmatrix}$$

This matrix may be reduced to

$$\begin{bmatrix} 3 & 1 & 1 & 1 & 0 \\ 0 & 4 & 1 & 4 & 0 \end{bmatrix}$$

If we let $x_3 = 4s$ and $x_4 = t$, then Row 2 implies that

$$4x_2 = -4t - 4s \quad \text{or} \quad x_2 = -t - s$$

Now Row 1 implies that

$$3x_1 = -x_2 - 4s - t = t + s - 4s - t = -3s \quad \text{or} \quad x_1 = -s$$

Therefore the solution is

$$x_1 = -s$$
$$x_2 = -(t + s)$$
$$x_3 = 4s$$
$$x_4 = t$$

3. **(b)** The augmented matrix of the homogenous system is

$$\begin{bmatrix} 0 & 1 & 3 & -2 & 0 \\ 2 & 1 & -4 & 3 & 0 \\ 2 & 3 & 2 & -1 & 0 \\ -4 & -3 & 5 & -4 & 0 \end{bmatrix}$$

This matrix can be reduced to

$$\begin{bmatrix} 1 & 0 & -\frac{7}{2} & \frac{5}{2} & 0 \\ 0 & 1 & 3 & -2 & 0 \\ 0 & 0 & 0 & 0 & 0 \\ 0 & 0 & 0 & 0 & 0 \end{bmatrix}$$

If we let $w = 2s$ and $x = 2t$, then the above matrix yields the solution

$$u = 7s - 5t$$
$$v = -6s + 4t$$
$$w = 2s$$
$$x = 2t$$

4. The reduced row-echelon form of the augmented matrix is

$$\begin{bmatrix} 1 & 1 & 0 & 0 & 1 & 0 \\ 0 & 0 & 1 & 0 & 1 & 0 \\ 0 & 0 & 0 & 1 & 0 & 0 \\ 0 & 0 & 0 & 0 & 0 & 0 \end{bmatrix}$$

If we let $Z_2 = s$ and $Z_5 = t$, then we obtain the solution

$$\begin{aligned} Z_1 &= -s - t \\ Z_2 &= s \\ Z_3 &= -t \\ Z_4 &= 0 \\ Z_5 &= t \end{aligned}$$

5. If we treat the given system as linear in the variables $\sin\alpha$, $\cos\beta$, and $\tan\gamma$, then the augmented matrix is

$$\begin{bmatrix} 1 & 2 & 3 & 0 \\ 2 & 5 & 3 & 0 \\ -1 & -5 & 5 & 0 \end{bmatrix}$$

This reduces to

$$\begin{bmatrix} 1 & 0 & 0 & 0 \\ 0 & 1 & 0 & 0 \\ 0 & 0 & 1 & 0 \end{bmatrix}$$

so that the solution (for α, β, γ between 0 and 2π) is

$$\begin{aligned} \sin\alpha &= 0 \quad \Rightarrow \quad \alpha = 0,\ \pi,\ 2\pi \\ \cos\beta &= 0 \quad \Rightarrow \quad \beta = \pi/2,\ 3\pi/2 \\ \tan\gamma &= 0 \quad \Rightarrow \quad \gamma = 0,\ \pi,\ 2\pi \end{aligned}$$

That is, there are $3 \cdot 2 \cdot 3 = 18$ possible triples α, β, γ which satisfy the system of equations.

6. If (x,y) is a solution, then the first equation implies that $y = -(\lambda - 3)x$. If we then substitute this value for y into the second equation, we obtain $[1 - (\lambda - 3)^2]x = 0$. But $[1 - (\lambda - 3)^2]x = 0$ implies that either $x = 0$ or $1 - (\lambda - 3)^2 = 0$. Now if $x = 0$, then either of the given equations implies that $y = 0$; that is, if $x = 0$, then the system of equations has only the trivial solution. On the

other hand, if $1 - (\lambda - 3)^2 = 0$, then either $\lambda = 2$ or $\lambda = 4$. For either of these values of λ, the two given equations are identical—that is, they represent exactly the same line—and, hence, there are infinitely many solutions.

7. **(a)** If the system of equations has only the trivial solution, then the three lines all pass through (0,0) and at least two of them are distinct.

   (b) Nontrivial solutions can exist only if the three lines coincide; that is, the three equations represent the same line and that line passes through (0,0).

8. Using the given points, we obtain the equations

$$\begin{aligned} 41a - 4b + 5c + d &= 0 \\ 53a - 2b + 7c + d &= 0 \\ 25a + 4b - 3c + d &= 0 \end{aligned}$$

This homogeneous system with more unknowns than equations will have infinitely many solutions. Since we know that $a \neq 0$ (otherwise we'd have a line and not a circle), we can expect to solve for the other variables in terms of a. Since the augmented matrix reduces to

$$\begin{bmatrix} 29 & 0 & 0 & 1 & 0 \\ 4 & 0 & 1 & 0 & 0 \\ 2 & 1 & 0 & 0 & 0 \end{bmatrix}$$

we have $d = -29a$, $c = -4a$, and $b = -2a$. If we let $a = 1$, the equation of the circle becomes $x^2 + y^2 - 2x - 4y - 29 = 0$. Any other choice of a will yield a multiple of this equation.

9. **(a)** Suppose that (x_0, y_0) is a solution and k is a constant. Then

$$a(kx_0) + b(ky_0) = k(ax_0 + by_0) = 0$$
$$c(kx_0) + d(ky_0) = k(cx_0 + dy_0) = 0$$

Thus (kx_0, ky_0) is also a solution.

(b) Suppose that (x_0, y_0) and (x_1, y_1) are solutions. Then

$$a(x_0 + x_1) + b(y_0 + y_1) = (ax_0 + by_0) + (ax_1 + by_1) = 0$$
$$c(x_0 + x_1) + d(y_0 + y_1) = (cx_0 + dy_0) + (cx_1 + dy_1) = 0$$

Thus, $(x_0 + x_1, y_0 + y_1)$ is also a solution.

11. **(a)** The leading 1's which appear in the row-echelon form need not occur in the columns corresponding to x_1, x_2, ..., x_r, as shown in Part (b) below.

(b) The row-echelon form of the augmented matrix of this system is

$$\begin{bmatrix} 1 & 1 & 0 & 0 & 1 & 0 \\ 0 & 0 & 1 & 0 & 1 & 0 \\ 0 & 0 & 0 & 1 & 0 & 0 \end{bmatrix}$$

Thus, $r = 3$, $k_1 = 1$, $k_2 = 3$, $k_3 = 4$, and the three sums corresponding to x_1, x_3, and x_4 are $x_2 + x_5$, x_5, and 0, respectively.

EXERCISE SET 1.4

1. **(c)** The matrix AE is 4×4. Since B is 4×5, $AE + B$ is not defined.

   (e) The matrix $A + B$ is 4×5. Since E is 5×4, $E(A + B)$ is 5×5.

   (h) Since A^t is 5×4 and E is 5×4, their sum is also 5×4. Thus $(A^t + E)D$ is 5×2.

2. Since two matrices are equal if and only if their corresponding entries are equal, we have the system of equations

$$\begin{aligned} a - b \qquad\qquad &= 8 \\ b + c \qquad &= 1 \\ c + 3d &= 7 \\ 2a \qquad\quad - 4d &= 6 \end{aligned}$$

   The augmented matrix of this system may be reduced to

$$\begin{bmatrix} 1 & 0 & 0 & 0 & 5 \\ 0 & 1 & 0 & 0 & -3 \\ 0 & 0 & 1 & 0 & 4 \\ 0 & 0 & 0 & 1 & 1 \end{bmatrix}$$

   Hence, $a = 5$, $b = -3$, $c = 4$, and $d = 1$.

3. **(e)** Since $2B$ is a 2×2 matrix and C is a 2×3 matrix, $2B - C$ is not defined.

3. **(g)** We have

$$-3(D + 2E) = -3\left[\begin{bmatrix} 1 & 5 & 2 \\ -1 & 0 & 1 \\ 3 & 2 & 4 \end{bmatrix} + \begin{bmatrix} 12 & 2 & 6 \\ -2 & 2 & 4 \\ 8 & 2 & 6 \end{bmatrix}\right]$$

$$= -3\begin{bmatrix} 13 & 7 & 8 \\ -3 & 2 & 5 \\ 11 & 4 & 10 \end{bmatrix} = \begin{bmatrix} -39 & -21 & -24 \\ 9 & -6 & -15 \\ -33 & -12 & -30 \end{bmatrix}$$

(j) We have $\mathrm{tr}(D - 3E) = (1 - 3(6)) + (0 - 3(1)) + (4 - 3(3)) = -25$.

4. **(b)** We have

$$D^t - E^t = \begin{bmatrix} 1 & -1 & 3 \\ 5 & 0 & 2 \\ 2 & 1 & 4 \end{bmatrix} - \begin{bmatrix} 6 & -1 & 4 \\ 1 & 1 & 1 \\ 3 & 2 & 3 \end{bmatrix} = \begin{bmatrix} -5 & 0 & -1 \\ 4 & -1 & 1 \\ -1 & -1 & 1 \end{bmatrix}$$

(d) Since B^t is a 2×2 matrix and $5C^t$ is a 3×2 matrix, the indicated addition is impossible.

(h) We have

$$(2E^t - 3D^t)^t = \left[\begin{bmatrix} 12 & -2 & 8 \\ 2 & 2 & 2 \\ 6 & 4 & 6 \end{bmatrix} - \begin{bmatrix} 3 & -3 & 9 \\ 15 & 0 & 6 \\ 6 & 3 & 12 \end{bmatrix}\right]^t$$

$$= \begin{bmatrix} 9 & 1 & -1 \\ -13 & 2 & -4 \\ 0 & 1 & -6 \end{bmatrix} = \begin{bmatrix} 9 & -13 & 0 \\ 1 & 2 & 1 \\ -1 & -4 & -6 \end{bmatrix}$$

5. **(b)** Since B is a 2×2 matrix and A is a 3×2 matrix, BA is not defined (although AB is).

(d) We have

$$AB = \begin{bmatrix} 12 & -3 \\ -4 & 5 \\ 4 & 1 \end{bmatrix}$$

Hence

$$(AB)C = \begin{bmatrix} 3 & 45 & 9 \\ 11 & -11 & 17 \\ 7 & 17 & 13 \end{bmatrix}$$

(e) We have

$$A(BC) = \begin{bmatrix} 3 & 0 \\ -1 & 2 \\ 1 & 1 \end{bmatrix} \begin{bmatrix} 1 & 15 & 3 \\ 6 & 2 & 10 \end{bmatrix} = \begin{bmatrix} 3 & 45 & 9 \\ 11 & -11 & 17 \\ 7 & 17 & 13 \end{bmatrix}$$

(f) We have

$$CC^t = \begin{bmatrix} 1 & 4 & 2 \\ 3 & 1 & 5 \end{bmatrix} \begin{bmatrix} 1 & 3 \\ 4 & 1 \\ 2 & 5 \end{bmatrix} = \begin{bmatrix} 21 & 17 \\ 17 & 35 \end{bmatrix}$$

(j) We have $\text{tr}(4E^t - D) = \text{tr}(4E - D) = (4(6) - 1) + (4(1) - 0) + (4(3) - 4) = 35$.

6. **(a)** We have

$$(2D^t - E)A = \left(\begin{bmatrix} 2 & -2 & 6 \\ 10 & 0 & 4 \\ 4 & 2 & 8 \end{bmatrix} - \begin{bmatrix} 6 & 1 & 3 \\ -1 & 1 & 2 \\ 4 & 1 & 3 \end{bmatrix} \right) \begin{bmatrix} 3 & 0 \\ -1 & 2 \\ 1 & 1 \end{bmatrix}$$

$$= \begin{bmatrix} -4 & -3 & 3 \\ 11 & -1 & 2 \\ 0 & 1 & 5 \end{bmatrix} \Bigg) \begin{bmatrix} 3 & 0 \\ -1 & 2 \\ 1 & 1 \end{bmatrix}$$

$$= \begin{bmatrix} -6 & -3 \\ 36 & 0 \\ 4 & 7 \end{bmatrix}$$

6. **(b)** Since $(4B)C$ is a 2×3 matrix and $2B$ is a 2×2 matrix, then $(4B)C + 2B$ is undefined.

(d)
$$BA^t - 2C = \begin{bmatrix} 4 & -1 \\ 0 & 2 \end{bmatrix}\begin{bmatrix} 3 & -1 & 1 \\ 0 & 2 & 1 \end{bmatrix} - 2\begin{bmatrix} 1 & 4 & 2 \\ 3 & 1 & 5 \end{bmatrix}$$
$$= \begin{bmatrix} 12 & -6 & 3 \\ 0 & 4 & 2 \end{bmatrix} - \begin{bmatrix} 2 & 8 & 4 \\ 6 & 2 & 10 \end{bmatrix}$$
$$= \begin{bmatrix} 10 & -14 & -1 \\ -6 & 2 & -8 \end{bmatrix}$$

Thus,

$$(BA^t - 2C)^t = \begin{bmatrix} 10 & -6 \\ -14 & 2 \\ -1 & -8 \end{bmatrix}$$

(f) Each of the matrices D^tE^t and $(ED)^t$ is equal to

$$\begin{bmatrix} 14 & 4 & 12 \\ 36 & -1 & 26 \\ 25 & 7 & 21 \end{bmatrix}$$

Thus their difference is the 3×3 matrix with all zero entries.

7. **(a)** The first row of A is

$$A_1 = [\, 3 \quad -2 \quad 7 \,]$$

Thus, the first row of AB is

$$A_1B = [\, 3 \quad -2 \quad 7 \,]\begin{bmatrix} 6 & -2 & 4 \\ 0 & 1 & 3 \\ 7 & 7 & 5 \end{bmatrix}$$
$$= [\, 67 \quad 41 \quad 41 \,]$$

(c) The second column of B is

$$B_2 = \begin{bmatrix} -2 \\ 1 \\ 7 \end{bmatrix}$$

Thus, the second column of AB is

$$AB_2 = \begin{bmatrix} 3 & -2 & 7 \\ 6 & 5 & 4 \\ 0 & 4 & 9 \end{bmatrix} \begin{bmatrix} -2 \\ 1 \\ 7 \end{bmatrix} = \begin{bmatrix} 41 \\ 21 \\ 67 \end{bmatrix}$$

(e) The third row of A is

$$A_3 = [\ 0 \quad 4 \quad 9\ ]$$

Thus, the third row of AA is

$$A_3A = [\ 0 \quad 4 \quad 9\ ] \begin{bmatrix} 3 & -2 & 7 \\ 6 & 5 & 4 \\ 0 & 4 & 9 \end{bmatrix}$$

$$= [\ 24 \quad 56 \quad 97\ ]$$

8. Let f_{ij} denote the entry in the ith row and j^{th} column of $C(DE)$. We are asked to find f_{23}. In order to compute f_{23}, we must calculate the elements in the second row of C and the third column of DE. According to Example 9, we can find the elements in the third column of DE by computing DE_3 where E_3 is the third column of E. That is,

$$f_{23} = [\ 3 \quad 1 \quad 5\ ] \left(\begin{bmatrix} 1 & 5 & 2 \\ -1 & 0 & 1 \\ 3 & 2 & 4 \end{bmatrix} \begin{bmatrix} 3 \\ 2 \\ 3 \end{bmatrix} \right)$$

$$= [\ 3 \quad 1 \quad 5\ ] \begin{bmatrix} 19 \\ 0 \\ 25 \end{bmatrix} = 182$$

9. **(a)** Suppose that A is $m \times n$ and B is $r \times s$. If AB is defined, then $n = r$. That is, B is $n \times s$. On the other hand, if BA is defined, then $s = m$. That is, B is $n \times m$. Hence AB is $m \times m$ and BA is $n \times n$, so that both are square matrices.

(b) Suppose that A is $m \times n$ and B is $r \times s$. If BA is defined, then $s = m$ and BA is $r \times n$. If $A(BA)$ is defined, then $n = r$. Thus B is an $n \times m$ matrix.

12. **(b)** Denote the ij^{th} entry of DA by c_{ij}. Then

$$\begin{aligned} c_{ij} &= 0a_{1j} + 0a_{2j} + \cdots + d_i a_{ij} + \cdots + 0a_{mj} \\ &= d_i a_{ij} \end{aligned}$$

Denote the ij^{th} entry of AE by f_{ij}. Then

$$\begin{aligned} f_{ij} &= a_{i1}0 + a_{i2}0 + \cdots + a_{ij}e_j + \cdots + a_{in} \\ &= a_{ij}e_j \end{aligned}$$

Thus, we have two simple rules:

1. To multiply a matrix A on the left by a diagonal matrix D, multiply every element of the ith row of A by the ith diagonal entry of D to obtain the ith row of the product.
2. To multiply a matrix A on the right by a diagonal matrix E, multiply every element of the jth column of A by the jth diagonal element of E to obtain the jth column of the product.

13. Suppose that A and B are diagonal $n \times n$ matrices and that $C = AB$. Then

$$c_{ij} = a_{i1}b_{1j} + a_{i2}b_{2j} + \cdots + a_{in}b_{nj}$$

A typical term on the right is of the form $a_{ik}b_{kj}$ where at least one factor is zero unless $i = k = j$. Therefore $c_{ij} = 0$ if $i \neq h$. If $i = j$ then all the terms vanish except $a_{ii}b_{ii}$; thus

$$c_{ii} = a_{ii}b_{ii}$$

Therefore, the rule is this: The product of two square, diagonal matrices, A and B, is a diagonal matrix whose elements are the products of the corresponding elements on the diagonals of A and B.

14. **(a)** The entry in the ith row and j^{th} column of AB is

$$a_{i1}b_{1j} + a_{i2}b_{2j} + \cdots + a_{in}b_{nj}$$

Thus, if the ith row of A consists entirely of zeros, i.e., if $a_{ik} = 0$ for i fixed and for $k = 1, 2, \ldots, n$, then the entry in the ith row and the jth column of AB is zero for fixed i and for $j = 1, 2, \ldots, n$. That is, if the ith row of A consists entirely of zeros, then so does the ith row of AB.

15. The matrix A^t is obtained from A by interchanging rows and columns. Since a_{ij} appears in the ith row and the jth column of A, that same entry will appear in the jth row and the ith column of A^t.

17. The element in the ith row and j^{th} column of I is

$$\delta_{ij} = \begin{cases} 0 & \text{if} \quad i \neq j \\ 1 & \text{if} \quad i = j \end{cases}$$

The element in the ith row and j^{th} column of AI is

$$a_{i1}\delta_{1j} + a_{i2}\delta_{2j} + \cdots + a_{in}\delta_{nj}$$

But the above sum equals a_{ij} because $\delta_{kj} = 0$ if $k \neq j$ and $\delta_{jj} = 1$. This proves that $AI = A$.

The proof that $IA = A$ is similar.

18. **(b)** If $i > j$, then the entry a_{ij} has row number larger than column number; that is, it lies below the matrix diagonal. Thus $[a_{ij}]$ has all zero elements below the diagonal.

(d) If $|i - j| > 1$, then either $i - j > 1$ or $i - j < -1$; that is, either $i > j + 1$ or $j > i + 1$. The first of these inequalities says that the entry a_{ij} lies below the diagonal and also below the "subdiagonal" consisting of all entries immediately below the diagonal ones. The second inequality says that the entry a_{ij} lies above the diagonal and also above the entries immediately above the diagonal ones. Thus we have

$$[a_{ij}] = \begin{bmatrix} a_{11} & a_{12} & 0 & 0 & 0 & 0 \\ a_{21} & a_{22} & a_{23} & 0 & 0 & 0 \\ 0 & a_{32} & a_{33} & a_{34} & 0 & 0 \\ 0 & 0 & a_{43} & a_{44} & a_{45} & 0 \\ 0 & 0 & 0 & a_{54} & a_{55} & a_{56} \\ 0 & 0 & 0 & 0 & a_{65} & a_{66} \end{bmatrix}$$

19. Suppose that A is an $m \times n$ matrix and that B is an $n \times p$ matrix.

(a) Then B_j is the $n \times 1$ matrix

$$B_j = \begin{bmatrix} b_{1j} \\ b_{2j} \\ \vdots \\ b_{nj} \end{bmatrix}$$

The product AB_j is the $m \times 1$ matrix

$$AB_j = \begin{bmatrix} a_{11}b_{1j} + a_{12}b_{2j} + \cdots + a_{1n}b_{nj} \\ a_{21}b_{1j} + a_{22}b_{2j} + \cdots + a_{2n}b_{nj} \\ \vdots \\ a_{m1}b_{1j} + a_{m2}b_{2j} + \cdots + a_{mn}b_{nj} \end{bmatrix}$$

The entries in AB_j are precisely the entries in the jth column of AB.

EXERCISE SET 1.5

1. **(a)** We have

$$A + B = \begin{bmatrix} 10 & -4 & -2 \\ 0 & 5 & 7 \\ 2 & -6 & 10 \end{bmatrix}$$

Hence,

$$(A + B) + C = \begin{bmatrix} 10 & -4 & -2 \\ 0 & 5 & 7 \\ 2 & -6 & 10 \end{bmatrix} + \begin{bmatrix} 0 & -2 & 3 \\ 1 & 7 & 4 \\ 3 & 5 & 9 \end{bmatrix}$$

$$= \begin{bmatrix} 10 & -6 & 1 \\ 1 & 12 & 11 \\ 5 & -1 & 19 \end{bmatrix}$$

On the other hand,

$$B + C = \begin{bmatrix} 8 & -5 & -2 \\ 1 & 8 & 6 \\ 7 & -2 & 15 \end{bmatrix}$$

Hence,

$$A + (B + C) = \begin{bmatrix} 2 & -1 & 3 \\ 0 & 4 & 5 \\ -2 & 1 & 4 \end{bmatrix} + \begin{bmatrix} 8 & -5 & -2 \\ 1 & 8 & 6 \\ 7 & -2 & 15 \end{bmatrix}$$

$$= \begin{bmatrix} 10 & -6 & 1 \\ 1 & 12 & 11 \\ 5 & -1 & 19 \end{bmatrix}$$

1. **(c)** Since $a + b = -3$, we have

$$(a + b)C = (-3)\begin{bmatrix} 0 & -2 & 3 \\ 1 & 7 & 4 \\ 3 & 5 & 9 \end{bmatrix} = \begin{bmatrix} 0 & 6 & -9 \\ -3 & -21 & -12 \\ -9 & -15 & -27 \end{bmatrix}$$

Also

$$aC + bC = \begin{bmatrix} 0 & -8 & 12 \\ 4 & 28 & 16 \\ 12 & 20 & 36 \end{bmatrix} + \begin{bmatrix} 0 & 14 & -21 \\ -7 & -49 & -28 \\ -21 & -35 & -63 \end{bmatrix}$$

$$= \begin{bmatrix} 0 & 6 & -9 \\ -3 & -21 & -12 \\ -9 & -15 & -27 \end{bmatrix}$$

2. **(a)** We have

$$a(BC) = (4)\left[\begin{bmatrix} 8 & -3 & -5 \\ 0 & 1 & 2 \\ 4 & -7 & 6 \end{bmatrix}\begin{bmatrix} 0 & -2 & 3 \\ 1 & 7 & 4 \\ 3 & 5 & 9 \end{bmatrix}\right]$$

$$= (4)\begin{bmatrix} -18 & -62 & -33 \\ 7 & 17 & 22 \\ 11 & -27 & 38 \end{bmatrix} = \begin{bmatrix} -72 & -248 & -132 \\ 28 & 68 & 88 \\ 44 & -108 & 152 \end{bmatrix}$$

On the other hand

$$(aB)C = \begin{bmatrix} 32 & -12 & -20 \\ 0 & 4 & 8 \\ 16 & -28 & 24 \end{bmatrix} \begin{bmatrix} 0 & -2 & 3 \\ 1 & 7 & 4 \\ 3 & 5 & 9 \end{bmatrix} = \begin{bmatrix} -72 & -248 & -132 \\ 28 & 68 & 88 \\ 44 & -108 & 152 \end{bmatrix}$$

Finally,

$$B(aC) = \begin{bmatrix} 8 & -3 & 5 \\ 0 & 1 & 2 \\ 4 & -7 & 6 \end{bmatrix} \begin{bmatrix} 0 & -8 & 12 \\ 4 & 28 & 16 \\ 12 & 20 & 36 \end{bmatrix} = \begin{bmatrix} -72 & -248 & -132 \\ 28 & 68 & 88 \\ 44 & -108 & 152 \end{bmatrix}$$

(c) Since

$$(B + C)A = \begin{bmatrix} 8 & -5 & -2 \\ 1 & 8 & 6 \\ 7 & -2 & 15 \end{bmatrix} \begin{bmatrix} 2 & -1 & 3 \\ 0 & 4 & 5 \\ -2 & 1 & 4 \end{bmatrix}$$

$$= \begin{bmatrix} 20 & -30 & -9 \\ -10 & 37 & 67 \\ -16 & 0 & 71 \end{bmatrix}$$

and

$$BA + CA = \begin{bmatrix} 26 & -25 & -11 \\ -4 & 6 & 13 \\ -4 & -26 & 1 \end{bmatrix} + \begin{bmatrix} -6 & -5 & 2 \\ -6 & 31 & 54 \\ -12 & 26 & 70 \end{bmatrix}$$

$$= \begin{bmatrix} 20 & -30 & -9 \\ -10 & 37 & 67 \\ -16 & 0 & 71 \end{bmatrix}$$

the two matrices are equal.

3. **(b)** Since

$$(A + B)^t = \begin{bmatrix} 10 & -4 & -2 \\ 0 & 5 & 7 \\ 2 & -6 & 10 \end{bmatrix}^t = \begin{bmatrix} 10 & 0 & 2 \\ -4 & 5 & -6 \\ -2 & 7 & 10 \end{bmatrix}$$

and

$$A^t + B^t = \begin{bmatrix} 2 & 0 & -2 \\ -1 & 4 & 1 \\ 3 & 5 & 4 \end{bmatrix} + \begin{bmatrix} 8 & 0 & 4 \\ -3 & 1 & -7 \\ -5 & 2 & 6 \end{bmatrix} = \begin{bmatrix} 10 & 0 & 2 \\ -4 & 5 & -6 \\ -2 & 7 & 10 \end{bmatrix}$$

the two matrices are equal.

3. **(d)** Since

$$(AB)^t = \begin{bmatrix} 28 & -28 & 6 \\ 20 & -31 & 38 \\ 0 & -21 & 36 \end{bmatrix}^t = \begin{bmatrix} 28 & 20 & 0 \\ -28 & -31 & -21 \\ 6 & 38 & 36 \end{bmatrix}$$

and

$$B^tA^t = \begin{bmatrix} 8 & 0 & 4 \\ -3 & 1 & -7 \\ -5 & 2 & 6 \end{bmatrix}\begin{bmatrix} 2 & 0 & -2 \\ -1 & 4 & 1 \\ 3 & 5 & 4 \end{bmatrix} = \begin{bmatrix} 28 & 20 & 0 \\ -28 & -31 & -21 \\ 6 & 38 & 36 \end{bmatrix}$$

the two matrices are equal.

4. For A, we have $ad - bc = 1$. Hence,

$$A^{-1} = \begin{bmatrix} 2 & -1 \\ -5 & 3 \end{bmatrix}$$

For B, we have $ad - bc = 20$. Hence,

$$B^{-1} = \frac{1}{20}\begin{bmatrix} 4 & 3 \\ -4 & 2 \end{bmatrix}$$

For C, we have $ad - bc = 6$. Hence,

$$C^{-1} = \frac{1}{6}\begin{bmatrix} 3 & 0 \\ 0 & 2 \end{bmatrix}$$

6. By definition,

$$(AB)^2 = (AB)(AB) = ABAB$$

However, we are guaranteed that $ABAB = AABB$ only if $AB = BA$. Since, in general, $AB \neq BA$, we cannot conclude that $(AB)^2 = A^2B^2$. For example, let $A = \begin{bmatrix} 1 & 0 \\ 0 & 0 \end{bmatrix}$ and $B = \begin{bmatrix} 0 & 1 \\ 1 & 0 \end{bmatrix}$; it is easy to verify that $(AB)^2 \neq A^2B^2$.

8. We are given that $(7A)^{-1} = \begin{bmatrix} -3 & 7 \\ 1 & -2 \end{bmatrix}$. Therefore

$$7A = ((7A)^{-1})^{-1} = \begin{bmatrix} -3 & 7 \\ 1 & -2 \end{bmatrix}^{-1} = \begin{bmatrix} 2 & 7 \\ 1 & 3 \end{bmatrix}$$

Thus,

$$A = \begin{bmatrix} 2/7 & 1 \\ 1/7 & 3/7 \end{bmatrix}$$

10. Let

$$A^{-1} = \begin{bmatrix} x_{11} & x_{12} & x_{13} \\ x_{21} & x_{22} & x_{23} \\ x_{31} & x_{32} & x_{33} \end{bmatrix}$$

Then

$$AA^{-1} = \begin{bmatrix} 1 & 0 & 1 \\ 1 & 1 & 0 \\ 0 & 1 & 1 \end{bmatrix} \begin{bmatrix} x_{11} & x_{12} & x_{13} \\ x_{21} & x_{22} & x_{23} \\ x_{31} & x_{32} & x_{33} \end{bmatrix}$$

$$= \begin{bmatrix} x_{11}+x_{31} & x_{12}+x_{32} & x_{13}+x_{33} \\ x_{11}+x_{21} & x_{12}+x_{22} & x_{13}+x_{23} \\ x_{21}+x_{31} & x_{22}+x_{32} & x_{23}+x_{33} \end{bmatrix}$$

Since $AA^{-1} = I$, we equate corresponding entries to obtain the system of equations

$$\begin{aligned}
x_{11} \qquad\qquad\qquad\qquad\qquad\qquad + x_{31} \qquad\qquad &= 1 \\
x_{12} \qquad\qquad\qquad\qquad\qquad\qquad + x_{32} \qquad &= 0 \\
x_{13} \qquad\qquad\qquad\qquad\qquad\qquad + x_{33} &= 0 \\
x_{11} \qquad + x_{21} \qquad\qquad\qquad\qquad\qquad\qquad &= 0 \\
x_{12} \qquad + x_{22} \qquad\qquad\qquad\qquad\qquad &= 1 \\
x_{13} \qquad + x_{23} \qquad\qquad\qquad\qquad &= 0 \\
x_{21} \qquad + x_{31} \qquad\qquad &= 0 \\
x_{22} \qquad + x_{32} \qquad &= 0 \\
x_{23} \qquad + x_{33} &= 1
\end{aligned}$$

The solution to this system of equations gives

$$A^{-1} = \begin{bmatrix} 1/2 & 1/2 & -1/2 \\ -1/2 & 1/2 & 1/2 \\ 1/2 & -1/2 & 1/2 \end{bmatrix}$$

11. Call the matrix A. By the formula in Example 7,

$$A^{-1} = \frac{1}{\cos^2\theta + \sin^2\theta}\begin{bmatrix} \cos\theta & -\sin\theta \\ \sin\theta & \cos\theta \end{bmatrix}$$

$$= \begin{bmatrix} \cos\theta & -\sin\theta \\ \sin\theta & \cos\theta \end{bmatrix}$$

since $\cos^2\theta + \sin^2\theta = 1$.

12. **(a)** Any pair of matrices that do not commute will work. For example, if we let

$$A = \begin{bmatrix} 1 & 0 \\ 0 & 0 \end{bmatrix} \qquad B = \begin{bmatrix} 0 & 1 \\ 0 & 1 \end{bmatrix}$$

then

$$(A + B)^2 = \begin{bmatrix} 1 & 1 \\ 0 & 1 \end{bmatrix}^2 = \begin{bmatrix} 1 & 2 \\ 0 & 1 \end{bmatrix}$$

whereas

$$A^2 + 2AB + B^2 = \begin{bmatrix} 1 & 3 \\ 0 & 1 \end{bmatrix}$$

(c) In general,

$$(A + B)^2 = (A + B)(A + B) = A^2 + AB + BA + B^2$$

13. If $a_{11}a_{22}\cdots a_{nn} \neq 0$, then $a_{ii} \neq 0$, and hence $1/a_{ii}$ is defined for $i = 1,2,\ldots,n$. It is now easy to verify that

$$A^{-1} = \begin{bmatrix} 1/a_{11} & 0 & \cdots & 0 \\ 0 & 1/a_{22} & \cdots & 0 \\ \vdots & \vdots & & \vdots \\ 0 & 0 & \cdots & 1/a_{nn} \end{bmatrix}$$

14. We are given that $3A - A^2 = I$. This implies that $A(3I - A) = (3I - A)A = I$; hence $A^{-1} = 3I - A$.

15. Let A denote a matrix which has an entire row or an entire column of zeros. Then if B is any matrix, either AB has an entire row of zeros or BA has an entire column of zeros, respectively. (See Exercise 14, Section 1.4.) Hence, neither AB nor BA can be the identity matrix; therefore, A cannot have an inverse.

16. Not necessarily; for example, both I and $-I$ are invertible, but their sum is not.

17. Suppose that $AB = 0$ and A is invertible. Then $A^{-1}(AB) = A^{-1}0$ or $IB = 0$. Hence, $B = 0$.

18. We cannot necessarily write $A0 = 0 = 0A$ because the matrices $A0$ and $0A$ may be of different sizes. In fact, the two zero matrices in the equation $A0 = 0$ may be of different sizes.

19. If

$$A = \begin{bmatrix} a_{11} & 0 & 0 \\ 0 & a_{22} & 0 \\ 0 & 0 & a_{33} \end{bmatrix} \quad \text{then} \quad A^2 = \begin{bmatrix} a_{11}^2 & 0 & 0 \\ 0 & a_{22}^2 & 0 \\ 0 & 0 & a_{33}^2 \end{bmatrix}$$

Thus, $A^2 = I$ if an only if $a_{11}^2 = a_{22}^2 = a_{33}^2 = 1$, or $a_{11} = \pm 1$, $a_{22} = \pm 1$, and $a_{33} = \pm 1$. There are exactly eight possibilities:

$$\begin{bmatrix} 1 & 0 & 0 \\ 0 & 1 & 0 \\ 0 & 0 & 1 \end{bmatrix} \begin{bmatrix} 1 & 0 & 0 \\ 0 & 1 & 0 \\ 0 & 0 & -1 \end{bmatrix} \begin{bmatrix} 1 & 0 & 0 \\ 0 & -1 & 0 \\ 0 & 0 & 1 \end{bmatrix} \begin{bmatrix} 1 & 0 & 0 \\ 0 & -1 & 0 \\ 0 & 0 & -1 \end{bmatrix}$$

$$\begin{bmatrix} -1 & 0 & 0 \\ 0 & 1 & 0 \\ 0 & 0 & 1 \end{bmatrix} \begin{bmatrix} -1 & 0 & 0 \\ 0 & 1 & 0 \\ 0 & 0 & -1 \end{bmatrix} \begin{bmatrix} -1 & 0 & 0 \\ 0 & -1 & 0 \\ 0 & 0 & 1 \end{bmatrix} \begin{bmatrix} -1 & 0 & 0 \\ 0 & -1 & 0 \\ 0 & 0 & -1 \end{bmatrix}$$

20. Suppose that X_1 is a fixed matrix which satisfies the equation $AX_1 = B$. Further, let X be any matrix whatsoever which satisfies the equation $AX = B$. We must then show that there is a matrix X_0 which satisfies both of the equations $X = X_1 + X_0$ and $AX_0 = 0$.

Clearly, the first equation implies that

$$X_0 = X - X_1$$

This candidate for X_0 will satisfy the second equation because

$$AX_0 = A(X - X_1) = AX - AX_1 = B - B = 0$$

We must also show that if both $AX_1 = B$ and $AX_0 = 0$, then $A(X_1 + X_0) = B$. But

$$A(X_1 + X_0) = AX_1 + AX_0 = B + 0 = B$$

21. **(a)** If we let a_{ij} denote the ij^{th} entry of A, then the ij^{th} entry in A^t will be a_{ji}. Hence, the ij^{th} entry in $(A^t)^t$ is just a_{ij}; thus $(A^t)^t = A$.

(c) If we let $C = AB$, then the element in the ith row and k^{th} column of C is

$$c_{ik} = \sum_j a_{ij} b_{jk}$$

If d_{ik} denotes the element in the ith row and k^{th} column of C^t, then

$$d_{ik} = c_{ki} = \sum_j a_{kj} b_{ji}$$

Now let $a'_{jk} = a_{kj}$, $b'_{ij} = b_{ji}$, and $E = B^t A^t$. If e_{ik} denotes the element in the ith row and k^{th} column of E, then

$$\begin{aligned} e_{ik} &= \sum_j b'_{ij} a'_{jk} = \sum_j b_{ji} a_{kj} \\ &= \sum_j a_{kj} b_{ji} \\ &= d_{ik} \end{aligned}$$

Therefore, the entries in the ith row and k^{th} column of $(AB)^t$ and $B^t A^t$ are identical.

22. **(b)** One such example is

$$A = \begin{bmatrix} 1 & 0 & -1 \\ 0 & 2 & -2 \\ -1 & -2 & 3 \end{bmatrix}$$

In general, any matrix of the form

$$\begin{bmatrix} a & d & e \\ d & b & f \\ e & f & c \end{bmatrix}$$

will work.

(c) One such example is

$$A = \begin{bmatrix} 0 & 0 & -1 \\ 0 & 0 & -2 \\ 1 & 2 & 0 \end{bmatrix}$$

In general, any matrix of the form

$$\begin{bmatrix} 0 & b & c \\ -b & 0 & f \\ -c & -f & 0 \end{bmatrix}$$

will work.

23. We use the results of Exercise 21, which hold in general, not just for 3×3 matrices.

(a) If $A = BB^t$, then

$$A^t = (BB^t)^t = (B^t)^t B^t = BB^t = A$$

Thus A is symmetric. On the other hand, if $A = B + B^t$, then

$$A^t = (B + B^t)^t = B^t + (B^t)^t = B^t + B = B + B^t = A$$

Thus A is symmetric.

23. **(b)** If $A = B - B^t$, then

$$A^t = (B - B^t)^t = [B + (-1)B^t]^t = B^t + [(-1)B^t]^t$$

$$= B^t + (-1)(B^t)^t = B^t + (-1)B = B^t - B = -A$$

Thus A is skew-symmetric.

24. Since the results of Exercise 21 hold in general, and not just for 3×3 matrices, we can apply Part (c) to show that

$$(A^n)^t = (A(A^{n-1}))^t = (A^{n-1})^t A^t$$

If we repeat this process n-1 times, we get that

$$(A^n)^t = (A^t)^n$$

which justifies the equality.

25. We wish to show that $A(B - C) = AB - AC$. By Part (d) of Theorem 1.5.1, we have $A(B - C) = A(B + (-C)) = AB + A(-C)$. Finally by Part ($m$), we have $A(-C) = -AC$ and the desired result can be obtained by substituting this result in the above equation.

26. By Parts (b) and (c) of Theorem 1.5.8,

$$(A - B)^t = (A + (-1)B)^t = A^t + (-1)B^t = A^t - B^t$$

27. Suppose that A is invertible. Then A is a square matrix, so that A^t is also square. Moreover, $A^{-1}A = AA^{-1} = I$. Thus

$$(A^{-1}A)^t = (AA^{-1})^t = I^t = I$$

or

$$A^t(A^{-1})^t = (A^{-1})^tA^t = I$$

That is, A^t has inverse $(A^{-1})^t$.

28. If we use the notation introduced in the proof of (h) in the text, we have

$$\ell_{ij} = a_{ij} + (b_{ij} + c_{ij})$$

Since the associative rule holds for real numbers, this implies that

$$\ell_{ij} = (a_{ij} + b_{ij}) + c_{ij} = r_{ij}$$

29. Let a_{ij} denote the ij^{th} entry of the $m \times n$ matrix A. All the entries of 0 are 0. In the proofs below, we need only show that the ij^{th} entries of all the matrices involved are equal.

(a) $a_{ij} + 0 = 0 + a_{ij} = a_{ij}$

(b) $a_{ij} - a_{ij} = 0$

(c) $0 - a_{ij} = -a_{ij}$

(d) $a_{i1}0 + a_{i2}0 + \cdots + a_{in}0 = 0$ and $0a_{1j} + 0a_{2j} + \cdots + a_{mj} = 0$

30. Again we use the notation introduced in the proof of Theorem 1.5.1, Part (h) in the text. Thus $\ell_{ij} = [A(BC)]_{ij}$ and $r_{ij} = [(AB)C]_{ij}$, where $[D]_{ij}$ denotes the entry in the ith row and j^{th} column of any matrix D. Suppose that A is $m \times n$, B is $n \times p$, and C is $p \times q$. Then

$$\begin{aligned}
\ell_{ij} &= a_{i1}[BC]_{1j} + a_{i2}[BC]_{2j} + \cdots + a_{in}[BC]_{nj} \\
&= a_{i1}(b_{11}c_{1j} + \cdots + b_{1p}c_{pi}) + a_{i2}(b_{21}c_{1j} + \cdots + \\
&\qquad + b_{2p}c_{pj}) + \cdots + a_{in}(b_{n1}c_{1j} + \cdots + b_{np}c_{pj}) \\
&= (a_{i1}b_{11} + a_{i2}b_{21} + \cdots + a_{in}b_{n1})c_{1j} + \cdots + \\
&\qquad (a_{i1}b_{1p} + a_{i2}b_{2p} + \cdots + a_{in}b_{np})c_{pj} \\
&= [AB]_{i1}c_{1j} + \cdots + [AB]_{ip}c_{pj} \\
&= r_{ij}
\end{aligned}$$

For anyone familiar with sigma notation, this proof can be simplified as follows:

$$\begin{aligned}
\ell_{ij} &= \sum_{s=1}^{n} a_{is}[BC]_{sj} = \sum_{s=1}^{n} a_{is} \sum_{t=1}^{p} b_{st}c_{tj} \\
&= \sum_{t=1}^{p} \left[\sum_{s=1}^{n} a_{is}b_{st} \right] c_{tj} = \sum_{t=1}^{p} [AB]_{it}c_{tj} \\
&= r_{ij}
\end{aligned}$$

31. We are given that A is invertible. Thus,

$$\underbrace{A\cdots A}_{\substack{n\\ \text{factors}}}\ \underbrace{A^{-1}\cdots A^{-1}}_{\substack{n\\ \text{factors}}} = \underbrace{A\cdots A}_{\substack{n-1\\ \text{factors}}}(AA^{-1})\underbrace{A^{-1}\cdots A^{-1}}_{\substack{n-1\\ \text{factors}}}$$

$$= A^{n-1}I(A^{-1})^{n-1}$$

$$= A^{n-1}(A^{-1})^{n-1}$$

$$= \underbrace{A\cdots A}_{\substack{n-2\\ \text{factors}}}\ (AA^{-1})\ \underbrace{A^{-1}\cdots A^{-1}}_{\substack{n-2\\ \text{factors}}}$$

$$= A^{n-2}(A^{-1})^{n-2}$$

Eventually this process yields

$$\underbrace{A\cdots A}_{\substack{n\\ \text{factors}}}\ \underbrace{A^{-1}\cdots A^{-1}}_{\substack{n\\ \text{factors}}} = AA^{-1} = I$$

Hence $(A^n)^{-1} = (A^{-1})^n$ for $n = 1,2,\ldots$. In case $n = 0$, we have

$$(A^n)^{-1} = (A^0)^{-1} = I^{-1} = I = (A^{-1})^0 = (A^{-1})^n$$

Hence, the formula also holds if $n = 0$

32. **(a)** We have

$$A^rA^s = \underbrace{(AA\cdots A)}_{\substack{r\\ \text{factors}}}\ \underbrace{(AA\cdots A)}_{\substack{s\\ \text{factors}}}$$

$$= \underbrace{AA\cdots A}_{\substack{r+s\\ \text{factors}}} = A^{r+s}$$

On the other hand,

$$(A^r)^s = \underbrace{\underbrace{(AA\cdots A)}_{\substack{r\\ \text{factors}}}\ \underbrace{(AA\cdots A)}_{\substack{r\\ \text{factors}}}\ \cdots\ \underbrace{(AA\cdots A)}_{\substack{r\\ \text{factors}}}}_{\substack{s\\ \text{factors}}}$$

$$= \underbrace{AA\cdots A}_{\substack{rs\\ \text{factors}}}$$

(b) Suppose that $r < 0$ and $s < 0$; let $\rho = -r$ and $\sigma = -s$, so that

$$A^rA^s = A^{-\rho}A^{-\sigma}$$

$$= (A^{-1})^{\rho}(A^{-1})^{\sigma} \qquad \text{(by the definition)}$$

$$= (A^{-1})^{\rho+\sigma} \qquad \text{(by Part (a))}$$

$$= A^{-(\rho+\sigma)} \qquad \text{(by the definition)}$$

$$= A^{-\rho-\sigma}$$

$$= A^{r+s}$$

Also

$$(A^r)^s = (A^{-\rho})^{-\sigma}$$

$$= \left[(A^{-1})^{\rho}\right]^{-\sigma} \qquad \text{(by the definition)}$$

$$= \left[\left[(A^{-1})^{\rho}\right]^{-1}\right]^{\sigma} \qquad \text{(by the definition)}$$

$$= \left[\left[(A^{-1})^{-1}\right]^{\rho}\right]^{\sigma} \qquad \text{(by Theorem 1.5.7b)}$$

$$= \left[[A]^{\rho}\right]^{\sigma} \qquad \text{(by Theorem 1.5.7a)}$$

$$= A^{\rho\sigma} \qquad \text{(by Part (a))}$$

$$= A^{(-\rho)(-\sigma)}$$

$$= A^{rs}$$

34. **(a)** If $AB = AC$, then

$$A^{-1}(AB) = A^{-1}(AC)$$

or

$$(A^{-1}A)B = (A^{-1}A)C$$

or

$$B = C$$

(b) The matrix A in Example 3 is not invertible.

EXERCISE SET 1.6

1. **(a)** The matrix may be obtained from I_2 only by performing two elementary row operations such as multiplying Row 1 of I_2 by −1 and then adding 5 times Row 1 to Row 2. Thus this is not an elementary matrix.

   (c) The matrix may be obtained from I_2 by multiplying Row 2 of I_2 by $\sqrt{3}$. Thus it is elementary.

   (e) This is not an elementary matrix because it is not invertible.

   (g) The matrix may be obtained from I_4 only by performing two elementary row operations such as replacing Row 1 of I_4 by Row 1 plus Row 4, and then multiplying Row 1 by 2. Thus it is not an elementary matrix.

3. **(a)** If we interchange Rows 1 and 3 of A, then we obtain B. Therefore, E_1 must be the matrix obtained from I_3 by interchanging Rows 1 and 3 of I_3, i.e.,

$$E_1 = \begin{bmatrix} 0 & 0 & 1 \\ 0 & 1 & 0 \\ 1 & 0 & 0 \end{bmatrix}$$

3. **(c)** If we multiply Row 1 of A by -2 and add it to Row 3, then we obtain C. Therefore, E_3 must be the matrix obtained from I_3 by replacing its third row by -2 times Row 1 plus Row 3, i.e.,

$$E_3 = \begin{bmatrix} 1 & 0 & 0 \\ 0 & 1 & 0 \\ -2 & 0 & 1 \end{bmatrix}$$

5. **(a)**

$$\left[\begin{array}{cc|cc} 1 & 4 & 1 & 0 \\ 2 & 7 & 0 & 1 \end{array}\right]$$

$$\left[\begin{array}{cc|cc} 1 & 4 & 1 & 0 \\ 0 & -1 & -3 & 1 \end{array}\right]$$

Add -2 times Row 1 to Row 2.

$$\left[\begin{array}{cc|cc} 1 & 0 & -7 & 4 \\ 0 & 1 & 2 & -1 \end{array}\right]$$

Multiply Row 2 by -1; then multiply Row 2 by -4 and add to Row 1.

Therefore

$$\begin{bmatrix} 1 & 4 \\ 2 & 7 \end{bmatrix}^{-1} = \begin{bmatrix} -7 & 4 \\ 2 & -1 \end{bmatrix}$$

(c)

$$\left[\begin{array}{cc|cc} 6 & -4 & 1 & 0 \\ -3 & 2 & 0 & 1 \end{array}\right]$$

$$\left[\begin{array}{cc|cc} 1 & -2/3 & 1/6 & 0 \\ -3 & 2 & 0 & 1 \end{array}\right]$$

Divide Row 1 by 6.

$$\left[\begin{array}{cc|cc} 1 & -2/3 & 1/6 & 0 \\ 0 & 0 & 1/2 & 1 \end{array}\right]$$

Multiply Row 1 by 3 and add to Row 2.

Thus, the given matrix is not invertible because we have obtained a row of zeros on the left side.

6\. **(a)**

$$\left[\begin{array}{ccc|ccc} 3 & 4 & -1 & 1 & 0 & 0 \\ 1 & 0 & 3 & 0 & 1 & 0 \\ 2 & 5 & -4 & 0 & 0 & 1 \end{array}\right]$$

$$\left[\begin{array}{ccc|ccc} 1 & 0 & 3 & 0 & 1 & 0 \\ 3 & 4 & -1 & 1 & 0 & 0 \\ 2 & 5 & -4 & 0 & 0 & 1 \end{array}\right]$$

Interchange Rows 1 and 2.

$$\left[\begin{array}{ccc|ccc} 1 & 0 & 3 & 0 & 1 & 0 \\ 0 & 4 & -10 & 1 & -3 & 0 \\ 0 & 5 & -10 & 0 & -2 & 1 \end{array}\right]$$

Add -3 times Row 1 to Row 2 and -2 times Row 1 to Row 3.

$$\left[\begin{array}{ccc|ccc} 1 & 0 & 3 & 0 & 1 & 0 \\ 0 & 4 & -10 & 1 & -3 & 0 \\ 0 & 1 & 0 & -1 & 1 & 1 \end{array}\right]$$

Add -1 times Row 2 to Row 3.

$$\left[\begin{array}{ccc|ccc} 1 & 0 & 3 & 0 & 1 & 0 \\ 0 & 1 & 0 & -1 & 1 & 1 \\ 0 & 0 & -10 & 5 & -7 & -4 \end{array}\right]$$

Add -4 times Row 3 to Row 2 and interchange Rows 2 and 3.

$$\left[\begin{array}{ccc|ccc} 1 & 0 & 0 & \frac{3}{2} & -\frac{11}{10} & -\frac{6}{5} \\ 0 & 1 & 0 & -1 & 1 & 1 \\ 0 & 0 & 1 & -\frac{1}{2} & \frac{7}{10} & \frac{2}{5} \end{array}\right]$$

Multiply Row 3 by -1/10. Then add -3 times Row 3 to Row 1.

Thus, the desired inverse is

$$\begin{bmatrix} \frac{3}{2} & -\frac{11}{10} & -\frac{6}{5} \\ -1 & 1 & 1 \\ -\frac{1}{2} & \frac{7}{10} & \frac{2}{5} \end{bmatrix}$$

6. (c)

$$\left[\begin{array}{ccc|ccc} 1 & 0 & 1 & 1 & 0 & 0 \\ 0 & 1 & 1 & 0 & 1 & 0 \\ 1 & 1 & 0 & 0 & 0 & 1 \end{array}\right]$$

$$\left[\begin{array}{ccc|ccc} 1 & 0 & 1 & 1 & 0 & 0 \\ 0 & 1 & 1 & 0 & 1 & 0 \\ 0 & 1 & -1 & -1 & 0 & 1 \end{array}\right]$$

Subtract Row 1 from Row 3.

$$\left[\begin{array}{ccc|ccc} 1 & 0 & 1 & 1 & 0 & 0 \\ 0 & 1 & 1 & 0 & 1 & 0 \\ 0 & 0 & 1 & \frac{1}{2} & \frac{1}{2} & -\frac{1}{2} \end{array}\right]$$

Subtract Row 2 from Row 3 and multiply Row 3 by $-\frac{1}{2}$.

$$\left[\begin{array}{ccc|ccc} 1 & 0 & 0 & \frac{1}{2} & -\frac{1}{2} & \frac{1}{2} \\ 0 & 1 & 0 & -\frac{1}{2} & \frac{1}{2} & \frac{1}{2} \\ 0 & 0 & 1 & \frac{1}{2} & \frac{1}{2} & -\frac{1}{2} \end{array}\right]$$

Subtract Row 3 from Rows 1 and 2.

Thus

$$\begin{bmatrix} 1 & 0 & 1 \\ 0 & 1 & 1 \\ 1 & 1 & 0 \end{bmatrix}^{-1} = \begin{bmatrix} \frac{1}{2} & -\frac{1}{2} & \frac{1}{2} \\ -\frac{1}{2} & \frac{1}{2} & \frac{1}{2} \\ \frac{1}{2} & \frac{1}{2} & -\frac{1}{2} \end{bmatrix}$$

(e)

$$\left[\begin{array}{ccc|ccc} 1 & 0 & 1 & 1 & 0 & 0 \\ -1 & 1 & 1 & 0 & 1 & 0 \\ 0 & 1 & 0 & 0 & 0 & 1 \end{array}\right]$$

$$\left[\begin{array}{ccc|ccc} 1 & 0 & 1 & 1 & 0 & 0 \\ 0 & 1 & 2 & 1 & 1 & 0 \\ 0 & 0 & -2 & -1 & -1 & 1 \end{array}\right]$$

Add Row 1 to Row 2 and subtract the new Row 2 from Row 3.

$$\left[\begin{array}{ccc|ccc} 1 & 0 & 1 & 1 & 0 & 0 \\ 0 & 1 & 0 & 0 & 0 & 1 \\ 0 & 0 & 1 & \frac{1}{2} & \frac{1}{2} & -\frac{1}{2} \end{array}\right]$$

Add Row 3 to Row 2 and then multiply Row 3 by $-\frac{1}{2}$.

$$\left[\begin{array}{ccc|ccc} 1 & 0 & 0 & \frac{1}{2} & -\frac{1}{2} & \frac{1}{2} \\ 0 & 1 & 0 & 0 & 0 & 1 \\ 0 & 0 & 1 & \frac{1}{2} & \frac{1}{2} & -\frac{1}{2} \end{array}\right]$$

Subtract Row 3 from Row 1.

Thus

$$\begin{bmatrix} 1 & 0 & 1 \\ -1 & 1 & 1 \\ 0 & 1 & 0 \end{bmatrix}^{-1} = \begin{bmatrix} \frac{1}{2} & -\frac{1}{2} & \frac{1}{2} \\ 0 & 0 & 1 \\ \frac{1}{2} & \frac{1}{2} & -\frac{1}{2} \end{bmatrix}$$

7. **(a)**

$$\left[\begin{array}{ccc|ccc} \frac{1}{5} & \frac{1}{5} & -\frac{2}{5} & 1 & 0 & 0 \\ \frac{1}{5} & \frac{1}{5} & \frac{1}{10} & 0 & 1 & 0 \\ \frac{1}{5} & -\frac{4}{5} & \frac{1}{10} & 0 & 0 & 1 \end{array}\right]$$

$$\left[\begin{array}{ccc|ccc} 1 & 1 & -2 & 5 & 0 & 0 \\ 1 & 1 & \frac{1}{2} & 0 & 5 & 0 \\ 1 & -4 & \frac{1}{2} & 0 & 0 & 5 \end{array}\right]$$

Multiply all three Rows by 5.

$$\left[\begin{array}{ccc|ccc} 1 & 1 & -2 & 5 & 0 & 0 \\ 0 & 0 & \frac{5}{2} & -5 & 5 & 0 \\ 0 & -5 & \frac{5}{2} & -5 & 0 & 5 \end{array}\right]$$

Subtract Row 1 from both Row 2 and Row 3.

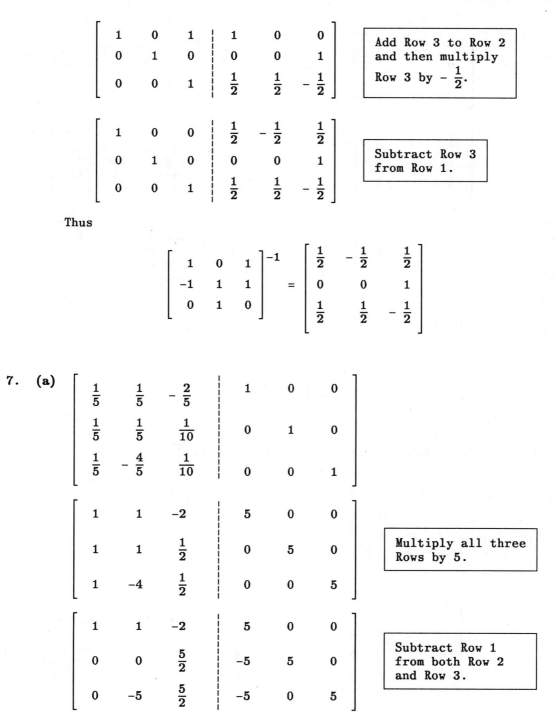

$$\left[\begin{array}{ccc|ccc} 1 & 1 & -2 & 5 & 0 & 0 \\ 0 & 0 & 1 & -2 & 2 & 0 \\ 0 & 1 & -\frac{1}{2} & 1 & 0 & -1 \end{array}\right]$$

Multiply Row 2 by 2/5 and Row 3 by -1/5.

$$\left[\begin{array}{ccc|ccc} 1 & 1 & 0 & 1 & 4 & 0 \\ 0 & 0 & 1 & -2 & 2 & 0 \\ 0 & 1 & 0 & 0 & 1 & -1 \end{array}\right]$$

Add 2 times Row 2 to Row 1 and 1/2 times Row 2 to Row 3.

$$\left[\begin{array}{ccc|ccc} 1 & 0 & 0 & 1 & 3 & 1 \\ 0 & 1 & 0 & 0 & 1 & -1 \\ 0 & 0 & 1 & -2 & 2 & 0 \end{array}\right]$$

Subtract Row 3 from Row 1 and interchange Rows 2 and 3.

Thus the desired inverse is

$$\begin{bmatrix} 1 & 3 & 1 \\ 0 & 1 & -1 \\ -2 & 2 & 0 \end{bmatrix}$$

7. **(b)** Hint: first multiply Rows 1 and 2 by $1/\sqrt{2}$.

(c)

$$\left[\begin{array}{cccc|cccc} 1 & 0 & 0 & 0 & 1 & 0 & 0 & 0 \\ 1 & 3 & 0 & 0 & 0 & 1 & 0 & 0 \\ 1 & 3 & 5 & 0 & 0 & 0 & 1 & 0 \\ 1 & 3 & 5 & 7 & 0 & 0 & 0 & 1 \end{array}\right]$$

$$\left[\begin{array}{cccc|cccc} 1 & 0 & 0 & 0 & 1 & 0 & 0 & 0 \\ 0 & 3 & 0 & 0 & -1 & 1 & 0 & 0 \\ 0 & 3 & 5 & 0 & -1 & 0 & 1 & 0 \\ 0 & 3 & 5 & 7 & -1 & 0 & 0 & 1 \end{array}\right]$$

Subtract Row 1 from Rows 2–4.

$$\left[\begin{array}{cccc|cccc} 1 & 0 & 0 & 0 & 1 & 0 & 0 & 0 \\ 0 & 1 & 0 & 0 & -\frac{1}{3} & \frac{1}{3} & 0 & 0 \\ 0 & 0 & 5 & 0 & 0 & -1 & 1 & 0 \\ 0 & 0 & 5 & 7 & 0 & -1 & 0 & 1 \end{array}\right]$$

Subtract Row 2 from Rows 3 and 4 and multiply Row 2 by 1/3.

$$\left[\begin{array}{cccc|cccc} 1 & 0 & 0 & 0 & 1 & 0 & 0 & 0 \\ 0 & 1 & 0 & 0 & -\frac{1}{3} & \frac{1}{3} & 0 & 0 \\ 0 & 0 & 1 & 0 & 0 & -\frac{1}{5} & \frac{1}{5} & 0 \\ 0 & 0 & 0 & 1 & 0 & 0 & -\frac{1}{7} & \frac{1}{7} \end{array}\right]$$

Subtract Row 3 from Row 4 and multiply Row 3 by 1/5 and Row 4 by 1/7.

Thus the desired inverse is

$$\begin{bmatrix} 1 & 0 & 0 & 0 \\ -\frac{1}{3} & \frac{1}{3} & 0 & 0 \\ 0 & -\frac{1}{5} & \frac{1}{5} & 0 \\ 0 & 0 & -\frac{1}{7} & \frac{1}{7} \end{bmatrix}$$

(d) Since the third row consists entirely of zeros, this matrix is not invertible.

8. **(b)** Multiplying Row i of

$$\left[\begin{array}{cccc|cccc} 0 & 0 & 0 & k_1 & 1 & 0 & 0 & 0 \\ 0 & 0 & k_2 & 0 & 0 & 1 & 0 & 0 \\ 0 & k_3 & 0 & 0 & 0 & 0 & 1 & 0 \\ k_4 & 0 & 0 & 0 & 0 & 0 & 0 & 1 \end{array}\right]$$

by $1/k_i$ for $i = 1,2,3,4$ and then reversing the order of the rows yields I_4 on the left and the desired inverse,

$$\begin{bmatrix} 0 & 0 & 0 & 1/k_4 \\ 0 & 0 & 1/k_3 & 0 \\ 0 & 1/k_2 & 0 & 0 \\ 1/k_1 & 0 & 0 & 0 \end{bmatrix}$$

on the right.

8. **(c)** To reduce

$$\left[\begin{array}{cccc|cccc} k & 0 & 0 & 0 & 1 & 0 & 0 & 0 \\ 1 & k & 0 & 0 & 0 & 1 & 0 & 0 \\ 0 & 1 & k & 0 & 0 & 0 & 1 & 0 \\ 0 & 0 & 1 & k & 0 & 0 & 0 & 1 \end{array}\right]$$

we multiply Row i by $1/k$ and then subtract Row i from Row $(i+1)$ for $i = 1,2,3$. Then multiply Row 4 by $1/k$. This produces I_4 on the left and the inverse,

$$\begin{bmatrix} 1/k & 0 & 0 & 0 \\ -1/k^2 & 1/k & 0 & 0 \\ 1/k^3 & -1/k^2 & 1/k & 0 \\ -1/k^4 & 1/k^3 & -1/k^2 & 1/k \end{bmatrix}$$

on the right.

9. **(a)** It takes 2 elementary row operations to convert A to I: first multiply Row 1 by 5 and add to Row 2, then multiply Row 2 by 1/2. If we apply each of these operations to I, we obtain the 2 elementary matrices

$$E_1 = \begin{bmatrix} 1 & 0 \\ 5 & 1 \end{bmatrix} \quad \text{and} \quad E_2 = \begin{bmatrix} 1 & 0 \\ 0 & 1/2 \end{bmatrix}$$

From Theorem 1.6.1, we conclude that

$$E_2 E_1 A = I$$

(b) Since the inverse of a matrix is unique, the above equation guarantees that (with E_1 and E_2 as defined above)

$$A^{-1} = E_2 E_1$$

(c) From the equation $E_2 E_1 A = I$, we conclude that

$$E_2^{-1} E_2 E_1 A = E_2^{-1} I$$

or

$$E_1 A = E_2^{-1}$$

or

$$E_1^{-1} E_1 A = E_1^{-1} E_2^{-1}$$

or

$$A = E_1^{-1} E_2^{-1}$$

so that

$$A = \begin{bmatrix} 1 & 0 \\ -5 & 1 \end{bmatrix} \begin{bmatrix} 1 & 0 \\ 0 & 2 \end{bmatrix}$$

11. We first reduce A to row-echelon form, keeping track as we do so of the elementary row operations which are required. Here

$$A = \begin{bmatrix} 0 & 1 & 7 & 8 \\ 1 & 3 & 3 & 8 \\ -2 & -5 & 1 & -8 \end{bmatrix}$$

1. If we interchange Rows 1 and 2, we obtain the matrix

$$\begin{bmatrix} 1 & 3 & 3 & 8 \\ 0 & 1 & 7 & 8 \\ -2 & -5 & 1 & -8 \end{bmatrix}$$

2. If we add twice Row 1 to Row 3, we obtain the matrix

$$\begin{bmatrix} 1 & 3 & 3 & 8 \\ 0 & 1 & 7 & 8 \\ 0 & 1 & 7 & 8 \end{bmatrix}$$

3. If we subtract Row 2 from Row 3, we obtain the matrix

$$R = \begin{bmatrix} 1 & 3 & 3 & 8 \\ 0 & 1 & 7 & 8 \\ 0 & 0 & 0 & 0 \end{bmatrix}$$

Note that the above matrix R is in row-echelon form. Thus, Theorem 1.6.1 implies that $E_3E_2E_1A = R$, where

$$E_1 = \begin{bmatrix} 0 & 1 & 0 \\ 1 & 0 & 0 \\ 0 & 0 & 1 \end{bmatrix}, \; E_2 = \begin{bmatrix} 1 & 0 & 0 \\ 0 & 1 & 0 \\ 2 & 0 & 1 \end{bmatrix}, \text{ and } E_3 = \begin{bmatrix} 1 & 0 & 0 \\ 0 & 1 & 0 \\ 0 & -1 & 1 \end{bmatrix}$$

Thus, $A = E_1^{-1}E_2^{-1}E_3^{-1}R$. We need only let

$$E = E_1^{-1} = \begin{bmatrix} 0 & 1 & 0 \\ 1 & 0 & 0 \\ 0 & 0 & 1 \end{bmatrix}$$

$$F = E_2^{-1} = \begin{bmatrix} 1 & 0 & 0 \\ 0 & 1 & 0 \\ -2 & 0 & 1 \end{bmatrix}$$

and

$$G = E_3^{-1} = \begin{bmatrix} 1 & 0 & 0 \\ 0 & 1 & 0 \\ 0 & 1 & 1 \end{bmatrix}$$

and we're done.

12. If A is an elementary matrix, then it can be obtained from the identity matrix I by a single elementary row operation. If we start with I and multiply Row 3 by a nonzero constant, then $a = b = 0$. If we interchange Row 1 or Row 2 with Row 3, then $c = 0$. If we add a nonzero multiple of Row 1 or Row 2 to Row 3, then either $b = 0$ or $a = 0$. Finally, if we operate only on the first two rows, then $a = b = 0$. Thus at least one entry in Row 3 must equal zero.

14. Every $m \times n$ matrix A can be transformed into reduced row-echelon form B by a sequence of row operations. From Theorem 1.6.1,

$$B = E_k E_{k-1} \cdots E_1 A$$

where $E_1, E_2, \ldots, E_k$ are the elementary matrices corresponding to the row operations. If we take $C = E_k E_{k-1} \cdots E_1$, then C is invertible by Theorem 1.6.2 and the rule following Theorem 1.5.5

15. The hypothesis that B is row equivalent to A can be expressed as follows:

$$B = E_k E_{k-1} \cdots E_1 A$$

where $E_1, E_2, \ldots, E_k$ are elementary matrices. Since A is invertible, then (by the rule following Theorem 1.5.5) B is invertible and

$$B^{-1} = A^{-1} E_1^{-1} E_2^{-1} \cdots E_k^{-1}$$

EXERCISE SET 1.7

1. This system of equations is of the form $AX = B$, where

$$A = \begin{bmatrix} 1 & 1 \\ 5 & 6 \end{bmatrix} \qquad X = \begin{bmatrix} x_1 \\ x_2 \end{bmatrix} \quad \text{and} \quad B = \begin{bmatrix} 2 \\ 9 \end{bmatrix}$$

By the rule given in Example 7,

$$A^{-1} = \begin{bmatrix} 6 & -1 \\ -5 & 1 \end{bmatrix}$$

Thus

$$X = A^{-1}B = \begin{bmatrix} 6 & -1 \\ -5 & 1 \end{bmatrix} \begin{bmatrix} 2 \\ 9 \end{bmatrix} = \begin{bmatrix} 3 \\ -1 \end{bmatrix}$$

That is,

$$x_1 = 3 \quad \text{and} \quad x_2 = -1$$

3. This system is of the form $AX = B$, where

$$A = \begin{bmatrix} 1 & 3 & 1 \\ 2 & 2 & 1 \\ 2 & 3 & 1 \end{bmatrix} \qquad X = \begin{bmatrix} x_1 \\ x_2 \\ x_3 \end{bmatrix} \quad \text{and} \quad B = \begin{bmatrix} 4 \\ -1 \\ 3 \end{bmatrix}$$

By direct computation we obtain

$$A^{-1} = \begin{bmatrix} -1 & 0 & 1 \\ 0 & -1 & 1 \\ 2 & 3 & -4 \end{bmatrix}$$

so that

$$X = A^{-1}B = \begin{bmatrix} -1 \\ 4 \\ -7 \end{bmatrix}$$

That is,

$$x_1 = -1, \; x_2 = 4, \text{ and } x_3 = -7$$

5. The system is of the form $AX = B$, where

$$A = \begin{bmatrix} 1 & 1 & 1 \\ 1 & 1 & -4 \\ -4 & 1 & 1 \end{bmatrix} \quad X = \begin{bmatrix} x_1 \\ x_2 \\ x_3 \end{bmatrix} \quad \text{and} \quad B = \begin{bmatrix} 5 \\ 10 \\ 0 \end{bmatrix}$$

By direct computation, we obtain

$$A^{-1} = \left(\frac{1}{5}\right) \begin{bmatrix} 1 & 0 & -1 \\ 3 & 1 & 1 \\ 1 & -1 & 0 \end{bmatrix}$$

Thus,

$$X = A^{-1}B = \begin{bmatrix} 1 \\ 5 \\ -1 \end{bmatrix}$$

That is,

$$x_1 = 1, \; x_2 = 5, \text{ and } x_3 = -1$$

7. The system is of the form $AX = B$ where

$$A = \begin{bmatrix} 3 & 5 \\ 1 & 2 \end{bmatrix} \quad X = \begin{bmatrix} x_1 \\ x_2 \end{bmatrix} \quad \text{and} \quad B = \begin{bmatrix} b_1 \\ b_2 \end{bmatrix}$$

By Example 7, we have

$$A^{-1} = \begin{bmatrix} 2 & -5 \\ -1 & 3 \end{bmatrix}$$

Thus

$$X = A^{-1}B = \begin{bmatrix} 2b_1 - 5b_2 \\ -b_1 + 3b_2 \end{bmatrix}$$

That is,

$$x_1 = 2b_1 - 5b_2 \quad \text{and} \quad x_2 = -b_1 + 3b_2$$

9. The system is of the form $AX = B$, where

$$A = \begin{bmatrix} 1 & 2 & 1 \\ 1 & -1 & 1 \\ 1 & 1 & 0 \end{bmatrix} \quad X = \begin{bmatrix} x_1 \\ x_2 \\ x_3 \end{bmatrix} \quad \text{and} \quad B = \begin{bmatrix} b_1 \\ b_2 \\ b_3 \end{bmatrix}$$

We compute

$$A^{-1} = \begin{bmatrix} -1/3 & 1/3 & 1 \\ 1/3 & -1/3 & 0 \\ 2/3 & 1/3 & -1 \end{bmatrix}$$

so that

$$X = A^{-1}B = \begin{bmatrix} -(1/3)b_1 + (1/3)b_2 + b_3 \\ (1/3)b_1 - (1/3)b_2 \\ (2/3)b_1 + (1/3)b_2 - b_3 \end{bmatrix}$$

9. **(a)** In this case, we let

$$B = \begin{bmatrix} -1 \\ 3 \\ 4 \end{bmatrix}$$

Then

$$X = A^{-1}B = \begin{bmatrix} 16/3 \\ -4/3 \\ -11/3 \end{bmatrix}$$

That is, $x_1 = 16/3$, $x_2 = -4/3$, and $x_3 = -11/3$.

(c) In this case, we let

$$B = \begin{bmatrix} -1 \\ -1 \\ 3 \end{bmatrix}$$

Then

$$X = A^{-1}B = \begin{bmatrix} 3 \\ 0 \\ -4 \end{bmatrix}$$

That is, $x_1 = 3$, $x_2 = 0$, and $x_3 = -4$.

10. We apply the method of Example 2 to the two systems (a) and (c) solved above. The coefficient matrix augmented by the two B matrices yields

$$\left[\begin{array}{ccc|c|c} 1 & 2 & 1 & -1 & -1 \\ 1 & -1 & 1 & 3 & -1 \\ 1 & 1 & 0 & 4 & 3 \end{array}\right]$$

The reduced row-echelon form may be obtained as follows:

$$\left[\begin{array}{ccc|c|c} 1 & 2 & 1 & -1 & -1 \\ 0 & -3 & 0 & 4 & 0 \\ 0 & 2 & -1 & 1 & 4 \end{array}\right]$$

Add -1 times Row 2 to Row 3 and -1 times Row 1 to Row 2.

$$\left[\begin{array}{ccc|c|c} 1 & 0 & 2 & -2 & -5 \\ 0 & 1 & 0 & -4/3 & 0 \\ 0 & 0 & -1/2 & 11/6 & 2 \end{array}\right]$$

Add -1 times Row 3 to Row 2, divide Row 2 by -3 and Row 3 by 2, add -1 times Row 2 to Row 3.

$$\left[\begin{array}{ccc|c|c} 1 & 0 & 0 & 16/3 & 3 \\ 0 & 1 & 0 & -4/3 & 0 \\ 0 & 0 & 1 & -11/3 & -4 \end{array}\right]$$

Multiply Row 3 by -2 and then add -2 times Row 3 to Row 1.

This, fortunately, yields the same results as in Exercise 9.

11. The coefficient matrix, augmented by the two B matrices, yields

$$\left[\begin{array}{cc|c|c} 1 & -5 & 1 & -2 \\ 3 & 2 & 4 & 5 \end{array}\right]$$

This reduces to

$$\left[\begin{array}{cc|c|c} 1 & -5 & 1 & -2 \\ 0 & 17 & 1 & 11 \end{array}\right]$$

Add -3 times Row 1 to Row 2.

or

$$\left[\begin{array}{cc|c|c} 1 & 0 & 22/17 & 21/17 \\ 0 & 1 & 1/17 & 11/17 \end{array}\right]$$

Divide Row 2 by 17 and add 5 times Row 2 to Row 1.

Thus the solution to Part (a) is $x_1 = 22/17$, $x_2 = 1/17$, and to Part (b) is $x_1 = 21/17$, $x_2 = 11/17$.

15. As above, we set up the matrix

$$\left[\begin{array}{ccc|c|c} 1 & -2 & 1 & -2 & 1 \\ 2 & -5 & 1 & 1 & -1 \\ 3 & -7 & 2 & -1 & 0 \end{array}\right]$$

This reduces to

$$\left[\begin{array}{ccc|c|c} 1 & -2 & 1 & -2 & 1 \\ 0 & -1 & -1 & 5 & -3 \\ 0 & -1 & -1 & 5 & -3 \end{array}\right]$$

Add appropriate multiples of Row 1 to Rows 2 and 3.

or

$$\left[\begin{array}{ccc|c|c} 1 & -2 & 1 & -2 & 1 \\ 0 & 1 & 1 & -5 & 3 \\ 0 & 0 & 0 & 0 & 0 \end{array}\right]$$

Add −1 times Row 2 to Row 3 and multiply Row 2 by −1.

or

$$\left[\begin{array}{ccc|c|c} 1 & 0 & 3 & -12 & 7 \\ 0 & 1 & 1 & -5 & 3 \\ 0 & 0 & 0 & 0 & 0 \end{array}\right]$$

Add twice Row 2 to Row 3.

Thus if we let $x_3 = t$, we have for Part (a) $x_1 = -12 - 3t$ and $x_2 = -5 - t$, while for Part (b) $x_1 = 7 - 3t$ and $x_2 = 3 - t$.

16. The augmented matrix for this sytem is

$$\left[\begin{array}{ccc} 6 & -4 & b_1 \\ 3 & -2 & b_2 \end{array}\right]$$

which, when reduced to row-echelon form, becomes

$$\begin{bmatrix} 1 & -2/3 & b_2/3 \\ 0 & 0 & b_1 - 2b_2 \end{bmatrix}$$

The system is consistent if and only if $b_1 - 2b_2 = 0$ or $b_1 = 2b_2$.

Thus $AX = B$ is consistent if and only if B has the form

$$B = \begin{bmatrix} 2b_2 \\ b_2 \end{bmatrix}$$

17. The augmented matrix for this system of equations is

$$\begin{bmatrix} 1 & -2 & 5 & b_1 \\ 4 & -5 & 8 & b_2 \\ -3 & 3 & -3 & b_3 \end{bmatrix}$$

If we reduce this matrix to row-echelon form, we obtain

$$\begin{bmatrix} 1 & -2 & 5 & b_1 \\ 0 & 1 & -4 & \frac{1}{3}(b_2-4b_1) \\ 0 & 0 & 0 & -b_1+b_2+b_3 \end{bmatrix}$$

The 3rd row implies that $b_3 = b_1 - b_2$. Thus, $AX = B$ is consistent if and only if B has the form

$$B = \begin{bmatrix} b_1 \\ \frac{1}{3}(b_2 - 4b_1) \\ b_1 - b_2 \end{bmatrix}$$

20. **(a)** In general, $AX = X$ if and only if $AX - X = 0$ if and only if $(A - I)X = 0$. In this case,

$$A = \begin{bmatrix} 2 & 1 & 2 \\ 2 & 2 & -2 \\ 3 & 1 & 1 \end{bmatrix}$$

and hence,

$$A - I = \begin{bmatrix} 1 & 1 & 2 \\ 2 & 1 & -2 \\ 3 & 1 & 0 \end{bmatrix}$$

The reduced row-echelon form of $A - I$ is

$$\begin{bmatrix} 1 & 0 & 0 \\ 0 & 1 & 0 \\ 0 & 0 & 1 \end{bmatrix}$$

Therefore, the system of equations $(A - I)X = 0$ has only the trivial solution $x_1 = x_2 = x_3 = 0$; that is, $X = 0$ is the only solution to $AX = X$.

22. **(a)** It is clear that $X = 0$ is the only solution to the equation $AX = 0$. We may then invoke Theorem 1.7.3 to conclude that A is invertible.

23. Since $AX = 0$ has only $X = 0$ as a solution, Theorem 1.7.3 guarantees that A is invertible. We now show that A^k is also invertible. In fact,

$$(A^k)^{-1} = (A^{-1})^k$$

since

$$\underbrace{A^{-1}A^{-1}\cdots A^{-1}}_{\substack{k \\ \text{factors}}}\ \underbrace{AA\cdots A}_{\substack{k \\ \text{factors}}} = I$$

Because A^k is invertible, Theorem 1.7.3 allows us to conclude that $A^kX = 0$ has only the trivial solution.

24. First, assume that $AX = 0$ has only the trivial solution. It then follows from Theorem 1.7.3 that A is invertible. Now Q is invertible by assumption. Thus, QA is invertible because the product of two invertible matrices is invertible. If we apply Theorem 1.7.3 again, we see that $(QA)X = 0$ has only the trivial solution.

Conversely, assume that $(QA)X = 0$ has only the trivial solution; thus, QA is invertible by Theorem 1.7.3. But if QA is invertible, then so is A because $A = Q^{-1}(QA)$ is the product of invertible matrices. But if A is invertible, then $AX = 0$ has only the trivial solution.

25. Suppose that A is invertible. From Theorem 1.7.3, we know that A is row-equivalent to I_n. If we let E_1, E_2, ..., E_k denote the elementary matrices which are associated with the elementary row operations necessary to convert I_n to A, then we know that we can write

$$A = E_kE_{k-1} \cdots E_1I = E_kE_{k-1} \cdots E_1$$

Conversely, suppose that A can be expressed as a product of elementary matrices. By Theorem 1.6.2 every elementary matrix is invertible, so A is a product of invertible matrices; thus, A is also invertible.

26. We are given that $AB = I$. By Part (*a*) of Theorem 1.7.2, A must be the inverse of B. But then

$$A = B^{-1} \Rightarrow A^{-1} = (B^{-1})^{-1} = B$$

That is, $A^{-1} = B$.

SUPPLEMENTARY EXERCISES 1

1.

$$\begin{bmatrix} \frac{3}{5} & -\frac{4}{5} & x \\ \frac{4}{5} & \frac{3}{5} & y \end{bmatrix}$$

$$\begin{bmatrix} 1 & -\frac{4}{3} & \frac{5}{3}x \\ \frac{4}{5} & \frac{3}{5} & y \end{bmatrix}$$

Multiply Row 1 by 5/3.

$$\begin{bmatrix} 1 & -\frac{4}{3} & \frac{5}{3}x \\ 0 & \frac{5}{3} & -\frac{4}{3}x + y \end{bmatrix}$$

Add -4/5 times Row 1 to Row 2.

$$\begin{bmatrix} 1 & -\frac{4}{3} & \frac{5}{3}x \\ 0 & 1 & -\frac{4}{5}x + \frac{3}{5}y \end{bmatrix}$$

Multiply Row 2 by 3/5.

$$\begin{bmatrix} 1 & 0 & \frac{3}{5}x + \frac{4}{5}y \\ 0 & 1 & -\frac{4}{5}x + \frac{3}{5}y \end{bmatrix}$$

Add 4/3 times Row 2 to Row 1.

Thus,

$$x' = \frac{3}{5}x + \frac{4}{5}y$$

$$y' = -\frac{4}{5}x + \frac{3}{5}y$$

3. We denote the system of equations by

$$\begin{aligned} a_{11}x_1 + a_{12}x_2 + a_{13}x_3 + a_{14}x_4 &= 0 \\ a_{21}x_1 + a_{22}x_2 + a_{23}x_3 + a_{24}x_4 &= 0 \end{aligned}$$

If we substitute both sets of values for x_1, x_2, x_3, and x_4 into the first equation, we obtain

$$\begin{aligned} a_{11} - a_{12} + a_{13} + 2a_{14} &= 0 \\ 2a_{11} \qquad + 3a_{13} - 2a_{14} &= 0 \end{aligned}$$

where a_{11}, a_{12}, a_{13}, and a_{14} are variables. If we substitute both sets of values for x_1, x_2, x_3, and x_4 into the second equation, we obtain

$$\begin{aligned} a_{21} - a_{22} + a_{23} + 2a_{24} &= 0 \\ 2a_{21} \qquad + 3a_{23} - a_{24} &= 0 \end{aligned}$$

where a_{21}, a_{22}, a_{23}, and a_{24} are again variables. The two systems above both yield the matrix

$$\begin{bmatrix} 1 & -1 & 1 & 2 & 0 \\ 2 & 0 & 3 & -1 & 0 \end{bmatrix}$$

which reduces to

$$\begin{bmatrix} 1 & 0 & 3/2 & -1/2 & 0 \\ 0 & 1 & 1/2 & -5/2 & 0 \end{bmatrix}$$

This implies that

$$\begin{aligned} a_{11} &= -(3/2)a_{13} + (1/2)a_{14} \\ a_{12} &= -(1/2)a_{13} + (5/2)a_{14} \end{aligned}$$

and similarly,

$$\begin{aligned} a_{21} &= (-3/2)a_{23} + (1/2)a_{24} \\ a_{22} &= (-1/2)a_{23} + (5/2)a_{24} \end{aligned}$$

As long as our choice of values for the numbers a_{ij} is consistent with the above, then the system will have a solution. For simplicity, and to insure that neither equation is a multiple of the other, we let $a_{13} = a_{14} = -1$ and $a_{23} = 0$, $a_{24} = 2$. This means that $a_{11} = 1$, $a_{12} = -2$, $a_{21} = 1$, and $a_{22} = 5$, so that the system becomes

$$\begin{aligned} x_1 - 2x_2 - x_3 - x_4 &= 0 \\ x_1 + 5x_2 \qquad + 2x_4 &= 0 \end{aligned}$$

Of course, this is just one of infinitely many possibilities.

4. Suppose that the box contains x pennies, y nickels, and z dimes. We know that

$$\begin{aligned} x + 5y + 10z &= 83 \\ x + y + z &= 13 \end{aligned}$$

The augmented matrix of this system is

$$\begin{bmatrix} 1 & 5 & 10 & 83 \\ 1 & 1 & 1 & 13 \end{bmatrix}$$

which can be reduced to the matrix

$$\begin{bmatrix} 1 & 0 & -\frac{5}{4} & -\frac{9}{2} \\ 0 & 1 & \frac{9}{4} & \frac{35}{2} \end{bmatrix}$$

by Gauss-Jordan elimination. Thus,

$$x = \frac{5}{4}z - \frac{9}{2} \quad \text{and} \quad y = -\frac{9}{4}z + \frac{35}{2}$$

Since x, y, and z must all be integers between 0 and 13, then z must be either 2, 6, or 10 because those are the only values of z for which x and y are both integers. However, $z = 2$ implies that $x < 0$ while $z = 10$ implies that $y < 0$. Therefore, the only solution is $x = 3$, $y = 4$, and $z = 6$.

5. As in Exercise 3, we reduce the system to the equations

$$x = \frac{1 + 5z}{4}$$

$$y = \frac{35 - 9z}{4}$$

Since x, y, and z must all be positive integers, we have $z > 0$ and $35 - 9z > 0$ or $4 > z$. Thus we need only check the three values $z = 1,2,3$ to see whether or not they produce integer solutions for x and y. This yields the unique solution $x = 4$, $y = 2$, $z = 3$.

6. The augmented matrix of this system is

$$\begin{bmatrix} 1 & 1 & 1 & 4 \\ 0 & 0 & 1 & 2 \\ 0 & 0 & a^2-4 & a-2 \end{bmatrix}$$

which can be reduced to the matrix

$$\begin{bmatrix} 1 & 1 & 0 & 2 \\ 0 & 0 & 1 & 2 \\ 0 & 0 & 0 & 2a^2-a-6 \end{bmatrix}$$

Row 3 of the above matrix implies that the system of equations will have a solution if and only if $2a^2 - a - 6 = 0$; that is, if and only if $a = 2$ or $a = -3/2$. For either of those values of a, the solution is $x_1 = 2 - t$, $x_2 = t$, and $x_3 = 2$. Hence, the system has infinitely many solutions provided $a = 2$ or $a = -3/2$. For all other values of a, there will be no solution.

8. Let $s = xy$, $t = \sqrt{y}$, and $u = zy$. Then we have a system of linear equations in s, t, and u with augmented matrix

$$\begin{bmatrix} 1 & -2 & 3 & 8 \\ 2 & -3 & 2 & 7 \\ -1 & 1 & 2 & 4 \end{bmatrix}$$

This reduces to

$$\begin{bmatrix} 1 & 0 & 0 & 5 \\ 0 & 1 & 0 & 3 \\ 0 & 0 & 1 & 3 \end{bmatrix}$$

so that

$$xy = 5$$

$$\sqrt{y} = 3$$

$$zy = 3$$

Thus from the second equation, $y = 9$, and from the first and third equations, $x = 5/9$ and $z = 3/9 = 1/3$.

9. Note that X must be a 2×2 matrix. Let

$$X = \begin{bmatrix} a & b \\ c & d \end{bmatrix}$$

Then

$$\begin{bmatrix} 1 & 4 \\ -2 & 3 \\ 1 & -2 \end{bmatrix} \begin{bmatrix} a & b \\ c & d \end{bmatrix} \begin{bmatrix} 2 & 0 & 0 \\ 0 & 1 & -1 \end{bmatrix} = \begin{bmatrix} 8 & 6 & -6 \\ 6 & -1 & 1 \\ -4 & 0 & 0 \end{bmatrix}$$

or

$$\begin{bmatrix} 1 & 4 \\ -2 & 3 \\ 1 & -2 \end{bmatrix} \begin{bmatrix} 2a & b & -b \\ 2c & d & -d \end{bmatrix} = \begin{bmatrix} 8 & 6 & -6 \\ 6 & -1 & 1 \\ -4 & 0 & 0 \end{bmatrix}$$

or

$$\begin{bmatrix} 2a+8c & b+4d & -b-4d \\ -4a+6c & -2b+3d & 2b-3d \\ 2a-4c & b-2d & -b+2d \end{bmatrix} = \begin{bmatrix} 8 & 6 & -6 \\ 6 & -1 & 1 \\ -4 & 0 & 0 \end{bmatrix}$$

Thus

$$\begin{aligned} 2a \quad\quad + 8c \quad\quad &= 8 \\ b \quad\quad + 4d &= 6 \\ -4a \quad\quad + 6c \quad\quad &= 6 \\ -2b \quad\quad + 3d &= -1 \\ 2a \quad\quad - 4c \quad\quad &= -4 \\ b \quad\quad - 2d &= 0 \end{aligned}$$

Note that we have omitted the 3 equations obtained by equating elements of the last columns of these matrices because the information so obtained would be just a repeat of that gained by equating elements of the second columns. The augmented matrix of the above system is

$$\begin{bmatrix} 2 & 0 & 8 & 0 & 8 \\ 0 & 1 & 0 & 4 & 6 \\ -4 & 0 & 6 & 0 & 6 \\ 0 & -2 & 0 & 3 & -1 \\ 2 & 0 & -4 & 0 & -4 \\ 0 & 1 & 0 & -2 & 0 \end{bmatrix}$$

The reduced row-echelon form of this matrix is

$$\begin{bmatrix} 1 & 0 & 0 & 0 & 0 \\ 0 & 1 & 0 & 0 & 2 \\ 0 & 0 & 1 & 0 & 1 \\ 0 & 0 & 0 & 1 & 1 \\ 0 & 0 & 0 & 0 & 0 \\ 0 & 0 & 0 & 0 & 0 \end{bmatrix}$$

Thus $a = 0$, $b = 2$, $c = 1$, and $d = 1$.

10. If we substitute the given values for x, y, and z into the system of equations, we obtain

$$\begin{aligned} a - b - 6 &= -3 \\ -2 + b + 2c &= -1 \\ a - 3 - 2c &= -3 \end{aligned}$$

or

$$\begin{aligned} a - b &= 3 \\ b + 2c &= 1 \\ a - 2c &= 0 \end{aligned}$$

This system of equations in a, b, and c yields the matrix

$$\begin{bmatrix} 1 & -1 & 0 & 3 \\ 0 & 1 & 2 & 1 \\ 1 & 0 & -2 & 0 \end{bmatrix}$$

which reduces to

$$\begin{bmatrix} 1 & 0 & 0 & 2 \\ 0 & 1 & 0 & -1 \\ 0 & 0 & 1 & 1 \end{bmatrix}$$

Hence, $a = 2$, $b = -1$, and $c = 1$.

11. The matrix X in Part (a) must be 2×3 for the operations to make sense. The matrices in Parts (b) and (c) must be 2×2.

(b) Let $X = \begin{bmatrix} x & y \\ z & w \end{bmatrix}$. Then

$$X \begin{bmatrix} 1 & -1 & 2 \\ 3 & 0 & 1 \end{bmatrix} = \begin{bmatrix} x+3y & -x & 2x+y \\ z+3w & -z & 2z+w \end{bmatrix}$$

If we equate matrix entries, this gives us the equations

$$\begin{aligned} x + 3y &= -5 \\ -x &= -1 \\ 2x + y &= 0 \end{aligned} \qquad\qquad \begin{aligned} z + 3w &= 6 \\ -z &= -3 \\ 2z + w &= 7 \end{aligned}$$

Thus $x = 1$ and $z = 3$, so that the top two equations give $y = -2$ and $w = 1$. Since these values are consistent with the bottom two equations, we have that

$$X = \begin{bmatrix} 1 & -2 \\ 3 & 1 \end{bmatrix}$$

11. (c) As above, let $X = \begin{bmatrix} x & y \\ z & w \end{bmatrix}$, so that the matrix equation becomes

$$\begin{bmatrix} 3x+z & 3y+w \\ -x+2z & -y+2w \end{bmatrix} - \begin{bmatrix} x+2y & 4x \\ z+2w & 4z \end{bmatrix} = \begin{bmatrix} 2 & -2 \\ 5 & 4 \end{bmatrix}$$

This yields the system of equations

$$\begin{aligned} 2x - 2y + z \phantom{{}- 2w} &= 2 \\ -4x + 3y \phantom{{}+ z} + w &= -2 \\ -x + z - 2w &= 5 \\ -y - 4z + 2w &= 4 \end{aligned}$$

with matrix

$$\begin{bmatrix} 2 & -2 & 1 & 0 & 2 \\ -4 & 3 & 0 & 1 & -2 \\ -1 & 0 & 1 & -2 & 5 \\ 0 & -1 & -4 & 2 & 4 \end{bmatrix}$$

which reduces to

$$\begin{bmatrix} 1 & 0 & 0 & 0 & -113/37 \\ 0 & 1 & 0 & 0 & -160/37 \\ 0 & 0 & 1 & 0 & -20/37 \\ 0 & 0 & 0 & 1 & -46/37 \end{bmatrix}$$

Hence, $x = -113/37$, $y = -160/37$, $z = -20/37$, and $w = -46/37$.

12. **(a)** By inspection

$$\begin{bmatrix} y_1 \\ y_2 \\ y_3 \end{bmatrix} = \begin{bmatrix} 1 & -1 & 1 \\ 3 & 1 & -4 \\ -2 & -2 & 3 \end{bmatrix} \begin{bmatrix} x_1 \\ x_2 \\ x_3 \end{bmatrix}$$

and

$$\begin{bmatrix} z_1 \\ z_2 \end{bmatrix} = \begin{bmatrix} 4 & -1 & 1 \\ -3 & 5 & -1 \end{bmatrix} \begin{bmatrix} y_1 \\ y_2 \\ y_3 \end{bmatrix}$$

Thus,

$$\begin{bmatrix} z_1 \\ z_2 \end{bmatrix} = \begin{bmatrix} 4 & -1 & 1 \\ -3 & 5 & -1 \end{bmatrix} \begin{bmatrix} 1 & -1 & 1 \\ 3 & 1 & -4 \\ -2 & -2 & 3 \end{bmatrix} \begin{bmatrix} x_1 \\ x_2 \\ x_3 \end{bmatrix}$$

$$= \begin{bmatrix} -1 & -7 & 11 \\ 14 & 10 & -26 \end{bmatrix} \begin{bmatrix} x_1 \\ x_2 \\ x_3 \end{bmatrix}$$

14. **(a)** If $A^4 = 0$, then

$$(I - A)(I + A + A^2 + A^3) = I + A + A^2 + A^3 - A - A^2 - A^3 - 0 = I$$

Therefore, $(I - A)^{-1} = I + A + A^2 + A^3$ by Theorem 1.7.2.

(b) If $A^{n+1} = 0$, then

$$(I - A)(I + A + A^2 + \cdots + A^n) = I - A^{n+1} = I + 0 = I$$

Thus Theorem 1.7.2 again supplies the desired result.

15. Since the coordinates of the given points must satisfy the polynomial, we have

$$\begin{aligned} p(1) &= 2 \implies a + b + c = 2 \\ p(-1) &= 6 \implies a - b + c = 6 \\ p(2) &= 3 \implies 4a + 2b + c = 3 \end{aligned}$$

The reduced row-echelon form of the augmented matrix of this system of equations is

$$\begin{bmatrix} 1 & 0 & 0 & 1 \\ 0 & 1 & 0 & -2 \\ 0 & 0 & 1 & 3 \end{bmatrix}$$

Thus, $a = 1$, $b = -2$, and $c = 3$.

16. If $p(x) = ax^2 + bx + c$, then $p'(x) = 2ax + b$. Hence,

$$\begin{aligned} p(-1) &= 0 \implies a - b + c = 0 \\ p(2) &= -9 \implies 4a + 2b + c = -9 \\ p'(2) &= 0 \implies 4a + b = 0 \end{aligned}$$

The augmented matrix of this system reduces to

$$\begin{bmatrix} 1 & 0 & 0 & 1 \\ 0 & 1 & 0 & -4 \\ 0 & 0 & 1 & -5 \end{bmatrix}$$

so that $a = 1$, $b = -4$, and $c = -5$.

17. We must show that $(I - J_n)\left(I - \frac{1}{n-1} J_n\right) = I$ or that $\left(I - \frac{1}{n-1} J_n\right)\left(I - J_n\right) = I$. (By virtue of Theorem 1.7.2, we need only demonstrate one of these equalities.) We have

$$(I - J_n)\left(I - \frac{1}{n-1} J_n\right) = I^2 - \frac{1}{n-1} IJ_n - J_nI + \frac{1}{n-1} J_n^2$$

$$= I - \frac{n}{n-1} J_n + \frac{1}{n-1} J_n^2$$

But $J_n^2 = nJ_n$ (think about actually squaring J_n), so that the right-hand side of the above equation is just I, as desired.

18. If $A^3 + 4A^2 - 2A + 7I = 0$, then $(A^3 + 4A^2 - 2A + 7I)^t = 0$ as well. Now repeated applications of Theorem 1.5.8 will finish the problem.

19. First suppose that $AB^{-1} = B^{-1}A$. Note that all matrices must be square and of the same size. Therefore

$$(AB^{-1})B = (B^{-1}A)B$$

or

$$A = B^{-1}AB$$

so that

$$BA = B(B^{-1}AB) = (BB^{-1})(AB) = AB$$

It remains to show that if $AB = BA$ then $AB^{-1} = B^{-1}A$. An argument similar to the one given above will serve, and we leave the details to you.

20. Suppose that A and $A + B$ are both invertible. Then there is a matrix C such that $(A + B)C = I$. We must show that $I + BA^{-1}$ is also invertible. Consider the matrix

$$\begin{aligned}(I + BA^{-1})(AC) &= AC + (BA^{-1})(AC)\\ &= AC + BIC\\ &= AC + BC\\ &= (A + B)C\\ &= I\end{aligned}$$

Thus AC is the inverse of $I + BA^{-1}$.

Now suppose that A is invertible but $A + B$ is not. We must show that $(I + BA^{-1})$ is also not invertible. Suppose that $(I + BA^{-1})D = I$ for some matrix D. Then

$$\begin{aligned}(A + B)A^{-1}D &= AA^{-1}D + BA^{-1}D \\ &= D + BA^{-1}D \\ &= (I + BA^{-1})D \\ &= I\end{aligned}$$

This implies that $A + B$ is invertible. Since this is contrary to our hypothesis, $I + BA^{-1}$ is not invertible.

21. **(b)** Let the ij^{th} entry of A be a_{ij}. Then $\text{tr}(A) = a_{11} + a_{22} + \cdots + a_{nn}$, so that

$$\begin{aligned}\text{tr}(kA) &= ka_{11} + ka_{22} + \cdots + ka_{nn} \\ &= k(a_{11} + a_{22} + \cdots + a_{nn}) \\ &= k\text{tr}(A)\end{aligned}$$

(d) Let the ij^{th} entries of A and B be a_{ij} and b_{ij}, respectively. Then

$$\begin{aligned}\text{tr}(AB) = {} & a_{11}b_{11} + a_{12}b_{21} + \cdots + a_{1n}b_{n1} \\ & + a_{21}b_{12} + a_{22}b_{22} + \cdots + a_{2n}b_{n2} \\ & + \cdots \\ & + a_{n1}b_{1n} + a_{n2}b_{2n} + \cdots + a_{nn}b_{nn}\end{aligned}$$

and

$$\begin{aligned}\text{tr}(BA) = {} & b_{11}a_{11} + b_{12}a_{21} + \cdots + b_{1n}a_{n1} \\ & + b_{21}a_{12} + b_{22}a_{22} + \cdots + b_{2n}a_{n2} \\ & + \cdots \\ & + b_{n1}a_{1n} + b_{n2}a_{2n} + \cdots + b_{nn}a_{nn}\end{aligned}$$

If we rewrite each of the terms $b_{ij}a_{ji}$ in the above expression as $a_{ji}b_{ij}$ and list the terms in the order indicated by the arrows below,

$$\begin{aligned}
\operatorname{tr}(BA) = {} & a_{11}b_{11} + a_{21}b_{12} + \cdots + a_{n1}b_{1n} \\
& + a_{12}b_{21} + a_{22}b_{22} + \cdots + a_{n2}b_{2n} \\
& + \cdots \\
& + a_{1n}b_{n1} + a_{2n}b_{n2} + \cdots + a_{nn}b_{nn}
\end{aligned}$$

then we have $\operatorname{tr}(AB) = \operatorname{tr}(BA)$.

22. Suppose that A and B are square matrices such that $AB - BA = I$. Then

$$\operatorname{tr}(AB - BA) = \operatorname{tr}(AB) - \operatorname{tr}(BA) = \operatorname{tr}(I).$$

But $\operatorname{tr}(I) = n$ for any $n \times n$ identity matrix I and $\operatorname{tr}(AB) - \operatorname{tr}(BA) = 0$ by Part (d) of Problem 21.

24. **(c)** Let $A = [a_{ij}(x)]$ be an $m \times n$ matrix and let $B = [b_{ij}(x)]$ be an $n \times s$ matrix. Then the ij^{th} entry of AB is

$$a_{i1}(x)b_{1j}(x) + \cdots + a_{in}(x)b_{nj}(x)$$

Thus the ij^{th} entry of $\frac{d}{dx}(AB)$ is

$$\begin{aligned}
&\frac{d}{dx}\left[a_{i1}(x)b_{1j}(x) + \cdots + a_{in}(x)b_{nj}(x)\right] \\
&= a'_{i1}(x)b_{1j}(x) + a_{i1}(x)b'_{1j}(x) + \cdots \\
&\qquad + a'_{in}(x)b_{nj}(x) + a_{in}(x)b'_{nj}(x) \\
&= [a'_{i1}(x)b_{1j}(x) + \cdots + a'_{in}(x)b_{nj}(x)] \\
&\qquad + [a_{i1}(x)b'_{1j}(x) + \cdots + a_{in}(x)b'_{nj}(x)]
\end{aligned}$$

This is just the ij^{th} entry of $\frac{dA}{dx}B + A\frac{dB}{dx}$.

25. Suppose that A is a square matrix whose entries are differentiable functions of x. Suppose also that A has an inverse, A^{-1}. Then we shall show that A^{-1} also has entries which are differentiable functions of x and that

$$\frac{dA^{-1}}{dx} = -A^{-1}\,\frac{dA}{dx}\,A^{-1}$$

Since we can find A^{-1} by the method used in Chapter 1, its entries are functions of x which are obtained from the entries of A by using only addition together with multiplication and division by constants or entries of A. Since sums, products, and quotients of differentiable functions are differentiable wherever they are defined, the resulting entries in the inverse will be differentiable functions except, perhaps, for values of x where their denominators are zero. (Note that we never have to divide by a function which is identically zero.) That is, the entries of A^{-1} are differentiable wherever they are defined. But since we are assuming that A^{-1} is defined, its entries must be differentiable. Moreover,

$$\frac{d}{dx}\,(AA^{-1}) = \frac{d}{dx}\,(I) = 0$$

or

$$\frac{dA}{dx}\,A^{-1} + A\,\frac{dA^{-1}}{dx} = 0$$

Therefore

$$A\,\frac{dA^{-1}}{dx} = -\,\frac{dA}{dx}\,A^{-1}$$

so that

$$\frac{dA^{-1}}{dx} = -A^{-1}\,\frac{dA}{dx}\,A^{-1}$$

26. Following the hint, we obtain the equation

$$x^2 + x - 2 = A(x^2 + 1) + (Bx + C)(3x - 1)$$

or

$$x^2 + x - 2 = (A + 3B)x^2 + (3C - B)x + (A - C)$$

If we equate coefficients of like powers of x, we obtain the system of equations

$$\begin{aligned} A + 3B &= 1 \\ - B + 3C &= 1 \\ A - C &= -2 \end{aligned}$$

the solution to this system is

$$A = -7/5, \qquad B = 4/5, \qquad C = 3/5$$

27. **(b)** Let H be a Householder matrix, so that $H = I - 2PP^t$ where P is an $n \times 1$ matrix. Then using Theorem 1.5.8,

$$\begin{aligned} H^t &= (I - 2PP^t)^t \\ &= I^t - (2PP^t)^t \\ &= I - 2(P^t)^t P^t \\ &= I - 2PP^t \\ &= H \end{aligned}$$

and (using Theorem 1.5.1)

$$\begin{aligned} H^tH &= H^2 && \text{(by the above result)} \\ &= (I - 2PP^t)^2 \\ &= I^2 - 2PP^t - 2PP^t + (-2PP^t)^2 \\ &= I - 4PP^t + 4PP^tPP^t \\ &= I - 4PP^t + 4PP^t && \text{(because } P^tP = 1\text{)} \\ &= I \end{aligned}$$

28. **(a)** We are asked to show that $(C^{-1} + D^{-1})^{-1} = C(C + D)^{-1}D$. Since all the matrices in sight are invertible and the inverse of a matrix is unique, it will suffice to show that the inverses of each side are equal, or that $(C(C + D)^{-1}D)^{-1} = C^{-1} + D^{-1}$. We have

$$\begin{aligned}(C(C + D)^{-1}D)^{-1} &= D^{-1}(C + D)C^{-1} \\ &= D^{-1}CC^{-1} + D^{-1}DC^{-1} \\ &= D^{-1} + C^{-1} \\ &= C^{-1} + D^{-1}\end{aligned}$$

(b) Since we don't know whether or not C is invertible, we can't use the technique of Part (a). Instead, we observe that

$$(*) \qquad (I + CD)C = C(I + DC)$$

since each side of (*) is just $C + CDC$. If we multiply both sides of (*) on the left by $(I + CD)^{-1}$ and on the right by $(I + DC)^{-1}$, we obtain the desired result.

(c) Start with $D + DD^tC^{-1}D$, write it in two different ways, and proceed more or less as in Part (b).

EXERCISE SET 2.1

1. **(a)** The number of inversions in (4,1,3,5,2) is 3 + 0 + 1 + 1 = 5.

   (d) The number of inversions in (5,4,3,2,1) is 4 + 3 + 2 + 1 = 10.

2. **(a)** The permutation is odd because 5 is odd.

   (d) The permutation is even because 10 is even.

3. $\begin{vmatrix} 3 & 5 \\ -2 & 4 \end{vmatrix} = 12 - (-10) = 22$

5. $\begin{vmatrix} -5 & 6 \\ -7 & -2 \end{vmatrix} = (-5)(-2) - (-7)(6) = 52$

7. $\begin{vmatrix} a-3 & 5 \\ -3 & a-2 \end{vmatrix} = (a - 3)(a - 2) - (-3)(5) = a^2 - 5a + 21$

9. $\begin{vmatrix} -2 & 1 & 4 \\ 3 & 5 & -7 \\ 1 & 6 & 2 \end{vmatrix} = (-20 - 7 + 72) - (20 + 84 + 6) = -65$

11. $\begin{vmatrix} 3 & 0 & 0 \\ 2 & -1 & 5 \\ 1 & 9 & -4 \end{vmatrix} = (12 + 0 + 0) - (0 + 135 + 0) = -123$

13. **(a)**

$$\det(A) = \begin{vmatrix} \lambda-2 & 1 \\ -5 & \lambda+4 \end{vmatrix} = (\lambda - 2)(\lambda + 4) + 5$$

$$= \lambda^2 + 2\lambda - 3 = (\lambda - 1)(\lambda + 3)$$

Hence, $\det(A) = 0$ if and only if $\lambda = 1$ or $\lambda = -3$.

14. The following permutations are even:

(1,2,3,4)	(1,3,4,2)	(1,4,2,3)
(2,1,4,3)	(2,3,1,4)	(2,4,3,1)
(3,1,2,4)	(3,2,4,1)	(3,4,1,2)
(4,1,3,2)	(4,2,1,3)	(4,3,2,1)

The following permutations are odd:

(1,2,4,3)	(1,3,2,4)	(1,4,3,2)
(2,1,3,4)	(2,3,4,1)	(2,4,1,3)
(3,1,4,2)	(3,2,1,4)	(3,4,2,1)
(4,1,2,3)	(4,2,3,1)	(4,3,1,2)

15. If A is a 4×4 matrix, then

$$\det(A) = \sum (-1)^p a_{1i_1} a_{2i_2} a_{3i_3} a_{4i_4}$$

where $p = 1$ if (i_1, i_2, i_3, i_4) is an odd permutation of $\{1,2,3,4\}$ and $p = 2$ otherwise. There are 24 terms in this sum.

17. **(a)** The only nonzero product in the expansion of the determinant is

$$a_{15}a_{24}a_{33}a_{42}a_{51} = (-3)(-4)(-1)(2)(5) = -120$$

Since (5,4,3,2,1) is even, $\det(A) = -120$.

(b) The only nonzero product in the expansion of the determinant is

$$a_{11}a_{25}a_{33}a_{44}a_{52} = (5)(-4)(3)(1)(-2) = 120$$

Since (1,5,3,4,2) is odd, $\det(A) = -120$.

18. We have

$$\begin{vmatrix} x & -1 \\ 3 & 1-x \end{vmatrix} = x(1 - x) - 3(-1) = 3 + x - x^2$$

and

$$\begin{vmatrix} 1 & 0 & -3 \\ 2 & x & -6 \\ 1 & 3 & x-5 \end{vmatrix} = x(x - 5) + 0 + (-3)(2)(3) - (-3x + (-6)(3) + 0) = x^2 - 2x$$

Thus $x^2 - 2x = 3 + x - x^2$ or $2x^2 - 3x - 3 = 0$. By the quadratic formula, $x = (3 \pm \sqrt{33})/4$.

19. The value of the determinant is

$$\sin^2\theta - (-\cos^2\theta) = \sin^2\theta + \cos^2\theta = 1$$

The identity $\sin^2\theta + \cos^2\theta = 1$ holds for all values of θ.

20. Since each elementary product in the expansion of the determinant contains a factor from each column, each elementary product must contain a factor from the column of zeros. Thus, each signed elementary product is zero and $\det(A) = 0$.

EXERCISE SET 2.2

1. **(a)** Since the matrix is upper triangular, its determinant equals the product of the diagonal elements. Hence, if we call the matrix A, then $\det(A) = -30$.

   (c) We replace Row 3 by Row 3 plus −1 times Row 1. By Theorem 2.2.3(c), the determinants of the two matrices are equal; that is

$$\det(A) = \begin{vmatrix} -2 & 1 & 3 \\ 1 & -7 & 4 \\ 0 & 0 & 0 \end{vmatrix}$$

   But the above matrix contains a row of zeros; hence, its determinant is zero. Thus $\det(A) = 0$.

   Alternatively, we may utilize the remark which appears in italics immediately above Example 8. The first and third rows of the given matrix are proportional – in fact, they are equal – and hence $\det(A) = 0$.

3. $$\det(A) = \begin{vmatrix} 0 & 3 & 1 \\ 1 & 1 & 2 \\ 3 & 2 & 4 \end{vmatrix} = (-1)\begin{vmatrix} 1 & 1 & 2 \\ 0 & 3 & 1 \\ 3 & 2 & 4 \end{vmatrix}$$ Interchange Row 1 and Row 2.

$$= (-1)\begin{vmatrix} 1 & 1 & 2 \\ 0 & 3 & 1 \\ 0 & -1 & -2 \end{vmatrix}$$ Add −3 times Row 1 to Row 3.

$$= (-1)(3)\begin{vmatrix} 1 & 1 & 2 \\ 0 & 1 & 1/3 \\ 0 & -1 & -2 \end{vmatrix}$$

Factor 3 from Row 2.

$$= -3\begin{vmatrix} 1 & 1 & 2 \\ 0 & 1 & 1/3 \\ 0 & 0 & -5/3 \end{vmatrix}$$

Add Row 2 to Row 3.

If we factor −5/3 from Row 3 and apply Theorem 2.2.2 we find that

$$\det(A) = -3(-5/3)(1) = 5$$

5.

$$\det(A) = \begin{vmatrix} 3 & -6 & 9 \\ -2 & 7 & -2 \\ 0 & 1 & 5 \end{vmatrix} = 3\begin{vmatrix} 1 & -2 & 3 \\ 0 & 3 & 4 \\ 0 & 1 & 5 \end{vmatrix}$$

Factor 3 from Row 1 and Add twice Row 1 to Row 2.

$$= (3)(3)\begin{vmatrix} 1 & -2 & 3 \\ 0 & 1 & 4/3 \\ 0 & 0 & 11/3 \end{vmatrix}$$

Factor 3 from Row 2 and subtract Row 2 from Row 3.

$$= 9\left[\frac{11}{3}\right]\begin{vmatrix} 1 & -2 & 3 \\ 0 & 1 & 4/3 \\ 0 & 0 & 1 \end{vmatrix}$$

Factor 11/3 from Row 3.

$$= 9(11/3)(1) = 33$$

7.

$$\det(A) = \begin{vmatrix} 2 & 1 & 3 & 1 \\ 1 & 0 & 1 & 1 \\ 0 & 2 & 1 & 0 \\ 0 & 1 & 2 & 3 \end{vmatrix} = (-1)\begin{vmatrix} 1 & 0 & 1 & 1 \\ 2 & 1 & 3 & 1 \\ 0 & 2 & 1 & 0 \\ 0 & 1 & 2 & 3 \end{vmatrix}$$

Interchange Row 1 and Row 2.

$$= (-1)\begin{vmatrix} 1 & 0 & 1 & 1 \\ 0 & 1 & 1 & -1 \\ 0 & 2 & 1 & 0 \\ 0 & 1 & 2 & 3 \end{vmatrix}$$

Add −2 times Row 1 to Row 2.

$$= (-1)\begin{vmatrix} 1 & 0 & 1 & 1 \\ 0 & 1 & 1 & -1 \\ 0 & 0 & -1 & 2 \\ 0 & 0 & 1 & 4 \end{vmatrix}$$

Add −2 times Row 2 to Row 3; subtract Row 2 from Row 4.

$$= (-1)\begin{vmatrix} 1 & 0 & 1 & 1 \\ 0 & 1 & 1 & -1 \\ 0 & 0 & -1 & 2 \\ 0 & 0 & 0 & 6 \end{vmatrix}$$

Add Row 3 to Row 4.

$$= (-1)(-1)(6)(1) = 6$$

9.

$$\det(A) = \begin{vmatrix} 1 & 3 & 1 & 5 & 3 \\ -2 & -7 & 0 & -4 & 2 \\ 0 & 0 & 1 & 0 & 1 \\ 0 & 0 & 2 & 1 & 1 \\ 0 & 0 & 0 & 1 & 1 \end{vmatrix}$$

$$= \begin{vmatrix} 1 & 3 & 1 & 5 & 3 \\ 0 & -1 & 2 & 6 & 8 \\ 0 & 0 & 1 & 0 & 1 \\ 0 & 0 & 0 & 1 & -1 \\ 0 & 0 & 0 & 1 & 1 \end{vmatrix}$$

Add 2 times Row 1 to Row 2; add −2 times Row 2 times Row 3 to Row 4

$$= \begin{vmatrix} 1 & 3 & 1 & 5 & 3 \\ 0 & -1 & 2 & 6 & 8 \\ 0 & 0 & 1 & 0 & 1 \\ 0 & 0 & 0 & 1 & -1 \\ 0 & 0 & 0 & 0 & 2 \end{vmatrix}$$

Add -1 times Row 4 to Row 5.

Hence, $\det(A) = (-1)(2)(1) = -2$.

10. Let

$$A = \begin{bmatrix} a & b & c \\ d & e & f \\ g & h & i \end{bmatrix}$$

We are given that $\det(A) = -6$.

(a)

$$\det(B) = \begin{vmatrix} d & e & f \\ g & h & i \\ a & b & c \end{vmatrix} = (-1)\begin{vmatrix} d & e & f \\ a & b & c \\ g & h & i \end{vmatrix}$$

Interchange Row 2 and Row 3.

$$= (-1)^2 \begin{vmatrix} a & b & c \\ d & e & f \\ g & h & i \end{vmatrix}$$

Interchange Row 1 and Row 2.

$$= \det(A) = -6$$

(b)

$$\det(B) = \begin{vmatrix} 3a & 3b & 3c \\ -d & -e & -f \\ 4g & 4h & 4i \end{vmatrix}$$

$$= (3)(-1)(4)\begin{vmatrix} a & b & c \\ d & e & f \\ g & h & i \end{vmatrix}$$

Factor 3 from Row 1, -1 from Row 2, and 4 from Row 3.

$$= -12 \det(A) = 72$$

(c) Let

$$D = \begin{bmatrix} a+g & b+h & c+i \\ d & e & f \\ g & h & i \end{bmatrix}$$

Then $\det(D) = \det(A) = -6$ because the matrix D may be obtained from A by replacing the first row by the sum of the first and third rows.

(d) Let

$$E = \begin{bmatrix} -3a & -3b & -3c \\ d & e & f \\ g-4d & h-4e & i-4f \end{bmatrix}$$

If we factor -3 from Row 1, we obtain

$$\det(E) = (-3) \begin{vmatrix} a & b & c \\ d & e & f \\ g-4d & h-4e & i-4f \end{vmatrix}$$

But the above matrix can be obtained from A by adding -4 times Row 2 to Row 3. Hence $\det(E) = (-3)\det(A) = 18$.

11.

$$\det(A) = \begin{vmatrix} 1 & 1 & 1 \\ a & b & c \\ a^2 & b^2 & c^2 \end{vmatrix}$$

$$= \begin{vmatrix} 1 & 1 & 1 \\ 0 & b-a & c-a \\ 0 & b^2-a^2 & c^2-a^2 \end{vmatrix}$$

Add $-a$ times Row 1 to Row 2; add $-a^2$ times Row 1 to Row 3.

Since $b^2 - a^2 = (b - a)(b + a)$, we add $-(b + a)$ times Row 2 to Row 3 to obtain

$$\det(A) = \begin{vmatrix} 1 & 1 & 1 \\ 0 & b-a & c-a \\ 0 & 0 & (c^2-a^2) - (c-a)(b+a) \end{vmatrix}$$

$$= (b - a)[(c^2 - a^2) - (c - a)(b + a)]$$

$$= (b - a)(c - a)[(c + a) - (b + a)]$$

$$= (b - a)(c - a)(c - b)$$

12. **(a)** Consider the elementary product $P = a_{1j_1}a_{2j_2}a_{3j_3}$. If $j_1 = 1$ or $j_1 = 2$, then $P = 0$ because $a_{11} = a_{12} = 0$. If $j_1 = 3$, then the only choices for j_2 are $j_2 = 1$ or $j_2 = 2$. But $a_{21} = 0$, so that $P = 0$ unless $j_2 = 2$. Finally, if $j_1 = 3$ and $j_2 = 2$, then the only choice for j_3 is $j_3 = 1$. Therefore, $a_{13}a_{22}a_{31}$ is the only elementary product which can be nonzero. Since the permutation (3,2,1) is odd, then

$$\det(A) = -a_{13}a_{22}a_{31}$$

(b) By an argument which is completely analogous to that given in Part (a), we can argue that $a_{14}a_{23}a_{32}a_{41}$ is the only nonzero elementary product. Since (4,3,2,1) is an even permutation, then

$$\det(A) = a_{14}a_{23}a_{32}a_{41}$$

13. Every elementary product contains a factor from every column. Thus if one column consists entirely of zeros, then every elementary product will be zero, and hence the determinant will be zero.

14. Each row in the row-echelon form of a square matrix is either all zeros or a string of zeros followed by a leading 1 and then, perhaps, some other numbers. Since the 1 cannot occur to the left of the main diagonal, clearly all entries below this diagonal are zero. Hence, the matrix is upper triangular.

15. In each case, D will denote the determinant on the left and, as usual, $\det(A) = \sum \pm a_{1j_1}a_{2j_2}a_{3j_3}$.

(a) $D = \sum \pm \left[ka_{1j_1}\right] a_{2j_2}a_{3j_3} = k \sum \pm a_{1j_1}a_{2j_2}a_{3j_3} = k\det(A)$

(b) $D = \sum \pm a_{2j_1}a_{1j_2}a_{3j_3} = \sum \pm a_{1j_2}a_{2j_1}a_{3j_3}$

The sign of a given elementary product on the left depends upon whether (j_1, j_2, j_3) is even or odd; the sign of the same elementary product appearing on the right depends upon whether (j_2, j_1, j_3) is even or odd. By counting the number of inversions in each of the six permutations of {1,2,3}, it can be verified that (j_1, j_2, j_3) is even if and only if (j_2, j_1, j_3) is odd. Therefore every elementary product in A appears in D with opposite sign, and $D = -\det(A)$.

(c) $D = \sum \pm \left[a_{1j_1} + ka_{2j_1}\right] a_{2j_2}a_{3j_3}$

$= \sum \pm a_{1j_1}a_{2j_2}a_{3j_3} + k \sum \pm a_{2j_1}a_{2j_2}a_{3j_3}$

$= \det(A) + kD_1$

where D_1 is a determinant with identical first and second rows. Thus $D_1 = 0$ and $D = \det(A)$.

EXERCISE SET 2.3

2. $\det(A) = \begin{vmatrix} 2 & 1 & 0 \\ 3 & 4 & 0 \\ 0 & 0 & 2 \end{vmatrix} = (2)\begin{vmatrix} 1 & 1/2 & 0 \\ 0 & 5/2 & 0 \\ 0 & 0 & 2 \end{vmatrix} = 10$

$$\det(B) = \begin{vmatrix} 1 & -1 & 3 \\ 7 & 1 & 2 \\ 5 & 0 & 1 \end{vmatrix} = \begin{vmatrix} -14 & -1 & 0 \\ -3 & 1 & 0 \\ 5 & 0 & 1 \end{vmatrix}$$

$$= \begin{vmatrix} -17 & 0 & 0 \\ -3 & 1 & 0 \\ 5 & 0 & 1 \end{vmatrix} = -17$$

Thus $\det(A) = 10$ and $\det(B) = -17$. Now by direct computation,

$$AB = \begin{bmatrix} 9 & -1 & 8 \\ 31 & 1 & 17 \\ 10 & 0 & 2 \end{bmatrix}$$

If we add 9 times Column 2 to Column 1, we find that

$$\det(AB) = \begin{vmatrix} 0 & -1 & 8 \\ 40 & 1 & 17 \\ 10 & 0 & 2 \end{vmatrix} = (10)\begin{vmatrix} 0 & -1 & 8 \\ 4 & 1 & 17 \\ 1 & 0 & 2 \end{vmatrix}$$

$$= (10)\begin{vmatrix} 0 & -1 & 8 \\ 0 & 1 & 9 \\ 1 & 0 & 2 \end{vmatrix} = -(10)\begin{vmatrix} 1 & 0 & 2 \\ 0 & 1 & 9 \\ 0 & 0 & 17 \end{vmatrix}$$

$$= -170 = \det(A)\det(B)$$

4. **(a)** $$\begin{vmatrix} 1 & 0 & -1 \\ 9 & -1 & 4 \\ 8 & 9 & -1 \end{vmatrix} = \begin{vmatrix} 1 & 0 & 0 \\ 9 & -1 & 13 \\ 8 & 9 & 7 \end{vmatrix} = \begin{vmatrix} 1 & 0 & 0 \\ 9 & -1 & 0 \\ 8 & 9 & 124 \end{vmatrix} = -124$$

Since the determinant of this matrix is not zero, the matrix is invertible.

(c) Since the first two rows of this matrix are proportional, its determinant is zero. Hence, it is not invertible.

5. **(a)** By Equation (1),

$$\det(3A) = 3^3 \det(A) = (27)(-7) = -189$$

(b) Again, by Equation (1), $\det(2A^{-1}) = 2^3 \det(A^{-1})$. By Corollary 2.3.5, we have

$$\det(2A^{-1}) = \frac{8}{\det(A)} = -\frac{8}{7}$$

(c) Again, by Equation (1), $\det(2A) = 2^3 \det(A) = -56$. By Corollary 2.3.5, we have

$$\det[(2A)^{-1}] = \frac{1}{\det(2A)} = -\frac{1}{56}$$

(d) $$\begin{vmatrix} a & g & d \\ b & h & e \\ c & i & f \end{vmatrix} = -\begin{vmatrix} a & d & g \\ b & e & h \\ c & f & i \end{vmatrix}$$

Interchange Columns 2 and 3.

$$= -\begin{vmatrix} a & b & c \\ d & e & f \\ g & h & i \end{vmatrix}$$

Take the transpose of the matrix.

$$= 7$$

7. If we replace Row 1 by Row 1 plus Row 2, we obtain

$$\begin{vmatrix} b+c & c+a & b+a \\ a & b & c \\ 1 & 1 & 1 \end{vmatrix} = \begin{vmatrix} a+b+c & b+c+a & c+b+a \\ a & b & c \\ 1 & 1 & 1 \end{vmatrix} = 0$$

because the first and third rows are proportional.

8. If we add -1 times Column 1 to Column 3 and -1 times Column 2 to Column 3 in the determinant on the left, we obtain the determinant on the right.

10. If we add $-t$ times Row 1 to Row 2 in the determinant on the left, we get

$$\begin{vmatrix} a_1+b_1t & a_2+b_2t & a_3+b_3t \\ (1-t^2)b_1 & (1-t^2)b_2 & (1-t^2)b_3 \\ c_1 & c_2 & c_3 \end{vmatrix}$$

or

$$(1-t^2)\begin{vmatrix} a_1+b_1t & a_2+b_2t & a_3+b_3t \\ b_1 & b_2 & b_3 \\ c_1 & c_2 & c_3 \end{vmatrix}$$

Adding $-t$ times Row 2 to Row 1 in the above determinant yields the desired result.

12. **(a)** We have

$$\det(A) = (k-3)(k-2) - 4$$
$$= k^2 - 5k + 2$$

From Theorem 2.3.4, A will be invertible if and only if $\det(A) \neq 0$. But $\det(A) = 0$ if and only if $k = (5 \pm \sqrt{17})/2$. Thus A is invertible unless $k = (5 \pm \sqrt{17})/2$.

13. By adding Row 1 to Row 2 and using the identity $\sin^2 x + \cos^2 x = 1$, we see that the determinant of the given matrix can be written as

$$\begin{vmatrix} \sin^2\alpha & \sin^2\beta & \sin^2\gamma \\ 1 & 1 & 1 \\ 1 & 1 & 1 \end{vmatrix}$$

But this is zero because two of its rows are identical. Therefore the matrix is not invertible.

14. Since A is invertible, $\det(A) \neq 0$. Thus

$$\det(A^{-1}BA) = \det(A^{-1})\det(B)\det(A)$$
$$= \frac{1}{\det(A)}\det(B)\det(A)$$
$$= \det(B)$$

15. **(a)** We have, for instance,

$$\begin{vmatrix} a_1+b_1 & c_1+d_1 \\ a_2+b_2 & c_2+d_2 \end{vmatrix} = \begin{vmatrix} a_1+b_1 & c_1+d_1 \\ a_2 & c_2 \end{vmatrix} + \begin{vmatrix} a_1+b_1 & c_1+d_1 \\ b_2 & d_2 \end{vmatrix}$$
$$= \begin{vmatrix} a_1 & c_1 \\ a_2 & c_2 \end{vmatrix} + \begin{vmatrix} b_1 & d_1 \\ a_2 & c_2 \end{vmatrix} + \begin{vmatrix} a_1 & c_1 \\ b_2 & d_2 \end{vmatrix} + \begin{vmatrix} b_1 & d_1 \\ b_2 & d_2 \end{vmatrix}$$

The answer is clearly not unique.

EXERCISE SET 2.4

1. For example,

$$M_{11} = \begin{vmatrix} 7 & -1 \\ 1 & 4 \end{vmatrix} = 29 \implies C_{11} = 29$$

$$M_{12} = \begin{vmatrix} 6 & -1 \\ -3 & 4 \end{vmatrix} = 21 \implies C_{12} = -21$$

$$M_{22} = \begin{vmatrix} 1 & 3 \\ -3 & 4 \end{vmatrix} = 13 \implies C_{22} = 13$$

2. **(a)**
$$M_{13} = \begin{vmatrix} 0 & 0 & 3 \\ 4 & 1 & 14 \\ 4 & 1 & 2 \end{vmatrix} = 0 \implies C_{13} = 0$$

(b)
$$M_{23} = \begin{vmatrix} 4 & -1 & 6 \\ 4 & 1 & 14 \\ 4 & 1 & 2 \end{vmatrix} = \begin{vmatrix} 8 & -1 & 6 \\ 0 & 1 & 14 \\ 0 & 1 & 2 \end{vmatrix} = 8(-12) = -96 \implies C_{23} = 96$$

3. **(a)**
$$\det(A) = (1)\begin{vmatrix} 7 & -1 \\ 1 & 4 \end{vmatrix} - (-2)\begin{vmatrix} 6 & -1 \\ -3 & 4 \end{vmatrix} + (3)\begin{vmatrix} 6 & 7 \\ -3 & 1 \end{vmatrix}$$

$$= (1)(29) + (2)(21) + (3)(27)$$

$$= 152$$

3. **(b)**

$$\det(A) = (1)\begin{vmatrix} 7 & -1 \\ 1 & 4 \end{vmatrix} - (6)\begin{vmatrix} -2 & 3 \\ 1 & 4 \end{vmatrix} + (-3)\begin{vmatrix} -2 & 3 \\ 7 & -1 \end{vmatrix}$$

$$= (1)(29) - (6)(-11) + (-3)(-19)$$

$$= 152$$

(c)

$$\det(A) = -(6)\begin{vmatrix} -2 & 3 \\ 1 & 4 \end{vmatrix} + (7)\begin{vmatrix} 1 & 3 \\ -3 & 4 \end{vmatrix} - (-1)\begin{vmatrix} 1 & -2 \\ -3 & 1 \end{vmatrix}$$

$$= -(6)(-11) + (7)(13) - (-1)(-5)$$

$$= 152$$

5. Since Column 2 contains 2 zeros, we expand using cofactors of that first column. Hence,

$$\det(A) = (5)\begin{vmatrix} -3 & 7 \\ -1 & 5 \end{vmatrix} = 5(-15 + 7) = -40$$

7. We factor k from the second column to obtain

$$\begin{vmatrix} 1 & k & k^2 \\ 1 & k & k^2 \\ 1 & k & k^2 \end{vmatrix} = k\begin{vmatrix} 1 & 1 & k^2 \\ 1 & 1 & k^2 \\ 1 & 1 & k^2 \end{vmatrix} = 0$$

because Column 1 and Column 2 are identical.

9. Let A denote the given matrix. Note that the third column of A has two zero entries. If we add Row 3 to Row 4, we won't change the value of $\det(A)$, but Column 3 will then have three zero entries. Thus

$$\det(A) = \begin{vmatrix} 3 & 3 & 0 & 5 \\ 2 & 2 & 0 & -2 \\ 4 & 1 & -3 & 0 \\ 6 & 11 & 0 & 2 \end{vmatrix}$$

$$= (-3)\begin{vmatrix} 3 & 3 & 5 \\ 2 & 2 & -2 \\ 6 & 11 & 2 \end{vmatrix}$$

Take cofactors of Column 3.

$$= (-3)\begin{vmatrix} 0 & 3 & 5 \\ 0 & 2 & -2 \\ -5 & 11 & 2 \end{vmatrix}$$

Subtract Column 2 from Column 1.

$$= (-3)(-5)\begin{vmatrix} 3 & 5 \\ 2 & -2 \end{vmatrix}$$

Take cofactors of Column 1.

$$= (-3)(-5)(-16)$$

$$= -240$$

11. The determinant of A is -1. The matrix of cofactors is

$$\begin{bmatrix} -3 & 3 & -2 \\ 5 & -4 & 2 \\ 5 & -5 & 3 \end{bmatrix}$$

Thus

$$-\operatorname{adj}(A) = A^{-1} = \begin{bmatrix} 3 & -5 & -5 \\ -3 & 4 & 5 \\ 2 & -2 & -3 \end{bmatrix}$$

13. The determinant of A is 4. The matrix of cofactors is

$$\begin{bmatrix} 2 & 0 & 0 \\ 6 & 4 & 0 \\ 4 & 6 & 2 \end{bmatrix}$$

Thus

$$\text{adj}(A) = \begin{bmatrix} 2 & 6 & 4 \\ 0 & 4 & 6 \\ 0 & 0 & 2 \end{bmatrix}$$

and

$$A^{-1} = \begin{bmatrix} 1/2 & 3/2 & 1 \\ 0 & 1 & 3/2 \\ 0 & 0 & 1/2 \end{bmatrix}$$

15. **(a)** We must first compute adj(A). By direct calculation, we find that the matrix of cofactors is

$$\begin{bmatrix} -4 & 2 & -7 & 6 \\ 3 & -1 & 0 & 0 \\ 0 & 0 & -1 & 1 \\ -1 & 0 & 8 & -7 \end{bmatrix}$$

Thus

$$\text{adj}(A) = \begin{bmatrix} -4 & 3 & 0 & -1 \\ 2 & -1 & 0 & 0 \\ -7 & 0 & -1 & 8 \\ 6 & 0 & 1 & -7 \end{bmatrix}$$

We must now compute det(A). We replace Row 2 by Row 2 plus (−2) times Row 1 to obtain

$$\det(A) = \begin{vmatrix} 1 & 3 & 1 & 1 \\ 0 & -1 & 0 & 0 \\ 1 & 3 & 8 & 9 \\ 1 & 3 & 2 & 2 \end{vmatrix}$$

$$= (-1)\begin{vmatrix} 1 & 1 & 1 \\ 1 & 8 & 9 \\ 1 & 2 & 2 \end{vmatrix} = (-1)(-1) = 1$$

Thus, $A^{-1} = \text{adj}(A)$.

(b)

$$\left[\begin{array}{cccc|cccc} 1 & 3 & 1 & 1 & 1 & 0 & 0 & 0 \\ 2 & 5 & 2 & 2 & 0 & 1 & 0 & 0 \\ 1 & 3 & 8 & 9 & 0 & 0 & 1 & 0 \\ 1 & 3 & 2 & 2 & 0 & 0 & 0 & 1 \end{array}\right]$$

$$\downarrow$$

$$\left[\begin{array}{cccc|cccc} 1 & 3 & 1 & 1 & 1 & 0 & 0 & 0 \\ 0 & -1 & 0 & 0 & -2 & 1 & 0 & 0 \\ 0 & 0 & 7 & 8 & -1 & 0 & 1 & 0 \\ 0 & 0 & 1 & 1 & -1 & 0 & 0 & 1 \end{array}\right]$$

$$\downarrow$$

$$\left[\begin{array}{cccc|cccc} 1 & 0 & 1 & 1 & -5 & 3 & 0 & 0 \\ 0 & 1 & 0 & 0 & 2 & -1 & 0 & 0 \\ 0 & 0 & 7 & 8 & -1 & 0 & 1 & 0 \\ 0 & 0 & 1 & 1 & -1 & 0 & 0 & 1 \end{array}\right]$$

$$\downarrow$$

$$\left[\begin{array}{cccc|cccc} 1 & 0 & 0 & 0 & -4 & 3 & 0 & -1 \\ 0 & 1 & 0 & 0 & 2 & -1 & 0 & 0 \\ 0 & 0 & 7 & 8 & -1 & 0 & 1 & 0 \\ 0 & 0 & 1 & 1 & -1 & 0 & 0 & 1 \end{array}\right]$$

$$\downarrow$$

$$\left[\begin{array}{cccc|cccc} 1 & 0 & 0 & 0 & -4 & 3 & 0 & -1 \\ 0 & 1 & 0 & 0 & 2 & -1 & 0 & 0 \\ 0 & 0 & 7 & 8 & -1 & 0 & 1 & 0 \\ 0 & 0 & 0 & 1 & 6 & 0 & 1 & -7 \end{array}\right]$$

$$\downarrow$$

$$\left[\begin{array}{cccc|cccc} 1 & 0 & 0 & 0 & -4 & 3 & 0 & -1 \\ 0 & 1 & 0 & 0 & 2 & -1 & 0 & 0 \\ 0 & 0 & 7 & 0 & -49 & 0 & -7 & 56 \\ 0 & 0 & 0 & 1 & 6 & 0 & 1 & -7 \end{array}\right]$$

$$\downarrow$$

$$\left[\begin{array}{cccc|cccc} 1 & 0 & 0 & 0 & -4 & 3 & 0 & -1 \\ 0 & 1 & 0 & 0 & 2 & -1 & 0 & 0 \\ 0 & 0 & 1 & 0 & -7 & 0 & -1 & 8 \\ 0 & 0 & 0 & 1 & 6 & 0 & 1 & -7 \end{array}\right]$$

Thus

$$A^{-1} = \begin{bmatrix} -4 & 3 & 0 & -1 \\ 2 & -1 & 0 & 0 \\ -7 & 0 & -1 & 8 \\ 6 & 0 & 1 & -7 \end{bmatrix}$$

and the answers to (a) and (b) agree.

17. By direct computation, we obtain

$$\det(A) = \begin{vmatrix} 4 & 5 & 0 \\ 11 & 1 & 2 \\ 1 & 5 & 2 \end{vmatrix} = -132$$

$$\det(A_1) = \begin{vmatrix} 2 & 5 & 0 \\ 3 & 1 & 2 \\ 1 & 5 & 2 \end{vmatrix} = -36$$

$$\det(A_2) = \begin{vmatrix} 4 & 2 & 0 \\ 11 & 3 & 2 \\ 1 & 1 & 2 \end{vmatrix} = -24$$

$$\det(A_3) = \begin{vmatrix} 4 & 5 & 2 \\ 11 & 1 & 3 \\ 1 & 5 & 1 \end{vmatrix} = 12$$

Thus, $x = 3/11$, $y = 2/11$, and $z = -1/11$.

19. By direct calculation, we have

$$\det(A) = \begin{vmatrix} 1 & -3 & 1 \\ 2 & -1 & 0 \\ 4 & 0 & -3 \end{vmatrix} = -11$$

$$\det(A_1) = \begin{vmatrix} 4 & -3 & 1 \\ -2 & -1 & 0 \\ 0 & 0 & -3 \end{vmatrix} = 30$$

$$\det(A_2) = \begin{vmatrix} 1 & 4 & 1 \\ 2 & -2 & 0 \\ 4 & 0 & -3 \end{vmatrix} = 38$$

$$\det(A_3) = \begin{vmatrix} 1 & -3 & 4 \\ 2 & -1 & -2 \\ 4 & 0 & 0 \end{vmatrix} = 40$$

Hence, $x_1 = -30/11$, $x_2 = -38/11$, and $x_3 = -40/11$.

21. By direct calculation, we have

$$\det(A) = \begin{vmatrix} 3 & -1 & 1 \\ -1 & 7 & -2 \\ 2 & 6 & -1 \end{vmatrix} = 0$$

Since $\det(A) = 0$, Cramer's Rule cannot be used to find the solution to this system of equations.

22. Since

$$\det(A) = \begin{vmatrix} \cos\theta & \sin\theta \\ -\sin\theta & \cos\theta \end{vmatrix} = \cos^2\theta + \sin^2\theta = 1$$

the matrix is invertible for all values of θ. The matrix of cofactors is

$$\begin{bmatrix} \cos\theta & \sin\theta & 0 \\ -\sin\theta & \cos\theta & 0 \\ 0 & 0 & 1 \end{bmatrix}$$

so that here

$$A^{-1} = \text{adj}(A) = \begin{bmatrix} \cos\theta & -\sin\theta & 0 \\ \sin\theta & \cos\theta & 0 \\ 0 & 0 & 1 \end{bmatrix}$$

24. **(a)** By direct computation, we have

$$\det(A) = -424$$

$$\det(A_1) = -424$$

$$\det(A_2) = 0$$

$$\det(A_3) = -848$$

$$\det(A_4) = 0$$

Hence, $x = 1$, $z = 2$, and $y = w = 0$.

(b) The augmented matrix is

$$\begin{bmatrix} 4 & 1 & 1 & 1 & 6 \\ 3 & 7 & -1 & 1 & 1 \\ 7 & 3 & -5 & 8 & -3 \\ 1 & 1 & 1 & 2 & 3 \end{bmatrix} \rightarrow \begin{bmatrix} 1 & 1 & 1 & 2 & 3 \\ 3 & 7 & -1 & 1 & 1 \\ 7 & 3 & -5 & 8 & -3 \\ 4 & 1 & 1 & 1 & 6 \end{bmatrix} \rightarrow$$

$$\begin{bmatrix} 1 & 1 & 1 & 2 & 3 \\ 0 & 4 & -4 & -5 & -8 \\ 0 & -4 & -12 & -6 & -24 \\ 0 & -3 & -3 & -7 & -6 \end{bmatrix} \rightarrow \begin{bmatrix} 1 & 1 & 1 & 2 & 3 \\ 0 & 1 & 3 & 3/2 & 6 \\ 0 & 4 & -4 & -5 & -8 \\ 0 & 3 & 3 & 7 & 6 \end{bmatrix} \rightarrow$$

$$\begin{bmatrix} 1 & 0 & -2 & 1/2 & -3 \\ 0 & 1 & 3 & 3/2 & 6 \\ 0 & 0 & -16 & -11 & -32 \\ 0 & 0 & -6 & 5/2 & -12 \end{bmatrix} \rightarrow \begin{bmatrix} 1 & 0 & -2 & 1/2 & -3 \\ 0 & 1 & 3 & 3/2 & 6 \\ 0 & 0 & 1 & -5/12 & 2 \\ 0 & 0 & 16 & 11 & 32 \end{bmatrix} \rightarrow$$

$$\begin{bmatrix} 1 & 0 & 0 & -1/3 & 1 \\ 0 & 1 & 0 & 11/4 & 0 \\ 0 & 0 & 1 & -5/12 & 2 \\ 0 & 0 & 0 & 53/3 & 0 \end{bmatrix} \rightarrow \begin{bmatrix} 1 & 0 & 0 & -1/3 & 1 \\ 0 & 1 & 0 & 11/4 & 0 \\ 0 & 0 & 1 & -5/12 & 2 \\ 0 & 0 & 0 & 1 & 0 \end{bmatrix} \rightarrow$$

$$\begin{bmatrix} 1 & 0 & 0 & 0 & 1 \\ 0 & 1 & 0 & 0 & 0 \\ 0 & 0 & 1 & 0 & 2 \\ 0 & 0 & 0 & 1 & 0 \end{bmatrix}$$

Hence, $x = 1$, $z = 2$, and $y = w = 0$.

25. In case $\det(A) = 1$, then A is invertible and, by Theorem 2.4.2, $A^{-1} = \text{adj}(A)$. If A has only integer entries, then each of its cofactors will be a sum of products of integers. Hence, each cofactor will be an integer. Since adj(A) is the transpose of a matrix with integer entries, then its entries will also be integers. Thus, A^{-1} can have only integer entries.

26. Since $\det(A) = 1$, then $x_i = \det(A_i)$. But since the coefficients and the constants are all integers, it follows that $\det(A_i)$ is an integer.

27. If A is an upper triangular matrix, then its ith row starts with at least $i - 1$ zeros and its $(i + 1)^{\text{st}}$ row starts with at least i zeros. If we delete the ith row of A and any column beyond the ith column, we are left with an upper triangular matrix M_{ij} whose i^{th} row is the old $i + 1^{\text{st}}$ row of A with the jth entry deleted. Thus M_{ij} has i zeros at the beginning of the ith row; that is, M_{ij} has a zero on the main diagonal. Since the determinant of an upper triangular matrix is just the product of the diagonal elements, then $\det(M_{ij}) = 0$. Because this is true whenever $i < j$, then the matrix of cofactors of A is lower triangular. Therefore, adj(A) and hence A^{-1} will be upper triangular.

28. To prove the first identity, we rearrange the terms in Equation (1) as follows:

$$\begin{aligned}\det(A) &= a_{11}(a_{22}a_{33} - a_{23}a_{32}) - a_{12}(a_{21}a_{33} - a_{23}a_{31}) \\ &\quad + a_{13}(a_{21}a_{32} - a_{22}a_{31}) \\ &= a_{11}C_{11} + a_{12}C_{12} + a_{13}C_{13}\end{aligned}$$

In order to prove the last identity, we rearrange the terms in Equation (1) as follows:

$$\begin{aligned}\det(A) &= a_{13}(a_{21}a_{32} - a_{22}a_{31}) - a_{23}(a_{11}a_{32} - a_{12}a_{31}) \\ &\quad + a_{33}(a_{11}a_{22} - a_{12}a_{21}) \\ &= a_{13}C_{13} + a_{23}C_{23} + a_{33}C_{33}\end{aligned}$$

29. The point (x,y) lies on the line determined by (a_1,b_1) and (a_2,b_2) if and only if there are constants α, β, and γ, not all zero, such that

$$\begin{aligned}\alpha x + \beta y + \gamma &= 0 \\ \alpha a_1 + \beta b_1 + \gamma &= 0 \\ \alpha a_2 + \beta b_2 + \gamma &= 0\end{aligned}$$

or equivalently, if and only if

$$\begin{bmatrix} x & y & 1 \\ a_1 & b_1 & 1 \\ a_2 & b_2 & 1 \end{bmatrix}\begin{bmatrix} \alpha \\ \beta \\ \gamma \end{bmatrix} = \begin{bmatrix} 0 \\ 0 \\ 0 \end{bmatrix}$$

has a nontrivial solution for α, β, and γ. This is so if and only if the determinant of the coefficient matrix vanishes, that is, if and only if

$$\begin{vmatrix} x & y & 1 \\ a_1 & b_1 & 1 \\ a_2 & b_2 & 1 \end{vmatrix} = 0$$

Alternatively, we can notice that the above equation is

$$(*) \qquad x(b_1 - b_2) + y(a_2 - a_1) + a_1b_2 - a_2b_1 = 0$$

If (a_1,b_1) and (a_2,b_2) are distinct points, then (*) is the equation of a line. Moreover, it is easy to verify that the coordinates of these two points satisfy (*).

30. The points (x_1,y_1), (x_2,y_2) and (x_3,y_3) are collinear if and only if for some α, β, and γ, not all zero, the equations

$$\alpha x_1 + \beta y_1 + \gamma = 0$$
$$\alpha x_2 + \beta y_2 + \gamma = 0$$
$$\alpha x_3 + \beta y_3 + \gamma = 0$$

are satisfied. If we use the same argument that we employed in the solution to Exercise 29, we see that the existence of a solution to the above system of equations for α, β, and γ not all zero is equivalent to the condition that

$$\begin{vmatrix} x_1 & y_1 & 1 \\ x_2 & y_2 & 1 \\ x_3 & y_3 & 1 \end{vmatrix} = 0$$

31. The point (x,y,z) lies in the plane determined by the given points if and only if there exist constants α, β, γ, and δ, not all zero, such that

$$\begin{aligned} \alpha x + \beta y + \gamma z + \delta &= 0 \\ \alpha a_1 + \beta b_1 + \gamma c_1 + \delta &= 0 \\ \alpha a_2 + \beta b_2 + \gamma c_2 + \delta &= 0 \\ \alpha a_3 + \beta b_3 + \gamma c_3 + \delta &= 0 \end{aligned}$$

or equivalently, if and only if

$$\begin{bmatrix} x & y & z & 1 \\ a_1 & b_1 & c_1 & 1 \\ a_2 & b_2 & c_2 & 1 \\ a_3 & b_3 & c_3 & 1 \end{bmatrix} \begin{bmatrix} \alpha \\ \beta \\ \gamma \\ \delta \end{bmatrix} = \begin{bmatrix} 0 \\ 0 \\ 0 \\ 0 \end{bmatrix}$$

has a nontrivial solution for α, β, γ, and δ. As in Problem 29, this is true if and only if the determinant of the coefficient matrix vanishes. Thus we have the desired result.

SUPPLEMENTARY EXERCISES 2

1.

$$x' = \frac{\begin{vmatrix} x & -\frac{4}{5} \\ y & \frac{3}{5} \end{vmatrix}}{\begin{vmatrix} \frac{3}{5} & -\frac{4}{5} \\ \frac{4}{5} & \frac{3}{5} \end{vmatrix}} = \frac{\frac{3}{5}x + \frac{4}{5}y}{\frac{9}{25} + \frac{16}{25}} = \frac{3}{5}x + \frac{4}{5}y$$

$$y' = \frac{\begin{vmatrix} \frac{3}{5} & x \\ \frac{4}{5} & y \end{vmatrix}}{\begin{vmatrix} \frac{3}{5} & -\frac{4}{5} \\ \frac{4}{5} & \frac{3}{5} \end{vmatrix}} = \frac{\frac{3}{5}y - \frac{4}{5}x}{1} = -\frac{4}{5}x + \frac{3}{5}y$$

3. The determinant of the coefficient matrix is

$$\begin{vmatrix} 1 & 1 & \alpha \\ 1 & 1 & \beta \\ \alpha & \beta & 1 \end{vmatrix} = \begin{vmatrix} 1 & 1 & \alpha \\ 0 & 0 & \beta-\alpha \\ \alpha & \beta & 1 \end{vmatrix} = -(\beta - \alpha)\begin{vmatrix} 1 & 1 \\ \alpha & \beta \end{vmatrix}$$

$$= -(\beta - \alpha)(\beta - \alpha)$$

The system of equations has a nontrivial solution if and only if this determinant is zero; that is, if and only if $\alpha = \beta$. (See Theorem 2.3.4 and Theorem 1.6.3.)

4. Recall that det(A) is the sum of 6 signed elementary products, 3 of which are positive and 3 of which are negative. Since each of the factors in these products is either 0 or 1, then the only values which the products can assume are 0, +1, or −1. If det(A) has 3 signed elementary products which are equal to +1, then the remaining 3 products must equal −1. (Why?) In this case, det(A) = 0. However, det(A) can have 2 signed elementary products which are equal to +1 and the remaining 4 products equal to 0. For instance, this is true of

$$\det\begin{bmatrix} 1 & 0 & 1 \\ 1 & 1 & 0 \\ 0 & 1 & 1 \end{bmatrix}$$

Thus, the largest possible value for det(A) is 2.

5. **(a)** If the perpendicular from the vertex of angle α to side a meets side a between angles β and γ, then we have the following picture:

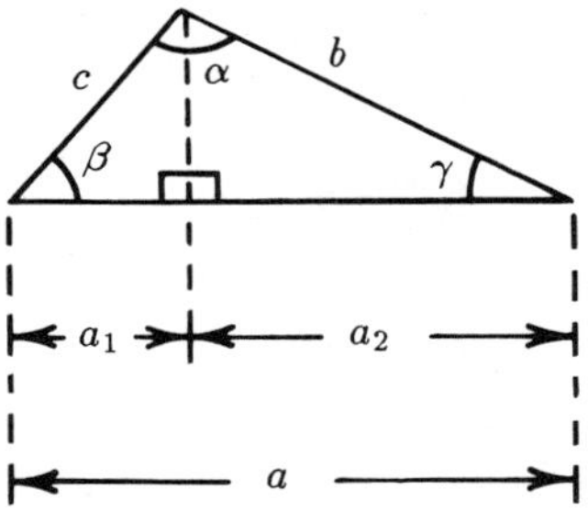

Thus $\cos\beta = \dfrac{a_1}{c}$ and $\cos\gamma = \dfrac{a_2}{b}$ and hence

$$a = a_1 + a_2 = c\cos\beta + b\cos\gamma$$

This is the first equation which you are asked to derive. If the perpendicular intersects side a outside of the triangle, the argument must be modified slightly, but the same result holds. Since there is nothing sacred about starting at angle α, the same argument starting at angles β and γ will yield the second and third equations.

Cramer's Rule applied to this system of equations yields the following results:

$$\cos \alpha = \frac{\begin{vmatrix} a & c & b \\ b & 0 & a \\ c & a & 0 \end{vmatrix}}{\begin{vmatrix} 0 & c & b \\ c & 0 & a \\ b & a & 0 \end{vmatrix}} = \frac{a(-a^2 + b^2 + c^2)}{2abc} = \frac{b^2 + c^2 - a^2}{2bc}$$

$$\cos \beta = \frac{\begin{vmatrix} 0 & a & b \\ c & b & a \\ b & c & 0 \end{vmatrix}}{2abc} = \frac{b(a^2 - b^2 + c^2)}{2abc} = \frac{a^2 + c^2 - b^2}{2ac}$$

$$\cos \gamma = \frac{\begin{vmatrix} 0 & c & a \\ c & 0 & b \\ b & a & c \end{vmatrix}}{2abc} = \frac{c(a^2 + b^2 - c^2)}{2abc} = \frac{a^2 + b^2 - c^2}{2ab}$$

6. The system of equations

$$\begin{aligned} (1 - \lambda)x - 2y &= 0 \\ x - (1 + \lambda)y &= 0 \end{aligned}$$

always has the trivial solution $x = y = 0$. There can only be a nontrivial solution in case the determinant of the coefficient matrix is zero. But this determinant is just

$$-(1 - \lambda)(1 + \lambda) + 2 = \lambda^2 + 1$$

and $\lambda^2 + 1$ can never equal zero for any real value of λ.

7. If A is invertible, then $A^{-1} = \frac{1}{\det(A)}\operatorname{adj}(A)$, or $\operatorname{adj}(A) = [\det(A)]A^{-1}$. Thus

$$\operatorname{adj}(A)\,\frac{A}{\det(A)} = I$$

That is, $\operatorname{adj}(A)$ is invertible and

$$[\operatorname{adj}(A)]^{-1} = \frac{1}{\det(A)}\,A$$

It remains only to prove that $A = \det(A)\operatorname{adj}(A^{-1})$. This follows from Theorem 2.4.2 and Corollary 2.3.5 as shown:

$$A = [A^{-1}]^{-1} = \frac{1}{\det(A^{-1})}\operatorname{adj}(A^{-1}) = \det(A)\operatorname{adj}(A^{-1})$$

8. From the proof of Theorem 2.4.2, we have

$$(*)\qquad A[\operatorname{adj}(A)] = [\det(A)]I$$

First, assume that A is invertible. Then $\det(A) \neq 0$. Notice that $[\det(A)]I$ is a diagonal matrix, each of whose diagonal elements is $\det(A)$. Thus, if we take the determinant of the matrices on the right- and left-hand sides of (*), we obtain

$$\det(A[\operatorname{adj}(A)]) = \det([\det(A)]I)$$

or

$$\det(A)\cdot\det(\operatorname{adj}(A)) = [\det(A)]^n$$

and the result follows.

Now suppose that A is not invertible. Then $\det(A) = 0$ and hence we must show that $\det(\operatorname{adj}(A)) = 0$. But if $\det(A) = 0$, then, by (*), $A[\operatorname{adj}(A)] = 0$. Certainly, if $A = 0$, then $\operatorname{adj}(A) = 0$; thus, $\det(\operatorname{adj}(A)) = 0$ and the result holds. On the other hand, if $A \neq 0$, then A must have a nonzero row X. Hence $A[\operatorname{adj}(A)] = 0$ implies that $X\operatorname{adj}(A) = 0$ or

$$[\operatorname{adj}(A)]^t X^t = 0^t$$

Since X^t is a nontrivial solution to the above system of homogeneous equations, it follows that the determinant of $[\text{adj}(A)]^t$ is zero. Hence, $\det(\text{adj}(A)) = 0$ and the result is proved.

9. We simply expand W. That is,

$$\frac{dW}{dx} = \frac{d}{dx}\begin{vmatrix} f_1(x) & f_2(x) \\ g_1(x) & g_2(x) \end{vmatrix}$$

$$= \frac{d}{dx}\,(f_1(x)g_2(x) - f_2(x)g_1(x))$$

$$= f_1'(x)g_2(x) + f_1(x)g_2'(x) - f_2'(x)g_1(x) - f_2(x)g_1'(x)$$

$$= \left[f_1'(x)g_2(x) - f_2'(x)g_1(x)\right] + \left[f_1(x)g_2'(x) - f_2(x)g_1'(x)\right]$$

$$= \begin{vmatrix} f_1'(x) & f_2'(x) \\ g_1(x) & g_2(x) \end{vmatrix} + \begin{vmatrix} f_1(x) & f_2(x) \\ g_1'(x) & g_2'(x) \end{vmatrix}$$

10. **(a)** As suggested in the figure and the hints,

area ABC = area $ADEC$ + area $CEFB$ − area $ADFB$

$$= \frac{1}{2}\,(x_3 - x_1)(y_1 + y_3) + \frac{1}{2}\,(x_2 - x_3)(y_3 + y_2)$$

$$- \frac{1}{2}\,(x_2 - x_1)(y_1 + y_2)$$

$$= \frac{1}{2}\Big[x_3y_1 - x_1y_1 + x_3y_3 - x_1y_3 + x_2y_3 - x_3y_3 + x_2y_2$$

$$- x_3y_2 - x_2y_1 + x_1y_1 - x_2y_2 + x_1y_2\Big]$$

$$= \frac{1}{2}\left[x_1y_2 - x_1y_3 - x_2y_1 + x_2y_3 + x_3y_1 - x_3y_2\right]$$

$$= \frac{1}{2}\begin{vmatrix} x_1 & y_1 & 1 \\ x_2 & y_2 & 1 \\ x_3 & y_3 & 1 \end{vmatrix}$$

10. (b) The area of the triangle is

$$\frac{1}{2}\begin{vmatrix} 3 & 3 & 1 \\ -2 & -1 & 1 \\ 4 & 0 & 1 \end{vmatrix} = \frac{1}{2}\,[3(-1) + 2(3) + 4(3 + 1)]$$

$$= \frac{19}{2}$$

11. Let A be an $n \times n$ matrix for which the entries in each row add up to zero and let X be the $n \times 1$ matrix each of whose entries is one. Then all of the entries in the $n \times 1$ matrix AX are zero since each of its entries is the sum of the entries of one of the rows of A. That is, the homogeneous system of linear equations

$$AX = \begin{bmatrix} 0 \\ \vdots \\ 0 \end{bmatrix}$$

has a nontrivial solution. Hence $\det(A) = 0$. (See Theorem 2.3.4 and Theorem 1.6.3.)

12. To obtain B from A, we first interchange Row 1 and Row n. This changes the sign of $\det(A)$. If $n = 2$ or 3, we are done. If $n > 3$, then we interchange Rows 2 and $n-1$, which again reverses the sign of the determinant. In general, we interchange Rows $k+1$ and $n-k$ until all of the rows have been interchanged.

If n is even, then we will have made $n/2$ interchanges and the sign of det(A) will have changed by a factor of $(-1)^{n/2}$. On the other hand, if n is odd, then we will have made $(n-1)/2$ interchanges and the sign of det(A) will have changed by a factor of $(-1)^{(n-1)/2}$. All of this can be summarized by the formula

$$\det(B) = (-1)^{\frac{n(n-1)}{2}} \det(A)$$

Can you see why?

13. **(a)** If we interchange the ith and j^{th} rows of A, then we claim that we must interchange the ith and j^{th} columns of A^{-1}. To see this, let

$$A = \begin{bmatrix} \text{Row } 1 \\ \text{Row } 2 \\ \vdots \\ \text{Row } n \end{bmatrix} \quad \text{and} \quad A^{-1} = [\ \text{Col.}1,\ \text{Col.}2,\ \ldots,\ \text{Col.}n\ ]$$

where $AA^{-1} = I$. Thus, the sum of the products of corresponding entries from Row s in A and from Column r in A^{-1} must be 0 unless $s = r$, in which case it is 1. That is, if Rows i and j are interchanged in A, then Columns i and j must be interchanged in A^{-1} in order to insure that only 1's will appear on the diagonal of the product AA^{-1}.

(b) If we multiply the ith row of A by a nonzero scalar c, then we must divide the ith column of A^{-1} by c. This will insure that the sum of the products of corresponding entries from the ith row of A and the ith column of A^{-1} will remain equal to 1.

13. **(c)** Suppose we add c times the ith row of A to the jth row of A. Call that matrix B. Now suppose that we add $-c$ times the jth column of A^{-1} to the ith column of A^{-1}. Call that matrix C. We claim that $C = B^{-1}$. To see that this is so, consider what happens when

$$\text{Row } j \longrightarrow \text{Row } j + c \text{ Row } i \qquad [\text{in } A]$$

$$\text{Column } i \longrightarrow \text{Column } i - c \text{ Column } j \qquad [\text{in } A^{-1}]$$

The sum of the products of corresponding entries from the jth row of B and any k^{th} column of C will clearly be 0 unless $k = i$ or $k = j$. If $k = i$, then the result will be $c - c = 0$. If $k = j$, then the result will be 1. The sum of the products of corresponding entries from any other row of B — say the rth row — and any column of C — say the kth column — will be 1 if $r = k$ and 0 otherwise. This follows because there has been no changes unless $k = i$. In case $k = i$, the result is easily checked.

14. Let C be the matrix which is the same as A except for its ith row which is the sum of the ith rows of B_1 and B_2, or twice the ith row of A. Then by Theorem 2.3.2,

$$\det(B_1) + \det(B_2) = \det(C)$$

But $\det(C) = 2\det(A)$, so that

$$\det(A) = \frac{1}{2}\left[\det(B_1) + \det(B_2)\right]$$

15. **(a)** We have

$$\det(\lambda I - A) = \begin{vmatrix} \lambda - a_{11} & -a_{12} & -a_{13} \\ -a_{21} & \lambda - a_{22} & -a_{23} \\ -a_{31} & -a_{32} & \lambda - a_{33} \end{vmatrix}$$

If we calculate this determinant by any method, we find that

$$\begin{aligned}\det(\lambda I - A) &= (\lambda - a_{11})(\lambda - a_{22})(\lambda - a_{33}) - a_{23}a_{32}(\lambda - a_{11}) \\ &\quad - a_{13}a_{31}(\lambda - a_{22}) - a_{12}a_{21}(\lambda - a_{33}) \\ &\quad - a_{13}a_{21}a_{32} - a_{12}a_{23}a_{31} \\ &= \lambda^3 + (-a_{11} - a_{22} - a_{33})\lambda^2 \\ &\quad + (a_{11}a_{22} + a_{11}a_{33} + a_{22}a_{33} - a_{12}a_{21} \\ &\qquad - a_{13}a_{31} - a_{23}a_{32})\lambda \\ &\quad + (a_{11}a_{23}a_{32} + a_{12}a_{21}a_{33} + a_{13}a_{22}a_{31} \\ &\qquad - a_{11}a_{22}a_{33} - a_{12}a_{23}a_{31} - a_{13}a_{21}a_{32})\end{aligned}$$

(b) From Part (a) we see that $b = -\mathrm{tr}(A)$ and $d = -\det(A)$. (It is less obvious that c is the trace of the matrix of minors of the entries of A; that is, the sum of the minors of the diagonal entries of A.)

16. Recall that $\sin(a + b) = \sin a \cos b + \cos a \sin b$. If we multiply the 1st column of the determinant by $\cos \delta$ and the 2nd column by $\sin \delta$ and then add the results, we get the 3rd column. Therefore the determinant is 0.

17. If we multiply Column 1 by 10^4, Column 2 by 10^3, Column 3 by 10^2, Column 4 by 10, and add the results to Column 5, we obtain a new Column 5 whose entries are just the 5 numbers listed in the problem. Since each is divisible by 19, so is the resulting determinant.

EXERCISE SET 3.1

1. **(a)**

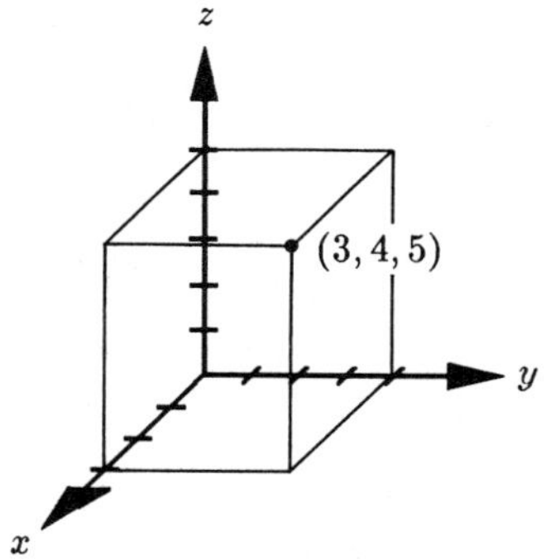

(c)

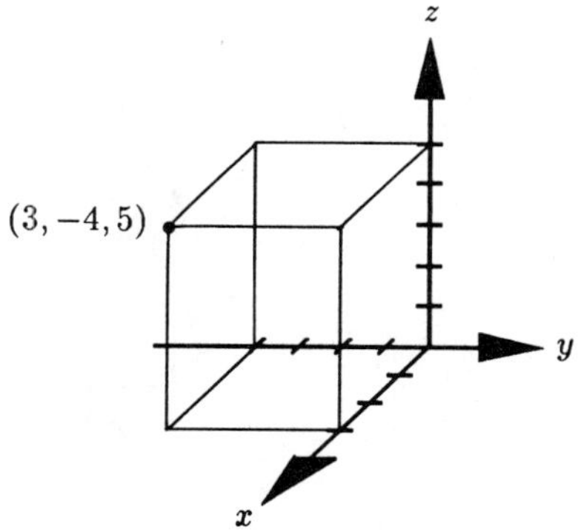

(e)

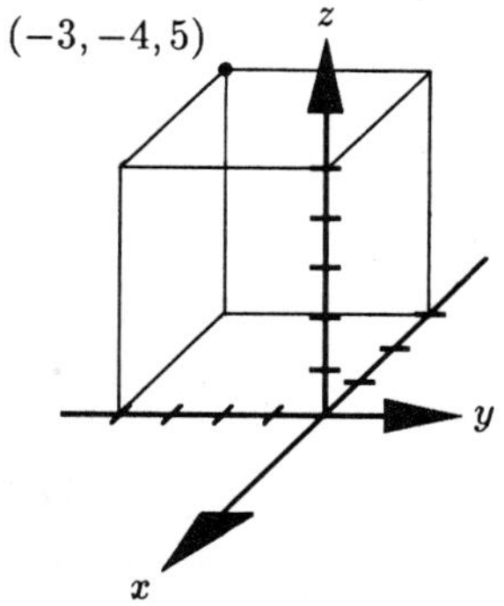

(j)

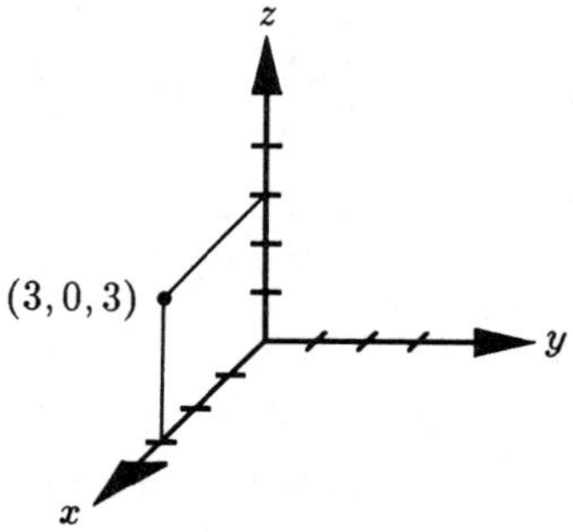

2. **(a)**

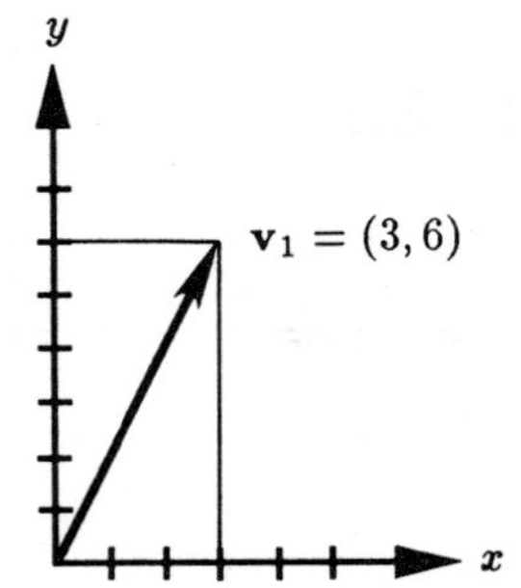

(c)

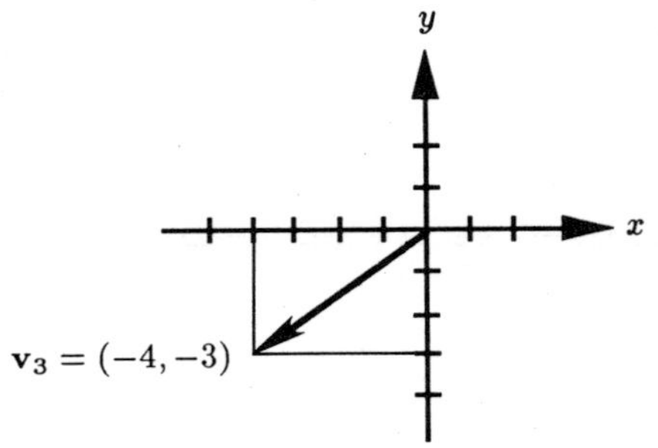

(g)

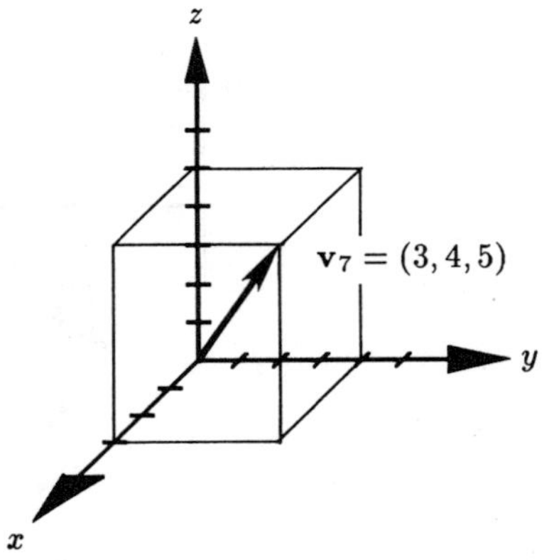

(i)

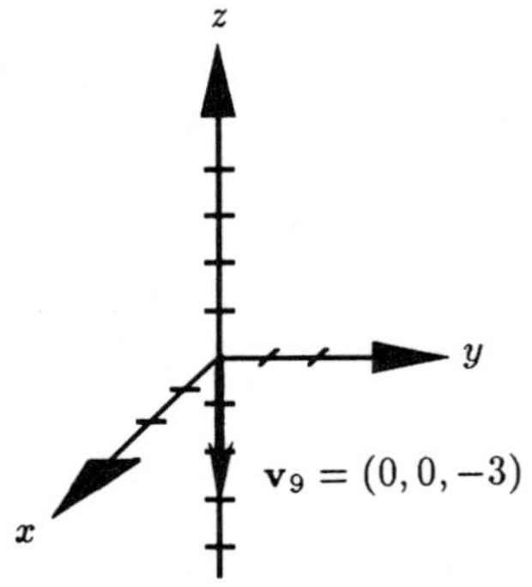

3. **(a)** $\overrightarrow{P_1P_2} = (3 - 4,\ 7 - 8) = (-1,-1)$

(e) $\overrightarrow{P_1P_2} = (-2 - 3,\ 5 + 7,\ -4 - 2) = (-5,12,-6)$

5. **(a)** Let $P = (x,y,z)$ be the initial point of the desired vector and assume that this vector has the same length as **v**. Since $\overrightarrow{PQ}$ has the same direction as $\mathbf{v} = (4,-2,-1)$, we have the equation

$$\overrightarrow{PQ} = (3 - x,\ 0 - y,\ -5 - z) = (4,-2,-1)$$

If we equate components in the above equation, we obtain

$$x = -1, \quad y = 2, \quad \text{and} \quad z = -4$$

Thus, we have found a vector $\overrightarrow{PQ}$ which satisfies the given conditions. Any positive multiple $k\,\overrightarrow{PQ}$ will also work provided the terminal point remains fixed at Q. Thus, P could be any point $(-k,2k,-4k)$ where $k > 0$.

(b) Let $P = (x,y,z)$ be the initial point of the desired vector and assume that this vector has the same length as **v**. Since $\overrightarrow{PQ}$ is oppositely directed to $\mathbf{v} = (4,-2,-1)$, we have the equation

$$\overrightarrow{PQ} = (3 - x,\ 0 - y,\ -5 - z) = (-4,2,1)$$

If we equate components in the above equation, we obtain

$$x = 7,\ y = -2,\ \text{and}\ z = -6$$

Thus, we have found a vector $\overrightarrow{PQ}$ which satisfies the given conditions. Any positive multiple $k\overrightarrow{PQ}$ will also work, provided the terminal point remains fixed at Q. Thus, P could be any point $(7k,-2k,-6k)$ where $k > 0$.

6. **(a)** $\mathbf{v} - \mathbf{w} = (4 - 6,\ 0 - (-1),\ -8 - (-4)) = (-2,1,-4)$

(c) $-\mathbf{v} + \mathbf{u} = (-4 - 3,\ 0 + 1,\ 8 + 2) = (-7,1,10)$

(e) $-3(\mathbf{v} - 8\mathbf{w}) = -3(4 - 48,\ 0 + 8,\ -8 + 32) = (132,-24,-72)$

7. Let $\mathbf{x} = (x_1,x_2,x_3)$. Then

$$\begin{aligned} 2\mathbf{u} - \mathbf{v} + \mathbf{x} &= (-6,2,4) - (4,0,-8) + (x_1,x_2,x_3) \\ &= (-10 + x_1,\ 2 + x_2,\ 12 + x_3) \end{aligned}$$

On the other hand,

$$\begin{aligned} 7\mathbf{x} + \mathbf{w} &= 7(x_1,x_2,x_3) + (6,-1,-4) \\ &= (7x_1 + 6,\ 7x_2 - 1,\ 7x_3 - 4) \end{aligned}$$

If we equate the components of these two vectors, we obtain

$$7x_1 + 6 = x_1 - 10$$
$$7x_2 - 1 = x_2 + 2$$
$$7x_3 - 4 = x_3 + 12$$

Hence, $\mathbf{x} = (-8/3, 1/2, 8/3)$.

9. Suppose there are scalars c_1, c_2, and c_3 which satisfy the given equation. If we equate components on both sides, we obtain the following system of equations:

$$\begin{aligned} -2c_1 - 3c_2 + c_3 &= 0 \\ 9c_1 + 2c_2 + 7c_3 &= 5 \\ 6c_1 + c_2 + 5c_3 &= 4 \end{aligned}$$

The augmented matrix of this system of equations can be reduced to

$$\begin{bmatrix} 2 & 3 & -1 & 0 \\ 0 & 2 & -2 & -1 \\ 0 & 0 & 0 & -1 \end{bmatrix}$$

The third row of the above matrix implies that $0c_1 + 0c_2 + 0c_3 = -1$. Clearly, there do not exist scalars c_1, c_2, and c_3 which satisfy the above equation, and hence the system is inconsistent.

10. If we equate components on both sides of the given equation, we obtain

$$\begin{aligned} c_1 + 2c_2 \qquad &= 0 \\ 2c_1 + c_2 + 3c_3 &= 0 \\ c_2 + c_3 &= 0 \end{aligned}$$

The augmented matrix of this system of equations can be reduced to

$$\begin{bmatrix} 1 & 0 & 0 & 0 \\ 0 & 1 & 0 & 0 \\ 0 & 0 & 1 & 0 \end{bmatrix}$$

Thus the only solution is $c_1 = c_2 = c_3 = 0$.

11. We work in the plane determined by the three points $O = (0,0,0)$, $P = (2,3,-2)$, and $Q = (7,-4,1)$. Let X be a point on the line through P and Q and let $t\overrightarrow{PQ}$ (where t is a positive, real number) be the vector with initial point P and terminal point X. Note that the length of $t\overrightarrow{PQ}$ is t times the length of $\overrightarrow{PQ}$. Referring to the figure below, we see that

$$\overrightarrow{OP} + t\overrightarrow{PQ} = \overrightarrow{OX} \qquad t$$

and

$$\overrightarrow{OP} + \overrightarrow{PQ} = \overrightarrow{OQ}$$
$$\overrightarrow{PQ} = (\overrightarrow{OQ} - \overrightarrow{OP})$$
$$+ \overrightarrow{PQ} = + (\overrightarrow{OQ} - \overrightarrow{OP})$$

Therefore,

$$\overrightarrow{OX} = \overrightarrow{OP} + t(\overrightarrow{OQ} - \overrightarrow{OP}) \quad = \overrightarrow{OP} + t\overrightarrow{OQ} - t\overrightarrow{OP}$$
$$= (1 - t)\overrightarrow{OP} + t\overrightarrow{OQ} \quad = \overrightarrow{OP} - t\overrightarrow{OP} + t\overrightarrow{OQ}$$
$$= \overrightarrow{OP}(1-t) + t\overrightarrow{OQ}$$

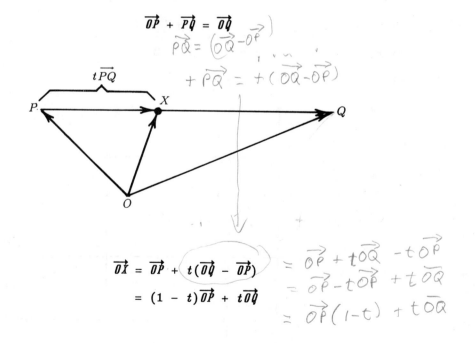

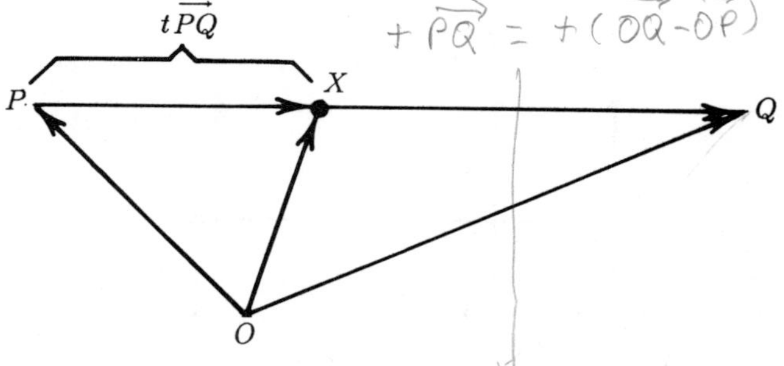

11. **(a)** To obtain the midpoint of the line segment connecting P and Q, we set $t = 1/2$. This gives

$$\overrightarrow{OX} = \frac{1}{2}\,\overrightarrow{OP} + \frac{1}{2}\,\overrightarrow{OQ}$$

$$= \frac{1}{2}\,(2,3,-2) + \frac{1}{2}\,(7,-4,1)$$

$$= \left(\frac{9}{2},\ -\frac{1}{2},\ -\frac{1}{2}\right)$$

(b) Now set $t = 3/4$. This gives

$$\overrightarrow{OX} = \frac{1}{4}\,(2,3,-2) + \frac{3}{4}\,(7,-4,1) = \left(\frac{23}{4},\ -\frac{9}{4},\ \frac{1}{4}\right)$$

12. The relationship between the xy-coordinate system and the $x'y'$-coordinate system is given by

$$x' = x - 2, \qquad y' = y - (-3) = y + 3$$

(a) $x' = 7 - 2 = 5$ and $y' = 5 + 3 = 8$

(b) $x = x' + 2 = -1$ and $y = y' - 3 = 3$

(c)

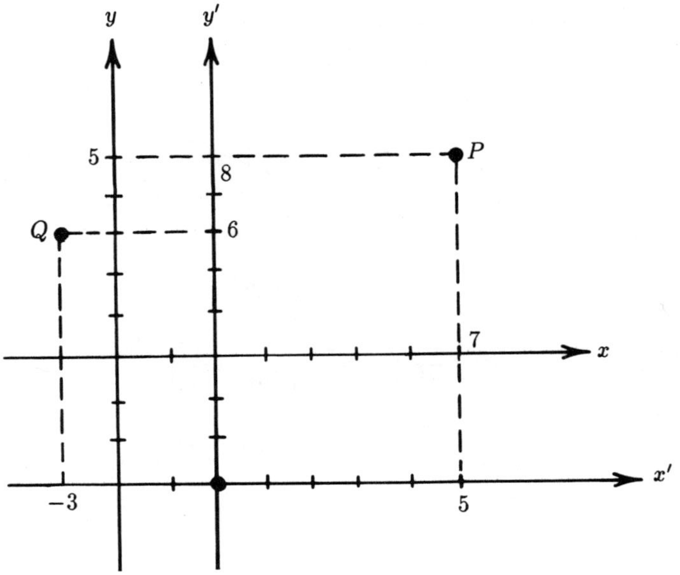

13. Let (x_0,y_0,z_0) denote the origin in the $x'y'z'$-coordinate system with respect to the xyz-coordinate system. Suppose that P_1 and P_2 are the initial and terminal points, respectively, for a vector **v**. Let (x_1,y_1,z_1) and (x_2,y_2,z_2) be the coordinates of P_1 and P_2 in the xyz-coordinate system, and let (x'_1,y'_1,z'_1) and (z'_2,y'_2,z'_2) be the coordinates of P_1 and P_2 in the $x'y'z'$-coordinate system. Further, let (v_1,v_2,v_3) and (v'_1,v'_2,v'_3) denote the coordinates of **v** with respect to the xyz- and $x'y'z'$-coordinate systems, respectively. Then

$$v_1 = x_2 - x_1, \quad v_2 = y_2 - y_1, \quad v_3 = z_2 - z_1$$

and

$$v'_1 = x'_2 - x'_1, \quad v'_2 = y'_2 - y'_1, \quad v'_3 = z'_2 - z'_1$$

However,

$$x'_1 = x_1 - x_0, \quad y'_1 = y_1 - y_0, \quad z'_1 = z_1 - z_0$$

and

$$x'_2 = x_2 - x_0, \quad y'_2 = y_2 - y_0, \quad z'_2 = z_2 - z_0$$

If we substitute the above expressions for x'_1 and x'_2 into the equation for v'_1, we obtain

$$v'_1 = (x_2 - x_0) - (x_1 - x_0) = x_2 - x_1 = v_1$$

In a similar way, we can show that $v'_2 = v_2$ and $v'_3 = v_3$.

15. Let $\mathbf{v} = (v_1, v_2)$ and $\mathbf{u} = k\mathbf{v} = (u_1, u_2)$. Let P and Q be the points (v_1, v_2) and (u_1, u_2), respectively. Since the triangles OPR and OQS are similar (see the diagram),

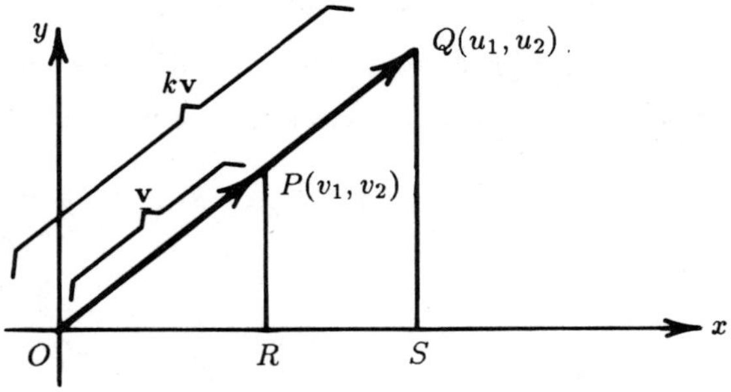

we have

$$\frac{u_1}{v_1} = \frac{\text{length of } \overrightarrow{OS}}{\text{length of } \overrightarrow{OR}} = \frac{\text{length of } \overrightarrow{OQ}}{\text{length of } \overrightarrow{OP}} = k$$

Thus, $u_1 = kv_1$. Similarly, $u_2 = kv_2$.

EXERCISE SET 3.2

1. **(a)** $\|\mathbf{v}\| = \left[4^2 + (-3)^2\right]^{1/2} = 5$

(c) $\|\mathbf{v}\| = \left[(-5)^2 + 0^2\right]^{1/2} = 5$

(e) $\|\mathbf{v}\| = \left[(-7)^2 + 2^2 + (-1)^2\right]^{1/2} = \sqrt{54}$

2. **(a)** $d = \left[(3 - 5)^2 + (4 - 7)^2\right]^{1/2} = \sqrt{13}$

(c) $d = \left[\ 7 - (-7)\ ^2 + \ -5 - (-2)\ ^2 + \ 1 - (-1)\ ^2\right]^{1/2} = \sqrt{209}$

3. **(a)** Since $\mathbf{u} + \mathbf{v} = (3,-5,7)$, then

$$\|\mathbf{u} + \mathbf{v}\| = \left[3^2 + (-5)^2 + 7^2\right]^{1/2} = \sqrt{83}$$

(c) Since

$$\|-2\mathbf{u}\| = \left[(-4)^2 + 4^2 + (-6)^2\right]^{1/2} = 2\sqrt{17}$$

and

$$2\|\mathbf{u}\| = 2\left[2^2 + (-2)^2 + 3^2\right]^{1/2} = 2\sqrt{17}$$

then

$$\|-2\mathbf{u}\| + 2\|\mathbf{u}\| = 4\sqrt{17}$$

3. **(e)** Since $\|\mathbf{w}\| = [3^2 + 6^2 + (-4)^2]^{1/2} = \sqrt{61}$, then

$$\frac{1}{\|\mathbf{w}\|}\mathbf{w} = \left(\frac{3}{\sqrt{61}}, \frac{6}{\sqrt{61}}, \frac{-4}{\sqrt{61}}\right)$$

4. Since $k\mathbf{v} = (-k, 2k, 5k)$, then

$$\|k\mathbf{v}\| = \left[k^2 + 4k^2 + 25k^2\right]^{1/2} = |k|\sqrt{30}$$

If $\|k\mathbf{v}\| = 4$, it follows that $|k|\sqrt{30} = 4$ or $k = \pm 4/\sqrt{30}$.

6. **(b)** From Part (a), we know that the norm of $\mathbf{v}/\|\mathbf{v}\|$ is 1. But if $\mathbf{v} = (3,4)$, then $\|\mathbf{v}\| = 5$. Hence $\mathbf{u} = \mathbf{v}/\|\mathbf{v}\| = (3/5, 4/5)$ has norm 1 and has the same direction as $\mathbf{v}$.

7. **(b)** By the result of Part (a), we have $4\mathbf{u} = 4(\|\mathbf{u}\| \cos 30^\circ, \|\mathbf{u}\| \sin 30^\circ)$ $= 4(3(\sqrt{3}/2), 3(1/2)) = (6\sqrt{3}, 6)$ and , also using Part (a), $5\mathbf{v} = 5(\|\mathbf{v}\| \cos (135^\circ), \|\mathbf{v}\| \sin (135^\circ)) = 5(2(-1/\sqrt{2}), 2(1/\sqrt{2}))$ $= (-10/\sqrt{2}, 10/\sqrt{2}) = (-5\sqrt{2}, 5\sqrt{2})$. Thus $4\mathbf{u} - 5\mathbf{v}$ $= (6\sqrt{3} + 5\sqrt{2}, 6 - 5\sqrt{2})$.

8. Note that $\|\mathbf{p} - \mathbf{p}_0\| = 1$ if and only if $\|\mathbf{p} - \mathbf{p}_0\|^2 = 1$. Thus

$$(x - x_0)^2 + (y - y_0)^2 + (z - z_0)^2 = 1$$

The points (x,y,z) which satisfy these equations are just the points on the sphere of radius 1 with center (x_0, y_0, z_0); that is, they are all the points whose distance from (x_0, y_0, z_0) is 1.

9. First, suppose that **u** and **v** are neither similarly nor oppositely directed and that neither is the zero vector.

If we place the initial point of **v** at the terminal point of **u**, then the vectors **u**, **v**, and **u** + **v** form a triangle, as shown in (i) below.

(i)

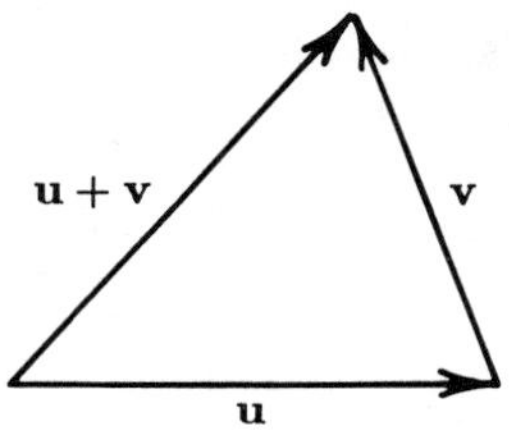

(ii)

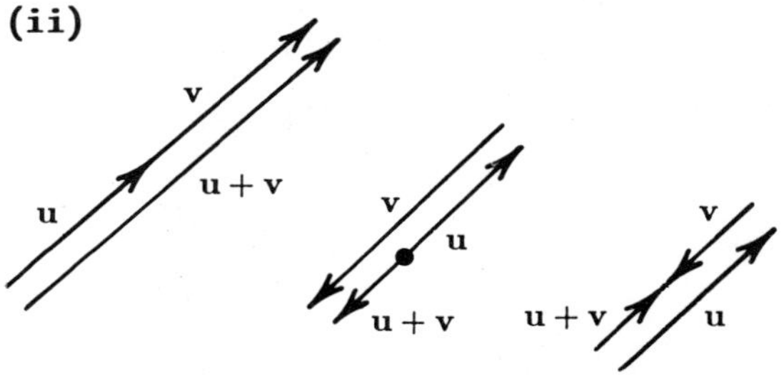

Since the length of one side of a triangle – say $\|\mathbf{u} + \mathbf{v}\|$ – cannot exceed the sum of the lengths of the other two sides, then $\|\mathbf{u} + \mathbf{v}\| < \|\mathbf{u}\| + \|\mathbf{v}\|$.

Now suppose that **u** and **v** have the same direction. From diagram (ii), we see that $\|\mathbf{u} + \mathbf{v}\| = \|\mathbf{u}\| + \|\mathbf{v}\|$. If **u** and **v** have opposite directions, then (again, see diagram (ii)) $\|\mathbf{u} + \mathbf{v}\| < \|\mathbf{u}\| + \|\mathbf{v}\|$.

Finally, if either vector is zero, then $\|\mathbf{u} + \mathbf{v}\| = \|\mathbf{u}\| + \|\mathbf{v}\|$.

10. These proofs are for vectors in 3-space. To obtain proofs in 2-space, just ignore the 3rd component. Let $\mathbf{u} = (u_1, u_2, u_3)$ and $\mathbf{v} = (v_1, v_2, v_3)$. Then

(a) $\mathbf{u} + \mathbf{v} = (u_1 + v_1,\ u_2 + v_2,\ u_3 + v_3)$

$= (v_1 + u_1,\ v_2 + u_2,\ v_3 + u_3) = \mathbf{v} + \mathbf{u}$

(c) $\mathbf{u} + \mathbf{0} = (u_1 + 0,\ u_2 + 0,\ u_3 + 0)$

$= (0 + u_1,\ 0 + u_2,\ 0 + u_3)$

$= (u_1, u_2, u_3) = \mathbf{0} + \mathbf{u} = \mathbf{u}$

(e) $k(l\mathbf{u}) = k(lu_1, lu_2, lu_3) = (klu_1, klu_2, klu_3) = (kl)\mathbf{u}$

11. Again, we work in 3-space. Let $\mathbf{u} = (u_1, u_2, u_3)$.

(d) $\mathbf{u} + (-\mathbf{u}) = (u_1 + (-u_1),\ u_2 + (-u_2),\ u_3 + (-u_3))$

$= (0,0,0) = \mathbf{0}$

(g) $(k + l)\mathbf{u} = ((k + l)u_1,\ (k + l)u_2,\ (k + l)u_3)$

$= (ku_1 + lu_1,\ ku_2 + lu_2,\ ku_3 + lu_3)$

$= (ku_1, ku_2, ku_3) + (lu_1, lu_2, lu_3)$

$= k\mathbf{u} + l\mathbf{u}$

(h) $1\mathbf{u} = (1u_1, 1u_2, 1u_3) = (u_1, u_2, u_3) = \mathbf{u}$

12. First, assume that $k > 0$.

(i) (ii)

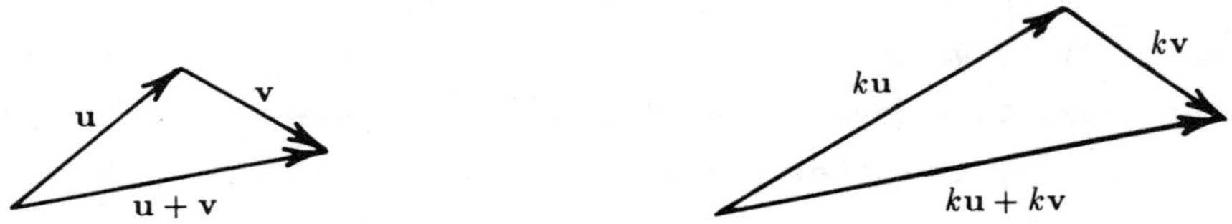

The two triangles pictured above are similar since $\mathbf{u}$ and $k\mathbf{u}$ and $\mathbf{v}$ and $k\mathbf{v}$ are parallel and their lengths are proportional; i.e., $\|k\mathbf{u}\| = k\|\mathbf{u}\|$ and $\|k\mathbf{v}\| = k\|\mathbf{v}\|$. Therefore $\|k\mathbf{u} + k\mathbf{v}\| = k\|\mathbf{u} + \mathbf{v}\|$. That is, the corresponding sides of the two triangles are all proportional, where k is the constant of proportionality. Thus, the vectors $k(\mathbf{u} + \mathbf{v})$ and $k\mathbf{u} + k\mathbf{v}$ have the same length. Moreover, $\mathbf{u} + \mathbf{v}$ and $k\mathbf{u} + k\mathbf{v}$ have the same direction because corresponding sides of the two triangles in (i) and (ii) are parallel. Hence, $k(\mathbf{u} + \mathbf{v})$ and $k\mathbf{u} + k\mathbf{v}$ have the same direction. Therefore, $k(\mathbf{u} + \mathbf{v}) = k\mathbf{u} + k\mathbf{v}$.

If $k < 0$, reverse the directions of all of the arrows in (ii). Now $\mathbf{u} + \mathbf{v}$ and $k\mathbf{u} + k\mathbf{v}$ will be oppositely directed, and we'll have $\|k\mathbf{u} + k\mathbf{v}\| = |k|\,\|\mathbf{u} + \mathbf{v}\|$. Thus the vectors $k(\mathbf{u} + \mathbf{v})$ and $(k\mathbf{u} + k\mathbf{v})$ will still have the same length. They will also have the same direction, since the direction of $k(\mathbf{u} + \mathbf{v})$ will be opposite to that of $\mathbf{u} + \mathbf{v}$.

The case where $k = 0$ is trivial.

EXERCISE SET 3.3

1. (a) $\mathbf{u} \cdot \mathbf{v} = (2)(5) + (3)(-7) = -11$

   (c) $\mathbf{u} \cdot \mathbf{v} = (1)(3) + (-5)(3) + (4)(3) = 0$

2. We have $\|\mathbf{u}\| = \left[2^2 + 3^2\right]^{1/2} = \sqrt{13}$ and $\|\mathbf{v}\| = \left[5^2 + (-7)^2\right]^{1/2} = \sqrt{74}$. From Problem 1.(a), we know that $\mathbf{u} \cdot \mathbf{v} = -11$. Hence,

$$\cos\theta = \frac{-11}{\sqrt{13}\,\sqrt{74}} \approx 110.8^{\circ}$$

   (c) From Problem 1.(c), we know that $\mathbf{u} \cdot \mathbf{v} = 0$. Since neither $\mathbf{u}$ nor $\mathbf{v}$ is the zero vector, this implies that $\cos\theta = 0$.

3. (a) $\mathbf{u} \cdot \mathbf{v} = (6)(2) + (1)(0) + (4)(-3) = 0$. Thus the vectors are orthogonal.

   (b) $\mathbf{u} \cdot \mathbf{v} = -1 < 0$. Thus θ is obtuse.

4. (a) Since $\mathbf{u} \cdot \mathbf{a} = 0$ and $\|\mathbf{a}\| \neq 0$, then $\mathbf{w}_1 = (0,0)$.

   (c) Since $\mathbf{u} \cdot \mathbf{a} = (3)(1) + (1)(0) + (-7)(5) = -32$ and $\|\mathbf{a}\|^2 = 1^2 + 0^2 + 5^2 = 26$, we have

$$\mathbf{w}_1 = -\frac{32}{26}(1,0,5) = \left(-\frac{16}{13},\ 0,\ -\frac{80}{13}\right)$$

5. **(a)** From Problem 4.(a), we have

$$\mathbf{w}_2 = \mathbf{u} - \mathbf{w}_1 = \mathbf{u} = (6,2)$$

(c) From Problem 4.(c), we have

$$\mathbf{w}_2 = (3, 1, -7) - (-16/13, 0, -80/13) = (55/13, 1, -11/13)$$

6. **(a)** Since $|\mathbf{u} \cdot \mathbf{a}| = |-4 + 6| = |2| = 2$ and $\|\mathbf{a}\| = 5$, it follows from Formula 10 that $\|\text{proj}_{\mathbf{a}}\mathbf{u}\| = 2/5$.

(d) Since $|\mathbf{u} \cdot \mathbf{a}| = |3 - 4 - 42| = |-43| = 43$ and $\|\mathbf{a}\| = \sqrt{54}$, it follows from Formula (10) that $\|\text{proj}_{\mathbf{a}}\mathbf{u}\| = 43/\sqrt{54}$.

8. **(c)** From Part (a) it follows that the vectors (-4,-3) and (4,3) are orthogonal to (-3,4). Divide each component by the norm, 5, to obtain the desired unit vectors.

10. **(a)** The inner product $\mathbf{x} \cdot \mathbf{y}$ is defined only if both $\mathbf{x}$ and $\mathbf{y}$ are vectors, but here $\mathbf{v} \cdot \mathbf{w}$ is a scalar.

(b) We can add two vectors or two scalars, but not one of each.

(c) The norm of $\mathbf{x}$ is defined only for $\mathbf{x}$ a vector, but $\mathbf{u} \cdot \mathbf{v}$ is a scalar.

(d) Again, the dot product of a scalar and a vector is undefined.

12. Since the three points are not collinear (why?), they do form a triangle. Since

$$\overrightarrow{AB} \cdot \overrightarrow{BC} = (1,3,-2) \cdot (4,-2,1) = 0$$

the right angle is at B.

13. If, for instance, $\mathbf{a} = (1,0,0)$, $\mathbf{b} = (0,1,0)$ and $\mathbf{c} = (0,0,1)$, we have $\mathbf{a} \cdot \mathbf{b} = \mathbf{a} \cdot \mathbf{c} = 0$, but $\mathbf{b} \neq \mathbf{c}$.

14. (a) We look for a value of k which will make $\mathbf{p}$ a nonzero multiple of $\mathbf{q}$. If $\mathbf{p} = c\mathbf{q}$, then $(2,k) = c(3,5)$, so that $k/5 = 2/3$ or $k = 10/3$.

(c) We want k such that

$$\mathbf{p} \cdot \mathbf{q} = \|\mathbf{p}\| \, \|\mathbf{q}\| \cos \frac{\pi}{3}$$

or

$$6 + 5k = \sqrt{4 + k^2} \sqrt{3^2 + 5^2} \left(\frac{1}{2}\right)$$

If we square both sides and use the quadratic formula, we find that

$$k = \frac{-60 \pm 34\sqrt{3}}{33}$$

15. (b) Here

$$D = \frac{|4(2) + 1(-5) - 2|}{\sqrt{(4)^2 + (1)^2}} = \frac{1}{\sqrt{17}}$$

16. From the definition of the norm, we have

$$\|\mathbf{u} + \mathbf{v}\|^2 = (\mathbf{u} + \mathbf{v}) \cdot (\mathbf{u} + \mathbf{v})$$

Using Theorem 3.3.2, we obtain

$$\|\mathbf{u} + \mathbf{v}\|^2 = (\mathbf{u} \cdot \mathbf{u}) + (\mathbf{u} \cdot \mathbf{v}) + (\mathbf{v} \cdot \mathbf{u}) + (\mathbf{v} \cdot \mathbf{v})$$

or

$$(*) \qquad \|\mathbf{u} + \mathbf{v}\|^2 = \|\mathbf{u}\|^2 + 2(\mathbf{u} \cdot \mathbf{v}) + \|\mathbf{v}\|^2$$

Similarly,

$$(**) \qquad \|\mathbf{u} - \mathbf{v}\|^2 = \|\mathbf{u}\|^2 - 2(\mathbf{u} \cdot \mathbf{v}) + \|\mathbf{v}\|^2$$

If we add Equations (*) and (**), we obtain the desired result.

17. If we subtract Equation (**) from Equation (*) in the solution to Problem 16, we obtain

$$\|\mathbf{u} + \mathbf{v}\|^2 - \|\mathbf{u} - \mathbf{v}\|^2 = 4(\mathbf{u} \cdot \mathbf{v})$$

If we then divide both sides by 4, we obtain the desired result.

18. Let $\mathbf{u}_1$, $\mathbf{u}_2$, and $\mathbf{u}_3$ be three sides of the cube as shown. The diagonal of the face of the cube determined by $\mathbf{u}_1$ and $\mathbf{u}_2$ is $\mathbf{b} = \mathbf{u}_1 + \mathbf{u}_2$; the diagonal of the cube itself is $\mathbf{a} = \mathbf{u}_1 + \mathbf{u}_2 + \mathbf{u}_3$. The angle θ between $\mathbf{a}$ and $\mathbf{b}$ is given by

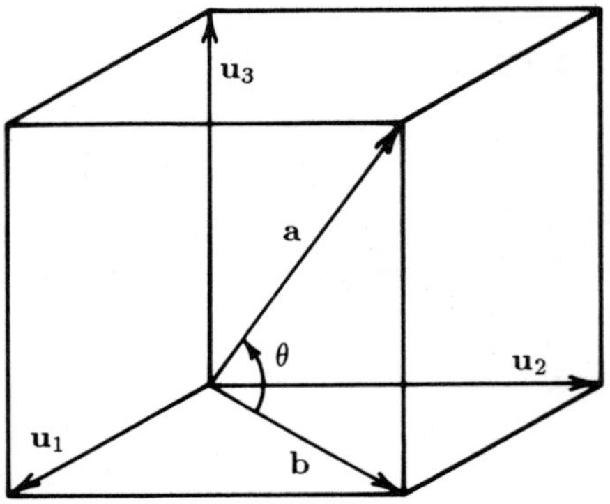

$$\cos\theta = \frac{\mathbf{a} \cdot \mathbf{b}}{\|\mathbf{a}\|\ \|\mathbf{b}\|}$$

$$= \frac{(\mathbf{u}_1 + \mathbf{u}_2 + \mathbf{u}_3) \cdot (\mathbf{u}_1 + \mathbf{u}_2)}{\sqrt{(\mathbf{u}_1 + \mathbf{u}_2 + \mathbf{u}_3)\cdot(\mathbf{u}_1 + \mathbf{u}_2 + \mathbf{u}_3)}\ \sqrt{(\mathbf{u}_1 + \mathbf{u}_2)\cdot(\mathbf{u}_1 + \mathbf{u}_2)}}$$

Because $\mathbf{u}_1$, $\mathbf{u}_2$, and $\mathbf{u}_3$ are mutually orthogonal, we have $\mathbf{u}_i \cdot \mathbf{u}_j = 0$ whenever $i \neq j$. Also $\mathbf{u}_i \cdot \mathbf{u}_i = \|\mathbf{u}_i\|^2$ and $\|\mathbf{u}_1\| = \|\mathbf{u}_2\| = \|\mathbf{u}_3\|$. Thus

$$\cos\theta = \frac{\|\mathbf{u}_1\|^2 + \|\mathbf{u}_2\|^2}{\sqrt{\|\mathbf{u}_1\|^2 + \|\mathbf{u}_2\|^2 + \|\mathbf{u}_3\|^2}\ \sqrt{\|\mathbf{u}_1\|^2 + \|\mathbf{u}_2\|^2}}$$

$$= \frac{2\|\mathbf{u}_1\|^2}{\sqrt{3\|\mathbf{u}_1\|^2}\ \sqrt{2\|\mathbf{u}_1\|^2}} = \frac{2}{\sqrt{6}}$$

That is, $\theta = \arccos(2/\sqrt{6})$.

19. (a) Let $\mathbf{i} = (1,0,0)$, $\mathbf{j} = (0,1,0)$, and $\mathbf{k} = (0,0,1)$ denote the unit vectors along the x, y, and z axes, respectively. If $\mathbf{v}$ is the arbitrary vector (a,b,c), then we can write $\mathbf{v} = a\mathbf{i} + b\mathbf{j} + c\mathbf{k}$. Hence, the angle α between $\mathbf{v}$ and $\mathbf{i}$ is given by

$$\cos\alpha = \frac{\mathbf{v}\cdot\mathbf{i}}{\|\mathbf{v}\|\ \|\mathbf{i}\|} = \frac{a}{\sqrt{a^2+b^2+c^2}} = \frac{a}{\|\mathbf{v}\|}$$

since $\|\mathbf{i}\| = 1$ and $\mathbf{i}\cdot\mathbf{j} = \mathbf{i}\cdot\mathbf{k} = 0$.

21. We can suppose that neither $\mathbf{v}_1$ nor $\mathbf{v}_2$ is the zero vector. By the results of Exercise 19, we have that if $\mathbf{v}_i = (a_i, b_i, c_i)$ for $i = 1,2$, then $\cos\alpha_i = a_i/\|\mathbf{v}_i\|$, $\cos\beta_i = b_i/\|\mathbf{v}_i\|$, and $\cos\gamma_i = c_i/\|\mathbf{v}_i\|$. Now $\mathbf{v}_1$ and $\mathbf{v}_2$ are orthogonal $\Longleftrightarrow \mathbf{v}_1\cdot\mathbf{v}_2 = 0 \Longleftrightarrow a_1a_2 + b_1b_2 + c_1c_2 = 0$

$$\Longleftrightarrow \frac{a_1a_2}{\|\mathbf{v}_1\|\ \|\mathbf{v}_2\|} + \frac{b_1b_2}{\|\mathbf{v}_1\|\ \|\mathbf{v}_2\|} + \frac{c_1c_2}{\|\mathbf{v}_1\|\ \|\mathbf{v}_2\|} = 0$$

$$\Longleftrightarrow \cos\alpha_1\ \cos\alpha_2 + \cos\beta_1\ \cos\beta_2 + \cos\gamma_1\ \cos\gamma_2 = 0$$

22. Note that

$$\mathbf{v}\cdot(k_1\mathbf{w}_1 + k_2\mathbf{w}_2) = k_1(\mathbf{v}\cdot\mathbf{w}_1) + k_2(\mathbf{v}\cdot\mathbf{w}_2) = 0$$

because, by hypothesis, $\mathbf{v}\cdot\mathbf{w}_1 = \mathbf{v}\cdot\mathbf{w}_2 = 0$. Therefore $\mathbf{v}$ is orthogonal to $k_1\mathbf{w}_1 + k_2\mathbf{w}_2$ for any scalars k_1 and k_2.

23. Note that $\mathbf{w}$ is a multiple of $\mathbf{u}$ plus a multiple of $\mathbf{v}$ and thus lies in the plain determined by $\mathbf{u}$ and $\mathbf{v}$. Let α be the angle between $\mathbf{u}$ and $\mathbf{w}$ and let β be the angle between $\mathbf{v}$ and $\mathbf{w}$. Then

$$\cos\alpha = \frac{\mathbf{u}\cdot\mathbf{w}}{\|\mathbf{u}\|\ \|\mathbf{v}\|} = \frac{[k\mathbf{u}\cdot\mathbf{v} + \ell\mathbf{u}\cdot\mathbf{u}]}{k\|\mathbf{w}\|}$$

$$= \frac{[k\mathbf{u}\cdot\mathbf{v} + \ell k^2]}{k\|\mathbf{w}\|}$$

$$= \frac{\mathbf{u}\cdot\mathbf{v} + \ell k}{\|\mathbf{w}\|}$$

Similarly, we can show that

$$\cos\beta = \frac{\mathbf{v}\cdot\mathbf{w}}{\|\mathbf{v}\|\ \|\mathbf{w}\|} = \frac{\mathbf{u}\cdot\mathbf{v} + kl}{\|\mathbf{w}\|}$$

Since both α and β lie between 0 and π, then $\cos\alpha = \cos\beta$ implies that $\alpha = \beta$.

EXERCISE SET 3.4

1. **(a)** $\mathbf{v} \times \mathbf{w} = \left(\begin{vmatrix} 2 & -3 \\ 6 & 7 \end{vmatrix}, -\begin{vmatrix} 0 & -3 \\ 2 & 7 \end{vmatrix}, \begin{vmatrix} 0 & 2 \\ 2 & 6 \end{vmatrix} \right) = (32,-6,-4)$

   (c) Since

$$\mathbf{u} \times \mathbf{v} = \left(\begin{vmatrix} 2 & -1 \\ 2 & -3 \end{vmatrix}, -\begin{vmatrix} 3 & -1 \\ 0 & -3 \end{vmatrix}, \begin{vmatrix} 3 & 2 \\ 0 & 2 \end{vmatrix} \right)$$

$$= (-4,9,6)$$

   we have

$$(\mathbf{u} \times \mathbf{v}) \times \mathbf{w} = \left(\begin{vmatrix} 9 & 6 \\ 6 & 7 \end{vmatrix}, -\begin{vmatrix} -4 & 6 \\ 2 & 7 \end{vmatrix}, \begin{vmatrix} -4 & 9 \\ 2 & 6 \end{vmatrix} \right)$$

$$= (27,40,-42)$$

   (e) Since

$$\mathbf{v} - 2\mathbf{w} = (0,2,-3) - (4,12,14) = (-4,-10,-17)$$

   we have

$$\mathbf{u} \times (\mathbf{v} - 2\mathbf{w}) = \left(\begin{vmatrix} 2 & -1 \\ -10 & -17 \end{vmatrix}, -\begin{vmatrix} 3 & -1 \\ -4 & -17 \end{vmatrix}, \begin{vmatrix} 3 & 2 \\ -4 & -10 \end{vmatrix} \right)$$

$$= (-44,55,-22)$$

2. **(a)** By Theorem 3.4.1, **u** × **v** will be orthogonal to both **u** and **v** where

$$\mathbf{u} \times \mathbf{v} = \left(\begin{vmatrix} 4 & 2 \\ 1 & 5 \end{vmatrix}, -\begin{vmatrix} -6 & 2 \\ 3 & 5 \end{vmatrix}, \begin{vmatrix} -6 & 4 \\ 3 & 1 \end{vmatrix} \right)$$

$$= (18,36,-18)$$

so (1,2,-1), for instance, is orthogonal to both **u** and **v**.

3. **(a)** We have $\mathbf{u} = \overrightarrow{PQ} = (-1,-5,2)$ and $\mathbf{v} = \overrightarrow{PR} = (2,0,3)$. Thus

$$\mathbf{u} \times \mathbf{v} = \left(\begin{vmatrix} -5 & 2 \\ 0 & 3 \end{vmatrix}, -\begin{vmatrix} -1 & 2 \\ 2 & 3 \end{vmatrix}, \begin{vmatrix} -1 & -5 \\ 2 & 0 \end{vmatrix} \right)$$

$$= (-15,7,10)$$

so that

$$\|\mathbf{u} \times \mathbf{v}\| = \left[(-15)^2 + 7^2 + 10^2\right]^{1/2} = \sqrt{374}$$

Thus the area of the triangle is $\sqrt{374}/2$.

4. **(a)** Since $\mathbf{u} \times \mathbf{v} = (-7,-1,3)$, the area of the parallelogram is $\|\mathbf{u} \times \mathbf{v}\| = \sqrt{59}$.

6. For Part (*a*), we have

$$\mathbf{u} \times \mathbf{v} = \left(\begin{vmatrix} 1 & 2 \\ 0 & -2 \end{vmatrix}, -\begin{vmatrix} 5 & 2 \\ 6 & -2 \end{vmatrix}, \begin{vmatrix} 5 & -1 \\ 6 & 0 \end{vmatrix} \right)$$

$$= (2,22,6)$$

and

$$\mathbf{v} \times \mathbf{u} = \left(\begin{vmatrix} 0 & -2 \\ -1 & 2 \end{vmatrix}, -\begin{vmatrix} 6 & -2 \\ 5 & 2 \end{vmatrix}, \begin{vmatrix} 6 & 0 \\ 5 & -1 \end{vmatrix} \right)$$

$$= (-2,-22,-6)$$

Hence

$$\mathbf{u} \times \mathbf{v} = -(\mathbf{v} \times \mathbf{u})$$

For Part (*b*), we have $\mathbf{v} + \mathbf{w} = (7,2,-3)$. Hence

$$\mathbf{u} \times (\mathbf{v} + \mathbf{w}) = \left(\begin{vmatrix} -1 & 2 \\ 2 & -3 \end{vmatrix}, -\begin{vmatrix} 5 & 2 \\ 7 & -3 \end{vmatrix}, \begin{vmatrix} 5 & -1 \\ 7 & 2 \end{vmatrix} \right)$$

$$= (-1,29,17)$$

On the other hand, from Part (*a*) we have

$$\mathbf{u} \times \mathbf{v} = (2,22,6)$$

Also

$$\mathbf{u} \times \mathbf{w} = \left(\begin{vmatrix} -1 & 2 \\ 2 & -1 \end{vmatrix}, -\begin{vmatrix} 5 & 2 \\ 1 & -1 \end{vmatrix}, \begin{vmatrix} 5 & -1 \\ 1 & 2 \end{vmatrix} \right)$$

$$= (-3,7,11)$$

Thus

$$(\mathbf{u} \times \mathbf{v}) + (\mathbf{u} \times \mathbf{w}) = (2,22,6) + (-3,7,11)$$
$$= (-1,29,17)$$

and we have that

$$\mathbf{u} \times (\mathbf{v} + \mathbf{w}) = (\mathbf{u} \times \mathbf{v}) + (\mathbf{u} \times \mathbf{w})$$

The proof of Part (*c*) is similar to the proof of Part (*b*).

For Part (*d*), we already have that $\mathbf{u} \times \mathbf{v} = (2,-22,6)$. Hence $k(\mathbf{u} \times \mathbf{v}) = (-10,110,-30)$. But $k\mathbf{u} = (-25,5,-10)$ and $k\mathbf{v} = (-30,0,10)$. Thus

$$(k\mathbf{u}) \times \mathbf{v} = \left(\begin{vmatrix} 5 & -10 \\ 0 & -2 \end{vmatrix}, -\begin{vmatrix} -25 & -10 \\ 6 & -2 \end{vmatrix}, \begin{vmatrix} -25 & 5 \\ 6 & 0 \end{vmatrix} \right)$$

$$= (-10,-110,-30)$$

and

$$\mathbf{u} \times (k\mathbf{v}) = \left(\begin{vmatrix} -1 & 2 \\ 0 & 10 \end{vmatrix}, -\begin{vmatrix} 5 & 2 \\ -30 & 10 \end{vmatrix}, \begin{vmatrix} 5 & -1 \\ -30 & 0 \end{vmatrix} \right)$$

$$= (-10,-110,-30)$$

Thus

$$k(\mathbf{u} \times \mathbf{v}) = (k\mathbf{u}) \times \mathbf{v} = \mathbf{u} \times (k\mathbf{v})$$

For Part (*e*), we have

$$\mathbf{u} \times \mathbf{0} = \left(\begin{vmatrix} -1 & 2 \\ 0 & 0 \end{vmatrix}, -\begin{vmatrix} 5 & 2 \\ 0 & 0 \end{vmatrix}, \begin{vmatrix} 5 & -1 \\ 0 & 0 \end{vmatrix} \right)$$

$$= (0,0,0) = \mathbf{0}$$

Similarly $\mathbf{0} \times \mathbf{u} = \mathbf{0}$.

Finally, for Part (*f*), we have

$$\mathbf{u} \times \mathbf{u} = \left(\begin{vmatrix} -1 & 2 \\ -1 & 2 \end{vmatrix}, -\begin{vmatrix} 5 & 2 \\ 5 & 2 \end{vmatrix}, \begin{vmatrix} 5 & -1 \\ 5 & -1 \end{vmatrix} \right)$$

$$= (0,0,0) = \mathbf{0}$$

8. **(a)** We have

$$\mathbf{u}\cdot(\mathbf{v}\times\mathbf{w}) = \begin{vmatrix} -1 & 2 & 4 \\ 3 & 4 & -2 \\ -1 & 2 & 5 \end{vmatrix}$$

$$= -(20 + 4) - 2(15 - 2) + 4(6 + 4)$$

$$= -10$$

9. **(e)** Since $(\mathbf{u}\times\mathbf{w})\cdot\mathbf{v} = \mathbf{v}\cdot(\mathbf{u}\times\mathbf{w})$ is a determinant whose rows are the components of **v**, **u**, and **w**, respectively, we interchange Rows 1 and 2 to obtain the determinant which represents $\mathbf{u}\cdot(\mathbf{v}\times\mathbf{w})$. Since the value of this determinant is 3, we have $(\mathbf{u}\times\mathbf{w})\cdot\mathbf{v} = -3$.

10. **(a)** Call this volume V. Then, since

$$\begin{vmatrix} 2 & -6 & 2 \\ 0 & 4 & -2 \\ 2 & 2 & -4 \end{vmatrix} = 2\begin{vmatrix} 4 & -2 \\ 2 & -4 \end{vmatrix} + 2\begin{vmatrix} -6 & 2 \\ 4 & -2 \end{vmatrix} = -16$$

we have $V = 16$.

11. **(a)** Since the determinant

$$\begin{vmatrix} -1 & -2 & 1 \\ 3 & 0 & -2 \\ 5 & -4 & 0 \end{vmatrix} = 16 \neq 0$$

the vectors do not lie in the same plane.

12. For a vector to be parallel to the yz-plane, it must be perpendicular to the vector (1,0,0), so we are looking for a vector which is perpendicular to both (1,0,0) and (3,−1,2). Such a vector is $(1,0,0)\times(3,-1,2) = (0,-2,-1)$. Since we want a unit vector, we divide through by the norm to obtain the vector $(0, -2/\sqrt{5}, -1/\sqrt{5})$. Obviously any vector parallel to this one, such as $(0, 2/\sqrt{5}, 1/\sqrt{5})$, will also work.

15. By Theorem 3.4.2, we have

$$\begin{aligned}(\mathbf{u} + \mathbf{v}) \times (\mathbf{u} - \mathbf{v}) &= \mathbf{u} \times (\mathbf{u} - \mathbf{v}) + \mathbf{v} \times (\mathbf{u} - \mathbf{v}) \\ &= (\mathbf{u} \times \mathbf{u}) + (\mathbf{u} \times (-\mathbf{v})) + (\mathbf{v} \times \mathbf{u}) + (\mathbf{v} \times (-\mathbf{v})) \\ &= 0 - (\mathbf{u} \times \mathbf{v}) - (\mathbf{u} \times \mathbf{v}) - (\mathbf{v} \times \mathbf{v}) \\ &= -2(\mathbf{u} \times \mathbf{v})\end{aligned}$$

17. **(a)** The area of the triangle with sides $\overrightarrow{AB}$ and $\overrightarrow{AC}$ is the same as the area of the triangle with sides (−1,2,2) and (1,1,−1) where we have "moved" A to the origin and translated B and C accordingly. This area is $\frac{1}{2}\,\|(-1,2,2) \times (1,1,-1)\| = \frac{1}{2}\,\|(-4,1,-3)\| = \sqrt{26}/2$.

18. If the vector **u** and the given line make an angle θ, then the distance D between the point and the line is given by

$$\begin{aligned}D &= \|\mathbf{u}\| \sin\theta && \text{(Why?)} \\ &= \|\mathbf{u} \times \mathbf{v}\|/\|\mathbf{v}\| && \text{(Why?)}\end{aligned}$$

19. **(a)** Let $\mathbf{u} = \overrightarrow{AP} = (-4,0,2)$ and $\mathbf{v} = \overrightarrow{AB} = (-3,2,-4)$. Then the distance we want is

$$\begin{aligned}\|(-4,0,2) \times (-3,2,-4)\|/\|(-3,2,-4)\| &= \|(-4,-22,-8)\|/\sqrt{29} \\ &= 2\sqrt{141}/\sqrt{29}\end{aligned}$$

20. We have that $\mathbf{u} \cdot \mathbf{v} = \|\mathbf{u}\|\,\|\mathbf{v}\| \cos\theta$ and $\|\mathbf{u} \times \mathbf{v}\| = \|\mathbf{u}\|\,\|\mathbf{v}\| \sin\theta$. Thus if $\mathbf{u} \cdot \mathbf{v} \neq 0$ then $\|\mathbf{u}\|$, $\|\mathbf{v}\|$, and $\cos\theta$ are all nonzero, and

$$\tan\theta = \frac{\sin\theta}{\cos\theta} = \frac{\|\mathbf{u} \times \mathbf{v}\|}{\|\mathbf{u}\|\,\|\mathbf{v}\|} \Big/ \frac{\mathbf{u} \cdot \mathbf{v}}{\|\mathbf{u}\|\,\|\mathbf{v}\|} = \frac{\|\mathbf{u} \times \mathbf{v}\|}{\mathbf{u} \cdot \mathbf{v}}$$

21. **(b)** Call the desired angle θ. Then we have that

$$|\text{proj}_{\mathbf{v}\times\mathbf{w}}\mathbf{u}| = \frac{|\mathbf{u} \cdot (\mathbf{v} \times \mathbf{w})|}{\|\mathbf{v} \times \mathbf{w}\|} = \|\mathbf{u}\| \sin\theta \qquad \text{(Why?)}$$

or

$$\frac{|(3,2,1) \cdot (-3,-1,2)|}{\sqrt{9+1+4}} = \sqrt{9+4+1}\ \sin\theta$$

or

$$\sin\theta = 9/14$$

so that

$$\theta = \sin^{-1}(9/14) \approx 40^\circ 19''$$

24. **(a)** By Theorems 3.4.2 and 3.2.1 and the definition of $k\mathbf{0}$,

$$\begin{aligned}(\mathbf{u} + k\mathbf{v}) \times \mathbf{v} &= (\mathbf{u} \times \mathbf{v}) + (k\mathbf{v} \times \mathbf{v}) \\ &= (\mathbf{u} \times \mathbf{v}) + k(\mathbf{v} \times \mathbf{v}) \\ &= (\mathbf{u} \times \mathbf{v}) + k\mathbf{0} \\ &= \mathbf{u} \times \mathbf{v}\end{aligned}$$

(b) Let $\mathbf{u} = (u_1, u_2, u_3)$, $\mathbf{v} = (v_1, v_2, v_3)$, and $\mathbf{z} = (z_1, z_2, z_3)$. Then we know from the text that

$$\mathbf{u} \cdot (\mathbf{v} \times \mathbf{z}) = \begin{vmatrix} u_1 & u_2 & u_3 \\ v_1 & v_2 & v_3 \\ z_1 & z_2 & z_3 \end{vmatrix}$$

Since $\mathbf{x} \cdot \mathbf{y} = \mathbf{y} \cdot \mathbf{x}$ for any vectors $\mathbf{x}$ and $\mathbf{y}$, we have

$$\begin{aligned}(\mathbf{u} \times \mathbf{z}) \cdot \mathbf{v} = \mathbf{v} \cdot (\mathbf{u} \times \mathbf{z}) &= \begin{vmatrix} v_1 & v_2 & v_3 \\ u_1 & u_2 & u_3 \\ z_1 & z_2 & z_3 \end{vmatrix} \\ &= -\begin{vmatrix} u_1 & u_2 & u_3 \\ v_1 & v_2 & v_3 \\ z_1 & z_2 & z_3 \end{vmatrix} \\ &= -\mathbf{u} \cdot (\mathbf{v} \times \mathbf{z})\end{aligned}$$

25. **(a)** By Theorem 3.4.1, we know that the vector $\mathbf{v} \times \mathbf{w}$ is perpendicular to both $\mathbf{v}$ and $\mathbf{w}$. Hence $\mathbf{v} \times \mathbf{w}$ is perpendicular to every vector in the plane determined by $\mathbf{v}$ and $\mathbf{w}$; moreover the only vectors perpendicular to $\mathbf{v} \times \mathbf{w}$ which share its initial point must be in this plane. But also by Theorem 3.4.1, $\mathbf{u} \times (\mathbf{v} \times \mathbf{w})$ is perpendicular to $\mathbf{v} \times \mathbf{w}$ for any vector $\mathbf{u} \neq \mathbf{0}$ and hence must lie in the plane determined by $\mathbf{v}$ and $\mathbf{w}$.

(b) The argument is completely similar to Part (a), above.

26. Let $\mathbf{x} = (x_1, x_2, x_3)$, $\mathbf{y} = (y_1, y_2, y_3)$, and $\mathbf{z} = (z_1, z_2, z_3)$. Following the hint, we find

$$\mathbf{y} \times \mathbf{i} = (0, y_3, -y_2)$$

and

$$\mathbf{x} \times (\mathbf{y} \times \mathbf{i}) = (-x_2y_2 - x_3y_3,\ x_1y_2,\ x_1y_3)$$

But

$$\begin{aligned}(\mathbf{x} \cdot \mathbf{i})\mathbf{y} - (\mathbf{x} \cdot \mathbf{y})\mathbf{i} &= x_1(y_1, y_2, y_3) - (x_1y_1 + x_2y_2 + x_3y_3)(1, 0, 0) \\ &= (x_1y_1, x_1y_2, x_1y_3) - (x_1y_1 + x_2y_2 + x_3y_3, 0, 0) \\ &= (-x_2y_2 - x_3y_3, x_1y_2, x_1y_3) \\ &= \mathbf{x} \times (\mathbf{y} \times \mathbf{i})\end{aligned}$$

Similarly,

$$\mathbf{x} \times (\mathbf{y} \times \mathbf{j}) = (\mathbf{x} \cdot \mathbf{j})\mathbf{y} - (\mathbf{x} \cdot \mathbf{y})\mathbf{j}$$

and

$$\mathbf{x} \times (\mathbf{y} \times \mathbf{k}) = (\mathbf{x} \cdot \mathbf{k})\mathbf{y} - (\mathbf{x} \cdot \mathbf{y})\mathbf{k}$$

Now write $\mathbf{z} = (z_1, z_2, z_3) = z_1\mathbf{i} + z_2\mathbf{j} + z_3\mathbf{k}$. Then

$$\begin{aligned}
\mathbf{x} \times (\mathbf{y} \times \mathbf{z}) &= \mathbf{x} \times (\mathbf{y} \times (z_1\mathbf{i} + z_2\mathbf{j} + z_3\mathbf{k})) \\
&= z_1[\mathbf{x} \times (\mathbf{y} \times \mathbf{i})] + z_2[\mathbf{x} \times (\mathbf{y} \times \mathbf{j})] + z_3[\mathbf{x} \times (\mathbf{y} \times \mathbf{k})] \\
&= z_1[(\mathbf{x} \cdot \mathbf{i})\mathbf{y} - (\mathbf{x} \cdot \mathbf{y})\mathbf{i}] + z_2[(\mathbf{x} \cdot \mathbf{j})\mathbf{y} - (\mathbf{x} \cdot \mathbf{y})\mathbf{j}] \\
&\quad + z_3[(\mathbf{x} \cdot \mathbf{k})\mathbf{y} - (\mathbf{x} \cdot \mathbf{y})\mathbf{k}] \\
&= (\mathbf{x} \cdot (z_1\mathbf{i} + z_2\mathbf{j} + z_3\mathbf{k}))\mathbf{y} - (\mathbf{x} \cdot \mathbf{y})(z_1\mathbf{i} + z_2\mathbf{j} + z_3\mathbf{k}) \\
&= (\mathbf{x} \cdot \mathbf{z})\mathbf{y} - (\mathbf{x} \cdot \mathbf{y})\mathbf{z}
\end{aligned}$$

28. If **a**, **b**, **c**, and **d** lie in the same plane, then $(\mathbf{a} \times \mathbf{b})$ and $(\mathbf{c} \times \mathbf{d})$ are both perpendicular to this plane, and are therefore parallel. Hence, their cross-product is zero.

29. Recall that $|\mathbf{a} \cdot (\mathbf{b} \times \mathbf{c})|$ represents the volume of the parallelepiped with sides **a**, **b**, and **c**, which is the area of its base times its height. The volume of the tetrahedron is just $\frac{1}{3}$ (area of its base) times (its height). The two heights are the same, but the area of the base of the tetrahedron is half of the area of the base of the parallelepiped. Hence, the volume of the tetrahedron is $\frac{1}{3}\left[\frac{1}{2}|\mathbf{a} \cdot (\mathbf{b} \times \mathbf{c})|\right]$.

30. **(a)** The required volume is

$$\begin{aligned}
\frac{1}{6}\,|(-1-3, 2+2, 0-3) \cdot ((2-3, 1+2, -3-3) \times (1-3, 0+2, 1-3))| \\
= \frac{1}{6}\,|(-4,4,-3) \cdot (6,10,4)| \\
= 2/3
\end{aligned}$$

32. Let $\mathbf{u} = (u_1, u_2, u_3)$, $\mathbf{v} = (v_1, v_2, v_3)$, and $\mathbf{w} = (w_1, w_2, w_3)$.

For Part (*c*), we have

$$\mathbf{u} \times \mathbf{w} = (u_2w_3 - u_3w_2,\ u_3w_1 - u_1w_3,\ u_1w_2 - u_2w_1)$$

and

$$\mathbf{v} \times \mathbf{w} = (v_2w_3 - v_3w_2,\ v_3w_1 - v_1w_3,\ v_1w_2 - v_2w_1)$$

Thus

$$(\mathbf{u} \times \mathbf{w}) + (\mathbf{v} \times \mathbf{w})$$

$$= \left([u_2+v_2]w_3 - [u_3+v_3]w_2,\ [u_3+v_3]w_1 - [u_1+v_1]w_3,\ [u_1+v_1]w_2 - [u_2+v_2]w_1\right)$$

But, by definition, this is just $(\mathbf{u} + \mathbf{v}) \times \mathbf{w}$.

For Part (*d*), we have

$$k(\mathbf{u} \times \mathbf{v}) = \left(k[u_2v_3 - u_3v_2],\ k[u_3v_1 - u_1v_3],\ k[u_1v_2 - u_2v_1]\right)$$

and

$$(k\mathbf{u}) \times \mathbf{v} = \left(ku_2v_3 - ku_3v_2,\ ku_3v_1 - ku_1v_3,\ ku_1v_2 - ku_2v_1\right)$$

Thus, $k(\mathbf{u} \times \mathbf{v}) = (k\mathbf{u}) \times \mathbf{v}$. The identity $k(\mathbf{u} \times \mathbf{v}) = \mathbf{u} \times (k\mathbf{v})$ may be proved in an analogous way.

EXERCISE SET 3.5

4. **(a)** We have

$$\overrightarrow{PQ} = (2,1,2) \quad \text{and} \quad \overrightarrow{PR} = (3,-1,-2)$$

Thus $\overrightarrow{PQ} \times \overrightarrow{PR} = (0,10,-5)$ is perpendicular to the plane determined by $\overrightarrow{PQ}$ and $\overrightarrow{PR}$ and P, say, is a point in that plane. Hence, an equation for the plane is

$$0(x + 4) + 10(y + 1) - 5(z + 1) = 0$$

or

$$2y - z + 1 = 0$$

5. **(a)** Normal vectors for the planes are (4,-1,2) and (7,-3,4). Since these vectors are not multiples of one another, the planes are not parallel.

   (b) The normal vectors are (1,-4,-3) and (3,-12,-9). Since one vector is three times the other, the planes are parallel.

6. **(a)** A normal vector for the plane is (1,2,3) and a direction vector for the line is (-4,-1,2). The inner product of these two vectors is $-4 - 2 + 6 = 0$, and therefore they are perpendicular. This guarantees that the line and the plane are parallel.

7. **(a)** Normal vectors for the planes are (3,−1,1) and (1,0,2). Since the inner product of these two vectors is not zero, the planes are not perpendicular.

8. **(a)** A normal vector for the plane is (2,1,−1) and a direction vector for the line is (−4,−2,2). Since one of these vectors is a multiple of the other, the line and the plane are perpendicular.

11. **(a)** Call the points P and Q and call the line ℓ. Then the vector $\overrightarrow{PQ} = (2,4,-8)$ is parallel to ℓ and the point $P = (5,-2,4)$ lies on ℓ. Hence, one set of parametric equations for ℓ is: $x = 5 + t$, $y = -2 + 2t$, $z = 4 - 4t$ where t is any real number.

12. **(a)** As in Example 5, we solve the two equations simultaneously. If we eliminate y, we have $x + 7z + 12 = 0$. Let, say, $z = t$, so that $x = -12 - 7t$, and substitute these values into the equation for either plane to get $y = -41 - 23t$.

Alternatively, recall that a direction vector for the line is just the cross-product of the normal vectors for the two planes, i.e.,

$$(7,-2,3) \times (-3,1,2) = (-7,-23,1)$$

Thus if we can find a point which lies on the line (that is, any point whose coordinates satisfy the equations for both planes), we are done. If we set $z = 0$ and solve the two equations simultaneously, we get $x = -12$ and $y = -41$, so that

$$x = -12 - 7t$$
$$y = -41 - 23t$$
$$z = 0 + t$$

is one set of equations for the line (see above).

13. **(a)** The symmetric equations for the line are

$$\frac{x - 7}{4} = \frac{y + 5}{2} = \frac{z - 5}{-1}$$

Thus, two of the planes are given by

$$\frac{x - 7}{4} = \frac{y + 5}{2} \quad \text{and} \quad \frac{x - 7}{4} = \frac{z - 5}{-1}$$

or, equivalently,

$$x - 2y - 17 = 0 \quad \text{and} \quad x + 4z - 27 = 0$$

14. **(a)** Since $6(0) + 4t - 4t = 0$ for all t, it follows that every point on the line also lies in the plane.

(b) The normal to the plane is $\mathbf{n} = (5,-3,3)$; the line is parallel to $\mathbf{v} = (0,1,1)$. But $\mathbf{n} \cdot \mathbf{v} = 0$, so $\mathbf{n}$ and $\mathbf{v}$ are perpendicular. Thus $\mathbf{v}$, and therefore the line, are parallel to the plane. To conclude that the line lies below the plane, simply note that $(0,0,1/3)$ is in the plane and $(0,0,0)$ is on the line.

(c) Here $\mathbf{n} = (6,2,-2)$ and $\mathbf{v}$ is the same as before. Again, $\mathbf{n} \cdot \mathbf{v} = 0$, so the line and the plane are parallel. Since $(0,0,0)$ lies on the line and $(0,0,-3/2)$ lies in the plane, then the line is above the plane.

15. Since the plane is perpendicular to a line with direction $(2,3,-5)$, we can use that vector as a normal to the plane. The point-normal form then yields the equation $2(x + 2) + 3(y - 1) - 5(z - 7) = 0$, or $2x + 3y - 5z + 36 = 0$.

17. **(a)** Since the vector $(0,0,1)$ is perpendicular to the xy-plane, we can use this as the normal for the plane. The point-normal form then yields the equation $z - z_0 = 0$. This equation could just as well have been derived by inspection, since it represents the set of all points with fixed z and x and y arbitrary.

18. The plane will have normal (7,4,−2), so the point-normal form yields $7x + 4y - 2z = 0$.

19. A normal to the plane is $\mathbf{n} = (5,-2,1)$ and the point (3,−6,7) is in the desired plane. Hence, an equation for the plane is $5(x - 3) - 2(y + 6) + (z - 7) = 0$ or $5x - 2y + z - 34 = 0$.

20. If we substitute $x = 9 - 5t$, $y = -1 - t$, and $z = 3 + t$ into the equation for the plane, we find that $t = 40/3$. Substituting this value for t into the parametric equations for the line yields $x = -173/3$, $y = -43/3$, and $z = 49/3$.

22. The two planes intersect at points given by $(-5 - 2t, -7 - 6t, t)$. Two such points are (−5,−7,0) and (−3,−1,−1). The plane $ax + by + cz + d = 0$ through these points and also through (2,4,−1) will satisfy the equations

$$\begin{aligned} -5a - 7b \phantom{{}- c} + d &= 0 \\ -3a - b - c + d &= 0 \\ 2a + 4b - c + d &= 0 \end{aligned}$$

This system of equations has the solution $a = (-1/2)t$, $b = (1/2)t$, $c = 2t$, $d = t$ where t is arbitrary. Thus, if we let $t = -2$, we obtain the equation

$$x - y - 4z - 2 = 0$$

for the desired plane.

Alternatively, note that the line of intersection of the two planes has direction given by the vector $\mathbf{v} = (-2,-6,1)$. It also contains the point (−5,−7,0). The direction of the line connecting this point and (2,4,−1) is given by $\mathbf{w} = (7,11,-1)$. Thus the vector $\mathbf{v} \times \mathbf{w} = (5,-5,-20)$ is normal to the desired plane, so that a point-normal form for the plane is

$$(x - 2) - (y - 4) - 4(z + 1) = 0$$

or

$$x - y - 4z - 2 = 0$$

23. Call the points A, B, C, and D, respectively. Since the vectors $\overrightarrow{AB} = (-1,2,4)$ and $\overrightarrow{BC} = (-2,-1,-2)$ are not parallel, then the points A, B, and C do determine a plane (and not just a line). The normal to this plane is $\overrightarrow{AB} \times \overrightarrow{BC} = (0,-10,5)$. Therefore an equation for the plane is

$$2y - z + 1 = 0$$

Since the coordinates of the point D satisfy this equation, all four points must lie in the same plane.

Alternatively, it would suffice to show that (for instance) $\overrightarrow{AB} \times \overrightarrow{BC}$ and $\overrightarrow{AD} \times \overrightarrow{DC}$ are parallel, so that the planes determined by A, B, and C and A, D, and C are parallel. Since they have points in common, they must coincide.

25. The normals to the two planes are $(4,-2,2)$ and $(3,3,-6)$ or $\mathbf{n}_1 = (2,-1,1)$ and $\mathbf{n}_2 = (1,1,-2)$. The normal $\mathbf{n}$ to a plane which is perpendicular to both of the given planes must be perpendicular to both $\mathbf{n}_1$ and $\mathbf{n}_2$. That is, $\mathbf{n} = \mathbf{n}_1 \times \mathbf{n}_2 = (1,5,3)$. The plane with this normal which passes through the point $(-2,1,5)$ has the equation

$$(x + 2) + 5(y - 1) + 3(z - 5) = 0$$

or

$$x + 5y + 3z - 18 = 0$$

26. The line of intersection of the given planes has the equations $x = -\frac{10}{9} - \frac{t}{3}$, $y = \frac{31}{18} - \frac{t}{3}$, $z = t$ and hence has direction $(1,1,-3)$. The plane with this normal vector which passes through the point $(2,-1,4)$ is the one we are looking for. Its equation is

$$(x - 2) + (y + 1) - 3(z - 4) = 0$$

Note that the normal vector for the desired plane can also be obtained by computing the cross product of the normal vectors of the two given planes.

28. The directions of the two lines are given by the vectors (-2,1,-1) and (2,-1,1). Since each is a multiple of the other, they represent the same direction. The first line passes (for example) through the point (3,4,1) and the second line passes through the points (5,1,7) and (7,0,8), among others. Either of the methods of Example 2 will yield an equation for the plane determined by these points.

29. If, for instance, we set $t = 0$ and $t = -1$ in the line equation, we obtain the points (0,1,-3) and (-1,0,-5). These, together with the given point and the methods of Example 2, will yield an equation for the desired plane.

31. The plane we are looking for is just the set of all points $P = (x,y,z)$ such that the distances from P to the two fixed points are equal. If we equate the squares of these distances, we have

$$(x + 1)^2 + (y + 4)^2 + (z + 2)^2 = (x - 0)^2 + (y + 2)^2 + (z - 2)^2$$

or

$$2x + 1 + 8y + 16 + 4z + 4 = 4y + 4 - 4z + 4$$

or

$$2x + 4y + 8z + 13 = 0$$

32. The vector $\mathbf{v} = (-1,2,-5)$ is parallel to the line and the vector $\mathbf{n} = (-3,1,1)$ is perpendicular to the plane. But $\mathbf{n} \cdot \mathbf{v} = 0$ and the result follows.

33. We change the parameter in the equations for the second line from t to s. The two lines will then intersect if we can find values of s and t such that the x, y, and z coordinates for the two lines are equal; that is, if there are values for s and t such that

$$\begin{aligned} 4t + 3 &= 12s - 1 \\ t + 4 &= 6s + 7 \\ 1 &= 3s + 5 \end{aligned}$$

This system of equations has the solution $t = -5$ and $s = -4/3$. If we then substitute $t = -5$ into the equations for the first line or $s = -4/3$ into the equations for the second line, we find that $x = -17$, $y = -1$, and $z = 1$ is the point of intersection.

34. The vector $\mathbf{v}_1 = (4,1,0)$ is parallel to the first line and $\mathbf{v}_2 = (4,2,1)$ is parallel to the second line. Hence $\mathbf{n} = \mathbf{v}_1 \times \mathbf{v}_2 = (1,-4,4)$ is perpendicular to both lines and is therefore a normal vector for the plane determined by the lines. If we substitute $t = 0$ into the parametric equations for the first line, we see that $(3,4,1)$ must lie in the plane. Hence, an equation for the plane is

$$(x - 3) - 4(y - 4) + 4(z - 1) = 0$$

or

$$x - 4y + 4z + 9 = 0$$

35. **(a)** If we set $z = t$ and solve for x and y in terms of z, then we find that

$$x = \frac{11}{23} + \frac{7}{23}t$$

$$y = -\frac{41}{23} - \frac{1}{23}t$$

$$z = t$$

36. Call the plane $Ax + By + Cz + D = 0$. Since the points $(a,0,0)$, $(0,b,0)$, and $(0,0,c)$ lie in this plane, we have

$$aA + D = 0$$
$$bB + D = 0$$
$$cC + D = 0$$

Thus $A = -D/a$, $B = -D/b$, and $C = -D/c$ and an equation for the plane is

$$\frac{x}{a} + \frac{y}{b} + \frac{z}{c} = 1$$

Alternatively, let P, Q, and R denote the points $(a,0,0)$, $(0,b,0)$, and $(0,0,c)$, respectively. Then

$$\overrightarrow{PQ} = (-a,b,0) \quad \text{and} \quad \overrightarrow{QR} = (0,-b,c)$$

So

$$\mathbf{n} = \overrightarrow{PQ} \times \overrightarrow{QR} = (bc,\ ac,\ ab)$$

is a normal vector for the plane. Since P, say, lies on the plane, we have

$$bc(x - a) + acy + abz = 0$$

Dividing the above equation by abc yields the desired result.

37. **(b)** By Theorem 3.5.1, the distance is

$$D = \frac{|2(-1) + 3(2) - 4(1) - 1|}{\sqrt{2^2 + 3^2 + (-4)^2}} = \frac{1}{\sqrt{29}}$$

38. **(a)** The point $(0,0,1)$ lies in the first plane. The distance between this point and the second plane is

$$D = \frac{|2 - 3|}{\sqrt{6^2 + (-8)^2 + (2)^2}} = \frac{1}{2\sqrt{26}}$$

(b) The two planes coincide, so the distance between them is zero.

39. **(a)** The normals to the two planes are (1,0,0) and (2,−1,1). The angle between them is given by

$$\cos\theta = \frac{(1,0,0)\cdot(2,-1,1)}{\sqrt{1}\sqrt{4+1+1}} = \frac{2}{\sqrt{6}}$$

Thus $\theta = \cos^{-1}(2/\sqrt{6}) \approx 35^\circ 15' 52''$.

EXERCISE SET 4.1

3. We must find numbers c_1, c_2, c_3, and c_4 such that

$$c_1(-1,3,2,0) + c_2(2,0,4,-1) + c_3(7,1,1,4) + c_4(6,3,1,2) = (0,5,6,-3)$$

If we equate vector components, we obtain the following system of equations:

$$\begin{aligned} -c_1 + 2c_2 + 7c_3 + 6c_4 &= 0 \\ 3c_1 \qquad + \ c_3 + 3c_4 &= 5 \\ 2c_1 + 4c_2 + \ c_3 + \ c_4 &= 6 \\ - \ c_2 + 4c_3 + 2c_4 &= 3 \end{aligned}$$

The augmented matrix of this system is

$$\begin{bmatrix} -1 & 2 & 7 & 6 & 0 \\ 3 & 0 & 1 & 3 & 5 \\ 2 & 4 & 1 & 1 & 6 \\ 0 & -1 & 4 & 2 & -3 \end{bmatrix}$$

This reduced row-echelon form of this matrix is

$$\begin{bmatrix} 1 & 0 & 0 & 0 & 1 \\ 0 & 1 & 0 & 0 & 1 \\ 0 & 0 & 1 & 0 & -1 \\ 0 & 0 & 0 & 1 & 1 \end{bmatrix}$$

Thus $c_1 = 1$, $c_2 = 1$, $c_3 = -1$, and $c_4 = 1$.

4. If we equate the second vector components in this equation, we find that $0c_1 + 0c_2 + 0c_3 = -2$. Hence, there do not exist scalars c_1, c_2, and c_3 which satisfy the given equation.

5. (c) $\|\mathbf{v}\| = \left[3^2 + 4^2 + 0^2 + (-12)^2\right]^{1/2} = \sqrt{169} = 13$

6. (a) $\|\mathbf{u} + \mathbf{w}\| = \|(4,4,10,1)\| = \left[4^2 + 4^2 + 10^2 + 1^2\right]^{1/2} = \sqrt{133}$

(c) $\|-2\mathbf{u}\| + 2\|\mathbf{u}\| = \left[(-8)^2 + (-2)^2 + (-4)^2 + (-6)^2\right]^{1/2}$

$$+ 2\left[4^2 + 1^2 + 2^2 + 3^2\right]^{1/2}$$

$$= 120^{1/2} + 2[30]^{1/2} = 4\sqrt{30}$$

(e) $\frac{1}{\|\mathbf{w}\|}\mathbf{w} = \frac{1}{[3^2 + 1^2 + 2^2 + 2^2]^{1/2}}(3,1,2,2) = \left(\frac{1}{\sqrt{2}}, \frac{1}{3\sqrt{2}}, \frac{\sqrt{2}}{3}, \frac{\sqrt{2}}{3}\right)$

8. $\|k\mathbf{v}\| = \left[(-2k)^2 + (3k)^2 + 0^2 + (6k)^2\right]^{1/2} = \left[49k^2\right]^{1/2} = 7|k|$

Thus $\|k\mathbf{v}\| = 5$ if and only if $k = \pm 5/7$.

9. (a) $(2,5) \cdot (-4,3) = (2)(-4) + (5)(3) = 7$

(c) $(3,1,4,-5) \cdot (2,2,-4,-3) = 6 + 2 - 16 + 15 = 7$

10. (a) Let $\mathbf{v} = (x,y)$ where $\|\mathbf{v}\| = 1$. We are given that $\mathbf{v} \cdot (3,-1) = 0$. Thus, $3x - y = 0$ or $y = 3x$. But $\|\mathbf{v}\| = 1$ implies that $x^2 + y^2 = x^2 + 9x^2 = 1$ or $x = \pm 1/\sqrt{10}$. Thus, the only possibilities are $\mathbf{v} = (1/\sqrt{10}, 3/\sqrt{10})$ or $\mathbf{v} = (-1/\sqrt{10}, -3/\sqrt{10})$. You should graph these two vectors and the vector $(3,-1)$ in an xy-coordinate system.

(b) Let $\mathbf{v} = (x,y,z)$ be a vector with norm 1 such that

$$x - 3y + 5z = 0$$

This equation represents a plane through (0,0,0) which is perpendicular to (1,-3,5). There are infinitely many vectors $\mathbf{v}$ which lie in this plane and have norm 1 and initial point (0,0,0).

11. **(a)** $d(\mathbf{u},\mathbf{v}) = \left[(1 - 2)^2 + (-2 - 1)^2\right]^{1/2} = \sqrt{10}$

(c) $d(\mathbf{u},\mathbf{v}) = \left[(0 + 3)^2 + (-2 - 2)^2 + (-1 - 4)^2 + (1 - 4)^2\right]^{1/2} = \sqrt{59}$

14. Use the ideas in the solution to Problem 16, Section 3.3, together with Theorem 4.1.2 and the definition of the norm.

15. See the solution to Problem 17, Section 3.3.

16. Let $\mathbf{u} = (u_1,\ldots,u_n)$ and $\mathbf{w} = (w_1,\ldots,w_n)$.

(a)
$$\begin{aligned}\mathbf{u} \cdot (k\mathbf{v}) &= (u_1,\ldots,u_n) \cdot (kv_1,\ldots,kv_n)\\ &= u_1kv_1 + \ldots + u_nkv_n\\ &= k(u_1v_1 + \ldots + u_nv_n)\\ &= k(\mathbf{u} \cdot \mathbf{v})\end{aligned}$$

(b)
$$\begin{aligned}\mathbf{u} \cdot (\mathbf{v} + \mathbf{w}) &= (u_1,\ldots,u_n) \cdot (v_1 + w_1,\ldots,v_n + w_n)\\ &= u_1\left[v_1 + w_1\right] + \ldots + u_n\left[v_n + w_n\right]\\ &= u_1v_1 + \ldots + u_nv_n + u_1w_1 + \ldots + u_nw_n\\ &= \mathbf{u} \cdot \mathbf{v} + \mathbf{u} \cdot \mathbf{w}\end{aligned}$$

19. Let $\mathbf{u} = (u_1,\ldots,u_n)$ and $\mathbf{v} = (v_1,\ldots,v_n)$.

(a) $\mathbf{u} \cdot \mathbf{v} = u_1v_1 + \cdots + u_nv_n = v_1u_1 + \cdots + v_nu_n$

$= \mathbf{v} \cdot \mathbf{u}$

(c) $(k\mathbf{u}) \cdot \mathbf{v} = (ku_1) + \cdots + (ku_n)v_n = k\left[u_1v_1 + \cdots + u_nv_n\right]$

$= k(\mathbf{u} \cdot \mathbf{v})$

EXERCISE SET 4.2

6. The pair (1,-2) is in the set but the pair (-1)(1,-2) = (-1,2) is not because the first component is negative; hence Axiom 6 fails. Axiom 5 also fails.

8. Axioms 1, 2, 3, 6, 9, and 10 are easily verified. Axiom 4 holds with $\mathbf{0} = (-1,-1)$ and Axiom 5 holds with $-(x,y) = (-x - 2, -y - 2)$. Axiom 7 fails because

$$k\Big[(x,y) + (x',y')\Big] = k(x' + x' + 1,\ y + y' + 1)$$
$$= (kx + kx' + k,\ ky + ky' + k)$$

while

$$k(x,y) + k(x',y') = (kx,ky) + (kx',\ ky')$$
$$= (kx + kx' + 1,\ ky + ky' + 1)$$

Hence, $k(\mathbf{u} + \mathbf{v}) = k\mathbf{u} + k\mathbf{v}$ only if $k = 1$. Axiom 8 also fails, since if $\mathbf{u} = (x,y)$, then

$$(k + \ell)\mathbf{u} = ((k + \ell)x, (k + \ell)y)$$

but

$$k\mathbf{u} + \ell\mathbf{u} = (kx,ky) + (\ell x,\ell y)$$
$$= ((k + \ell)x + 1,\ (k + \ell)y + 1)$$

9. This is a vector space. We must check all ten properties:

(1) If x and y are positive reals, so is $x + y = xy$.

(2) $x + y = xy = yx = y + x$

(3) $x + (y + z) = x(yz) = (xy)z = (x + y) + z$

(4) There is an object **0**, the positive real number 1, which is such that

$$1 + x = 1 \cdot x = x = x \cdot 1 = x + 1$$

for all positive real numbers x.

(5) For each positive real x, the positive real $1/x$ acts as a negative:

$$x + (1/x) = x(1/x) = 1 = \mathbf{0} = 1 = (1/x)x = (1/x) + x$$

(6) If k is a real and x is a positive real, then $kx = x^k$ is again a positive real.

(7) $k(x + y) = (xy)^k = x^k y^k = kx + ky$

(8) $(k + \ell)x = x^{k+\ell} = x^k x^\ell = kx + \ell x$

(9) $k(\ell x) = (\ell x)^k = (x^\ell)^k = x^{\ell k} = x^{k\ell} = (k\ell)x$

(10) $1x = x^1 = x$

11. This is a vector space. Axioms 2, 3, 7, 8, 9, and 10 follow from properties of matrix addition and scalar multiplication. We verify the remaining axioms.

(1) If we add two matrices of this form, the result will again be a matrix of this form:

$$(*) \quad \begin{bmatrix} a & 0 \\ 0 & b \end{bmatrix} + \begin{bmatrix} c & 0 \\ 0 & d \end{bmatrix} = \begin{bmatrix} a + c & 0 \\ 0 & b + d \end{bmatrix}$$

(4) The 2 × 2 zero matrix is of the appropriate form and has the desired properties.

(5) If **u** is a matrix of the given form, then

$$-\mathbf{u} = \begin{bmatrix} -a & 0 \\ 0 & -b \end{bmatrix}$$

is again of the desired form and $\mathbf{u} + (-\mathbf{u}) = (-\mathbf{u}) + \mathbf{u} = \mathbf{0}$

(6) If **u** is any matrix of this form, then $k\mathbf{u}$ is

$$(**) \qquad k\begin{bmatrix} a & 0 \\ 0 & b \end{bmatrix} = \begin{bmatrix} ka & 0 \\ 0 & kb \end{bmatrix}$$

and $k\mathbf{u}$ has the desired form.

12. This is a vector space. We shall check only four of the axioms because the others follow easily from various properties of the real numbers.

(1) If f and g are real-valued functions defined everywhere, then so is $f + g$. We must also check that if $f(1) = g(1) = 0$, then $(f+g)(1) = 0$. But $(f+g)(1) = f(1) + g(1) = 0 + 0 = 0$.

(4) The zero vector is the function z which is zero everywhere on the real line. In particular, $z(1) = 0$.

(5) If f is a function in the set, then $-f$ is also in the set since it is defined for all real numbers and $-f(1) = -0 = 0$. Moreover, $f + (-f) = (-f) + f = z$.

(6) If f is in the set and k is any real number, then kf is a real valued function defined everywhere. Moreover, $kf(1) = k0 = 0$.

13. This is a vector space and the proof is almost a direct repeat of that for Problem 11. In fact, we need only modify the two equations (*) and (**) in the following way:

$$\begin{bmatrix} a & a+b \\ a+b & b \end{bmatrix} + \begin{bmatrix} c & c+d \\ c+d & d \end{bmatrix}$$

$$= \begin{bmatrix} a+c & (a+c)+(b+d) \\ (a+c)+(b+d) & b+d \end{bmatrix}$$

$$k\begin{bmatrix} a & a+b \\ a+b & b \end{bmatrix} = \begin{bmatrix} ka & ka+kb \\ ka+kb & kb \end{bmatrix}$$

Note that

$$-\mathbf{u} = \begin{bmatrix} -a & -(a+b) \\ -(a+b) & -b \end{bmatrix}$$

15. Properties (2), (3), and (7) – (10) all hold because a line passing through the origin in 3-space is a collection of triples of real numbers and the set of all such triples with the usual operations is a vector space. To verify the remaining four properties, we need only check that

(1) If **u** and **v** lie on the line, so does **u** + **v**.

(4) The vector (0,0,0) lies on the line (which it does by hypothesis, since the line passes through the origin).

(5) If **u** lies on the line, so does −**u**.

(6) If **u** lies on the line, so does any real multiple $k\mathbf{u}$ of **u**.

We check (1), leaving (5) and (6) to you.

The line passes through the origin and therefore has the parametric equations $x = at$, $y = bt$, $z = ct$ where a, b, and c are fixed real numbers and t is the parameter. Thus, if **u** and **v** lie on the line, we have $\mathbf{u} = (at_1, bt_1, ct_1)$ and $\mathbf{v} = (at_2, bt_2, ct_2)$. Therefore $\mathbf{u} + \mathbf{v} = (a(t_1 + t_2), b(t_1 + t_2), c(t_1 + t_2))$, which is also on the line.

18. To prove Part (*b*) of Theorem 4.2.1, we write

$$k\mathbf{0} + k\mathbf{u} = k(\mathbf{0} + \mathbf{u}) \qquad \text{(Axiom 7)}$$
$$= k\mathbf{u} \qquad \text{(Axiom 4)}$$

By Axiom 5, the vector $k\mathbf{u}$ has a negative, $-(k\mathbf{u})$. We add this negative to both sides of the above equation to obtain

$$[k\mathbf{0} + k\mathbf{u}] + (-k\mathbf{u}) = k\mathbf{u} + (-k\mathbf{u})$$

or

$$k\mathbf{0} + [k\mathbf{u} + (-k\mathbf{u})] = k\mathbf{u} + (-k\mathbf{u}) \qquad \text{(Axiom 3)}$$

or

$$k\mathbf{0} + \mathbf{0} = \mathbf{0} \qquad \text{(Axiom 5)}$$

or

$$k\mathbf{0} = \mathbf{0} \qquad \text{(Axiom 4)}$$

19. We are given that $k\mathbf{u} = \mathbf{0}$. Suppose that $k \neq 0$. Then

$$\frac{1}{k}\,(k\mathbf{u}) = \left[\frac{1}{k}\,k\right]\mathbf{u} = (1)\mathbf{u} = \mathbf{u} \qquad \text{(Axioms 9 and 10)}$$

But

$$\frac{1}{k}\,(k\mathbf{u}) = \frac{1}{k}\,\mathbf{0} = \mathbf{0} \qquad \text{(By hypothesis and Part (b))}$$

Thus $\mathbf{u} = \mathbf{0}$. That is, either $k = 0$ or $\mathbf{u} = \mathbf{0}$.

20. Suppose that there are two zero vectors, $\mathbf{0}$ and $\mathbf{0}'$. If we apply Axiom 4 to both of these zero vectors, we have

$$\mathbf{0} = \mathbf{0} + \mathbf{0}' = \mathbf{0}'$$

Hence, the two zero vectors are identical.

21. Suppose that $\mathbf{u}$ has two negatives, $\mathbf{v}$ and $\mathbf{w}$. Then

$$\mathbf{v} = \mathbf{v} + \mathbf{0} = \mathbf{v} + (\mathbf{u} + \mathbf{w}) = (\mathbf{v} + \mathbf{u}) = \mathbf{0} + \mathbf{w} = \mathbf{w}$$

Axiom 5 guarantees that $\mathbf{u}$ must have at least one negative. We have proved that it has at most one.

EXERCISE SET 4.3

1. **(a)** The set is closed under vector addition because

$$(a,0,0) + (b,0,0) = (a + b,0,0)$$

It is closed under scalar multiplication because

$$k(a,0,0) = (ka,0,0)$$

Therefore it is a subspace of R^3.

(b) This set is not closed under either addition or scalar multiplication. (For example, $(a,1,1) + (b,1,1) = (a + b,2,2)$ and $(a + b,2,2)$ does not belong to the set.) Thus it is not a subspace.

2. **(a)** This set is closed under addition since the sum of two integers is again an integer. However, it is not closed under scalar multiplication since the product ku where k is real and a is an integer need not be an integer. Thus, the set is not a subspace.

(c) The elements of this set are 2 x 2 matrices whose off-diagonal elements are equal. The set is closed under addition because

$$\begin{bmatrix} a & b \\ b & c \end{bmatrix} + \begin{bmatrix} d & e \\ e & f \end{bmatrix} = \begin{bmatrix} a+d & b+e \\ b+g & c+f \end{bmatrix}$$

It is closed under scalar multiplication because

$$k\begin{bmatrix} a & b \\ b & c \end{bmatrix} = \begin{bmatrix} ka & kb \\ kb & kc \end{bmatrix}$$

Thus, this set forms a subspace.

3. **(a)** This is the set of all polynominals with degree ≤ 3 and with a constant term which is equal to zero. Certainly, the sum of any two such polynomials is a polynomial with degree ≤ 3 and with a constant term which is equal to zero. The same is true of a constant multiple of such a polynomial. Hence, this set is a subspace of P_3.

(c) The sum of two polynomials, each with degree ≤ 3 and each with integral coefficients, is again a polynomial with degree ≤ 3 and with integral coefficients. Hence, the subset is closed under vector addition. However, a constant multiple of such a polynomial will not necessarily have integral coefficients since the constant need not be an integer. Thus, the subset is not closed under scalar multiplication and is therefore not a subspace.

4. **(a)** The function $f(x) = -1$ for all x belongs to the set. However, the function $(-1)f(x) = 1$ for all x does not. Hence, the set is not closed under scalar multiplication and is therefore not a subspace.

(c) Suppose that f and g are in the set. Then

$$(f + g)(0) = f(0) + g(0) = 2 + 2 = 4$$

and

$$-2f(0) = (-2)(2) = -4$$

Thus this set is not closed under either operation.

(e) Let $f(x) = a + b\sin x$ and $g(x) = c + d\sin x$ be two functions in this set. Then

$$(f + g)(x) = (a + c) + (b + d)\sin x$$

and

$$k(f(x)) = ka + kb \sin x$$

Thus both closure properties are satisfied, and the set is a subspace.

5. **(a)** We look for constants a and b such that $a\mathbf{u} + b\mathbf{v} = (2,2,2)$, or

$$a(0,-2,2) + b(1,3,-1) = (2,2,2)$$

Equating corresponding vector components gives the following system of equations:

$$\begin{aligned} b &= 2 \\ -2a + 3b &= 2 \\ 2a - b &= 2 \end{aligned}$$

From the first equation, we see that $b = 2$. Substituting this value into the remaining equations yields $a = 2$. Thus $(2,2,2)$ is a linear combination of **u** and **v**.

(c) We look for constants a and b such that $a\mathbf{u} + b\mathbf{v} = (0,4,5)$, or

$$a(0,-2,2) + b(1,3,-1) = (0,4,5)$$

Equating corresponding components gives the following system of equations:

$$\begin{aligned} b &= 0 \\ -2a + 3b &= 4 \\ 2a - b &= 5 \end{aligned}$$

From the first equation, we see that $b = 0$. If we substitute this value into the remaining equations, we find that $a = -2$ and $a = 5/2$. Thus, the system of equations is inconsistent and therefore $(0,4,5)$ is not a linear combination of **u** and **v**.

6. (a) We look for constants a, b, and c such that $a\mathbf{u} + b\mathbf{v} + c\mathbf{w} =$ $(-9,-7,-15)$; that is, such that

$$a(2,1,4) + b(1,-1,3) + c(3,2,5) = (-9,-7,-15)$$

If we equate corresponding components, we obtain the system

$$\begin{aligned} 2a + b + 3c &= -9 \\ a - b + 2c &= -7 \\ 4a + 3b + 5c &= -15 \end{aligned}$$

The augmented matrix for this system is

$$\begin{bmatrix} 2 & 1 & 3 & -9 \\ 1 & -1 & 2 & -7 \\ 4 & 3 & 5 & -15 \end{bmatrix}$$

The reduced row-echelon form of this matrix is

$$\begin{bmatrix} 1 & 0 & 0 & -2 \\ 0 & 1 & 0 & 1 \\ 0 & 0 & 1 & -2 \end{bmatrix}$$

Thus $a = -2$, $b = 1$, and $c = -2$ and $(-9,-7,-15)$ is therefore a linear combination of $\mathbf{u}$, $\mathbf{v}$, and $\mathbf{w}$.

(c) This time we look for constants a, b, and c such that

$$a\mathbf{u} + b\mathbf{v} + c\mathbf{w} = (0,0,0)$$

If we choose $a = b = c = 0$, then it is obvious that $a\mathbf{u} + b\mathbf{v} + c\mathbf{w} = (0,0,0)$. We now proceed to show that $a = b = c = 0$ is the only choice. To this end, we equate components to obtain a system of equations whose augmented matrix is

$$\begin{bmatrix} 2 & 1 & 3 & 0 \\ 1 & -1 & 2 & 0 \\ 4 & 3 & 5 & 0 \end{bmatrix}$$

From Part (a), we know that this matrix can be reduced to

$$\begin{bmatrix} 1 & 0 & 0 & 0 \\ 0 & 1 & 0 & 0 \\ 0 & 0 & 1 & 0 \end{bmatrix}$$

Thus, $a = b = c = 0$ is the only solution.

7. **(a)** We look for constants a, b, and c such that

$$a\mathbf{p}_1 + b\mathbf{p}_2 + c\mathbf{p}_3 = -9 - 7x - 15x^2$$

If we substitute the expressions for $\mathbf{p}_1$, $\mathbf{p}_2$, and $\mathbf{p}_3$ into the above equation and equate corresponding coefficients, we find that we have exactly the same system of equations that we had in Problem 6(a), above. Thus, we know that $a = -2$, $b = 1$, and $c = -2$ and thus $-2\mathbf{p}_1 + 1\mathbf{p}_2 - 2\mathbf{p}_3 = -9 - 7x - 15x^2$.

(c) Just as Problem 7(a) was Problem 6(a) in disguise, Problem 7(c) is Problem 6(c) in different dress. The constants are the same, so that $\mathbf{0} = 0\mathbf{p}_1 + 0\mathbf{p}_2 + 0\mathbf{p}_3$.

8. **(a)** We ask if there are constants a, b, and c such that

$$a\begin{bmatrix} 4 & 0 \\ -2 & -2 \end{bmatrix} + b\begin{bmatrix} 1 & -1 \\ 2 & 3 \end{bmatrix} + c\begin{bmatrix} 0 & 2 \\ 1 & 4 \end{bmatrix} = \begin{bmatrix} 6 & -8 \\ -1 & -8 \end{bmatrix}$$

If we multiply, add, and equate corresponding matrix entries, we obtain the following system of equations:

$$\begin{aligned} 4a + b \phantom{{}+2c} &= 6 \\ -b + 2c &= -8 \\ -2a + 2b + c &= -1 \\ -2a + 3b + 4c &= -8 \end{aligned}$$

This system has the solution $a = 1$, $b = 2$, $c = -3$; thus, the matrix <u>is</u> a linear combination of the three given matrices.

8. **(b)** Clearly the zero matrix is a linear combination of any set of matrices since we can always choose the scalars to be zero.

9. **(a)** Given any vector (x,y,z) in R^3, we must determine whether or not there are constants a, b, and c such that

$$\begin{aligned}(x,y,z) &= a\mathbf{v}_1 + b\mathbf{v}_2 + c\mathbf{v}_3 \\ &= a(2,2,2) + b(0,0,3) + c(0,1,1) \\ &= (2a,\ 2a + c,\ 2a + 3b + c)\end{aligned}$$

or

$$\begin{aligned}x &= 2a \\ y &= 2a \qquad + c \\ z &= 2a + 3b + c\end{aligned}$$

This is a system of equations for a, b, and c. Since the determinant of the system is nonzero, the system of equations must have a solution for any values of x, y, and z, whatsoever. Therefore, $\mathbf{v}_1$, $\mathbf{v}_2$, and $\mathbf{v}_3$ do indeed span R^3.

Note that we can also show that the system of equations has a solution by solving for a, b, and c explicitly.

(c) We follow the same procedure that we used in Part (a). This time we obtain the system of equations

$$\begin{aligned}3a + 2b + 5c + d &= x \\ a - 3b - 2c + 4d &= y \\ 4a + 5b + 9c - d &= z\end{aligned}$$

The augmented matrix of this system is

$$\begin{bmatrix} 3 & 2 & 5 & 1 & x \\ 1 & -3 & -2 & 4 & y \\ 4 & 5 & 9 & -1 & z \end{bmatrix}$$

which reduces to

$$\begin{bmatrix} 1 & -3 & -2 & 4 & y \\ 0 & 1 & 1 & -1 & \frac{x-3y}{11} \\ 0 & 0 & 0 & 0 & \frac{z-4y}{17} - \frac{x-3y}{11} \end{bmatrix}$$

Thus the system is inconsistent unless the last entry in the last row of the above matrix is zero. Since this is not the case for all values of x, y, and z, the given vectors do not span R^3.

10. **(a)** Since $\cos(2x) = (1)\cos^2 x + (-1)\sin^2 x$ for all x, it follows that $\cos(2x)$ lies in the space spanned by $\cos^2 x$ and $\sin^2 x$.

(b) Suppose that $3 + x^2$ is in the space spanned by $\cos^2 x$ and $\sin^2 x$; that is, $3 + x^2 = a\cos^2 x + b\sin^2 x$ for some constants a and b. This equation must hold for all x. If we set $x = 0$, we find that $a = 3$. However, if we set $x = \pi$, we find $a = 3 + \pi^2$. Thus we have a contradiction, so $3 + x^2$ is not in the space spanned by $\cos^2 x$ and $\sin^2 x$.

11. Given an arbitrary polynomial $a_0 + a_1x + a_2x^2$ in P_2, we ask whether there are numbers a, b, c and d such that

$$a_0 + a_1x + a_2x^2 = a\mathbf{p}_1 + b\mathbf{p}_2 + c\mathbf{p}_3 + d\mathbf{p}_4$$

If we equate coefficients, we obtain the system of equations:

$$\begin{aligned} a_0 &= a + 3b + 5c - 2d \\ a_1 &= -a + b - c - 2d \\ a_2 &= 2a + 4c + 2d \end{aligned}$$

The row-echelon form of the augmented matrix of this system is

$$\begin{bmatrix} 1 & 3 & 5 & -2 & a_0 \\ 0 & 1 & 1 & -1 & \frac{a_0+a_1}{4} \\ 0 & 0 & 0 & 0 & -a_0+3a_1+2a_2 \end{bmatrix}$$

Thus the system is inconsistent whenever $-a_0 + 3a_1 + 2a_2 \neq 0$ (for example, when $a_0 = 0$, $a_1 = 0$, and $a_2 = 1$). Hence the given polynomials do not span P_2.

12. **(a)** As before, we look for constants a, b, and c such that

$$(2,3,-7,3) = a\mathbf{v}_1 + b\mathbf{v}_2 + c\mathbf{v}_3$$

If we equate components, we obtain the following system of equations:

$$\begin{aligned} 2a + 3b - c &= 2 \\ a - b \phantom{{}- c} &= 3 \\ 5b + 2c &= -7 \\ 3a + 2b + c &= 3 \end{aligned}$$

The augmented matrix of this system is

$$\begin{bmatrix} 2 & 3 & -1 & 2 \\ 1 & -1 & 0 & 3 \\ 0 & 5 & 2 & -7 \\ 3 & 2 & 1 & 3 \end{bmatrix}$$

This reduces to

$$\begin{bmatrix} 1 & 0 & 0 & 2 \\ 0 & 1 & 0 & -1 \\ 0 & 0 & 1 & -1 \\ 0 & 0 & 0 & 0 \end{bmatrix}$$

Thus $a = 2$, $b = -1$, and $c = -1$, and the existence of a solution guarantees that the given vector is in $\text{lin}\{\mathbf{v}_1,\mathbf{v}_2,\mathbf{v}_3\}$.

(c) Proceeding as in Part (a), we obtain the matrix

$$\begin{bmatrix} 2 & 3 & -1 & 1 \\ 1 & -1 & 0 & 1 \\ 0 & 5 & 2 & 1 \\ 3 & 2 & 1 & 1 \end{bmatrix}$$

This reduces to a matrix whose last row is $[0\ 0\ 0\ 1]$. Thus the system is inconsistent and hence the given vector is not in $\text{lin}\{\mathbf{v}_1,\mathbf{v}_2,\mathbf{v}_3\}$.

13. The plane has the vector $\mathbf{u} \times \mathbf{v} = (0,7,-7)$ as a normal and passes through the point $(0,0,0)$. Thus its equation is $y - z = 0$.

Alternatively, we look for conditions on a vector (x,y,z) which will insure that it lies in $\text{lin}\{\mathbf{u},\mathbf{v}\}$. That is , we look for numbers a and b such that

$$\begin{aligned}(x,y,z) &= a\mathbf{u} + b\mathbf{v} \\ &= a(-1,1,1) + b(3,4,4)\end{aligned}$$

If we expand and equate components, we obtain a system whose augmented matrix is

$$\begin{bmatrix} -1 & 3 & x \\ 1 & 4 & y \\ 1 & 4 & z \end{bmatrix}$$

This reduces to the matrix

$$\begin{bmatrix} 1 & -3 & -x \\ 0 & 1 & \frac{x+y}{7} \\ 0 & 0 & \frac{-y+z}{7} \end{bmatrix}$$

Thus the system is consistent if and only if $\frac{-y + z}{7} = 0$ or $y = z$.

15. Let the system have the form $AX = B$ where B has at least one nonzero entry. Suppose that X_1 and X_2 are two solutions of this system; that is, $AX_1 = B$ and $AX_2 = B$. Then

$$A(X_1 + X_2) = AX_1 + AX_2 = B + B \neq B$$

Thus the solution set is not closed under addition and so cannot form a subspace of R^n. Alternatively, we could show that it is not closed under scalar multiplication.

17. **(a)** We simply note that the sum of two continuous functions is a continuous function and that a constant times a continuous funtion is a continuous function.

(b) We recall that the sum of two differentiable functions is a differentiable functions and that a constant times a differentiable function is a differentiable function.

EXERCISE SET 4.4

2. (a) Clearly neither of these vectors is a multiple of the other. Thus they are linearly independent.

   (b) Following the technique used in Example 4, we consider the system of linear equations

$$\begin{aligned} -3k_1 + 5k_2 + k_3 &= 0 \\ - k_2 + k_3 &= 0 \\ 4k_1 + 2k_2 + 3k_3 &= 0 \end{aligned}$$

   Since the determinant of the coefficient matrix is nonzero, the system has <u>only</u> the trivial solution. Therefore, the three vectors are linearly independent.

   (d) By Theorem 4.4.3, any four vectors in R^3 are linearly dependent.

3. (a) Following the technique used in Example 4, we obtain the system of equations

$$\begin{aligned} 3k_1 + k_2 + 2k_3 + k_4 &= 0 \\ 8k_1 + 5k_2 - k_3 + 4k_4 &= 0 \\ 7k_1 + 3k_2 + 2k_3 &= 0 \\ -3k_1 - k_2 + 6k_3 + 3k_4 &= 0 \end{aligned}$$

Since the determinant of the coefficient matrix is nonzero, the system has only the trivial solution. Hence, the four vectors are linearly independent.

3. **(b)** Again following the technique of Example 4, we obtain the system of equations

$$\begin{aligned} 3k_2 + k_3 &= 0 \\ 3k_2 + k_3 &= 0 \\ 2k_1 \qquad\qquad &= 0 \\ 2k_1 \qquad - k_3 &= 0 \end{aligned}$$

The third equation, above, gives $k_1 = 0$. This implies that k_3 and hence k_2 must also equal zero. Thus the three vectors are linearly independent.

4. **(a)** We ask whether there exist constants a, b, and c such that

$$a(2 - x + 4x^2) + b(3 + 6x + 2x^2) + c(2 + 10x - 4x^2) = 0$$

If we equate all of the coefficients in the above polynomial to zero, we obtain the following system of equations:

$$\begin{aligned} 2a + 3b + 2c &= 0 \\ -a + 6b + 10c &= 0 \\ 4a + 2b - 4c &= 0 \end{aligned}$$

Since the coefficient matrix of this system is invertible, the trivial solution is the only solution. Hence, the polynomials are linearly independent.

(d) If we set up this problem in the same way we set up Part (a), above, we obtain four equations in three unknowns. Since this is equivalent to having four vectors in R^3, the vectors are linearly dependent by Theorem 4.4.3.

5. (a) Since $\sin^2 x + \cos^2 x = 1$, we observe that

$$2(3\sin^2 x) + 3(2\cos^2 x) + (-1)(6) = 0$$

Hence the vectors are linearly dependent.

(c) Suppose that there are constants a, b, and c such that

$$a(1) + b\sin x + c\sin 2x = 0$$

Setting $x = 0$ yields $a = 0$. Setting $x = \pi/2$ yields $b = 0$, and thus, since $\sin 2x \not\equiv 0$, we must also have $c = 0$. Therefore the vectors are linearly independent.

(e) Suppose that there are constants a, b, and c such that

$$a(3 - x)^2 + b(x^2 - 6x) + c(5) = 0$$

or

$$(9a + 5c) + (-6a - 6b)x + (a + b)x^2 = 0$$

Clearly $a = -b = -(5/9)c$. Thus $a = 5$, $b = -5$, $c = -9$ is one solution and the vectors are linearly dependent.

This conclusion may also be reached by noting that the determinant of the system of equations

$$\begin{aligned} 9a \quad\quad + 5c &= 0 \\ -6a - 6b \quad\quad &= 0 \\ a + b \quad\quad &= 0 \end{aligned}$$

is zero.

6. (a) The vectors lie in the same plane through the origin if and only if they are linearly dependent. Since the determinant of the matrix

$$\begin{bmatrix} 2 & 6 & 2 \\ -2 & 1 & 0 \\ 0 & 4 & -4 \end{bmatrix}$$

is not zero, the matrix is invertible and the vectors are linearly independent. Thus they do not lie in the same plane.

7. **(a)** Since $\mathbf{v}_2 = -2\mathbf{v}_1$, $\mathbf{v}_1$ and $\mathbf{v}_2$ lie on the same line. But since $\mathbf{v}_3$ is not a multiple of $\mathbf{v}_1$ or $\mathbf{v}_2$, the three vectors do not lie on the same line through the origin.

   (c) Since $\mathbf{v}_1 = 2\mathbf{v}_2 = -2\mathbf{v}_3$, these vectors all lie on the same line through the origin.

8. **(a)** Note that $7\mathbf{v}_1 - 2\mathbf{v}_2 + 3\mathbf{v}_3 = \mathbf{0}$.

   $v_3 = -\frac{7}{3}v_1 + \frac{2}{3}v_2$

   $\frac{7}{3}v_1 - \frac{2}{3}v_2 + v_3 = 0$

9. If there are constants a, b, and c such that

$$a(\lambda,-1/2,-1/2) + b(-1/2,\lambda,-1/2) + c(-1/2,-1/2,\lambda) = (0,0,0)$$

then

$$\begin{bmatrix} \lambda & -1/2 & -1/2 \\ -1/2 & \lambda & -1/2 \\ -1/2 & -1/2 & \lambda \end{bmatrix} \begin{bmatrix} a \\ b \\ c \end{bmatrix} = \begin{bmatrix} 0 \\ 0 \\ 0 \end{bmatrix}$$

The determinant of the coefficient matrix is

$$\lambda^3 - \frac{3}{4}\lambda - \frac{1}{4} = (\lambda - 1)\left[\lambda + \frac{1}{2}\right]^2$$

This equals zero if and only if $\lambda = 1$ or $\lambda = -1/2$. Thus the vectors are linearly dependent for these two values of λ and linearly independent for all other values.

11. Suppose that S has a linearly dependent subset T. Denote its vectors by $\mathbf{w}_1, \ldots, \mathbf{w}_m$. Then there exist constants k_i, not all zero, such that

$$k_1\mathbf{w}_1 + \cdots + k_m\mathbf{w}_m = \mathbf{0}$$

But if we let $\mathbf{u}_1, \ldots, \mathbf{u}_{n-m}$ denote the vectors which are in S but not in T, then

$$k_1\mathbf{w}_1 + \cdots + k_m\mathbf{w}_m + 0\mathbf{u}_1 + \cdots + 0\mathbf{u}_{n-m} = \mathbf{0}$$

Thus we have a linear combination of the vectors $\mathbf{v}_1, \ldots, \mathbf{v}_n$ which equals $\mathbf{0}$. Since not all of the constants are zero, it follows that S is not a linearly independent set of vectors, contrary to the hypothesis. That is, if S is a linearly independent set, then so is every non-empty subset T.

13. This is similar to Problem 11. Since $\{\mathbf{v}_1, \mathbf{v}_2, \ldots, \mathbf{v}_r\}$ is a linearly dependent set of vectors, there exist constants $c_1, c_2, \ldots, c_r$ not all zero such that

$$c_1\mathbf{v}_1 + c_2\mathbf{v}_2 + \cdots + c_r\mathbf{v}_r = \mathbf{0}$$

But then

$$c_1\mathbf{v}_1 + c_2\mathbf{v}_2 + \cdots + c_r\mathbf{v}_r + 0\mathbf{v}_{r+1} + \cdots + 0\mathbf{v}_n = \mathbf{0}$$

The above equation implies that the vectors $\mathbf{v}_1, \ldots, \mathbf{v}_n$ are linearly dependent.

15. Suppose that $\{\mathbf{v}_1, \mathbf{v}_2, \mathbf{v}_3\}$ is linearly dependent. Then there exist constants a, b, and c not all zero such that

$$(*) \qquad a\mathbf{v}_1 + b\mathbf{v}_2 + c\mathbf{v}_3 = \mathbf{0}$$

<u>Case 1</u>: $c = 0$. Then (*) becomes

$$a\mathbf{v}_1 + b\mathbf{v}_2 = \mathbf{0}$$

where not both a and b are zero. But then $\{\mathbf{v}_1, \mathbf{v}_2\}$ is linearly dependent, contrary to hypothesis.

<u>Case 2</u>: $c \neq 0$. Then solving (*) for $\mathbf{v}_3$ yields

$$\mathbf{v}_3 = -\frac{a}{c}\mathbf{v}_1 - \frac{b}{c}\mathbf{v}_2$$

This equation implies that $\mathbf{v}_3 \in \text{lin}\{\mathbf{v}_1, \mathbf{v}_2\}$, contrary to hypothesis. Thus, $\{\mathbf{v}_1, \mathbf{v}_2, \mathbf{v}_3\}$ is linearly independent.

16. Note that $(\mathbf{u} - \mathbf{v}) + (\mathbf{v} - \mathbf{w}) + (\mathbf{w} - \mathbf{u}) = \mathbf{0}$.

18. Suppose that $w(x) \not\equiv 0$. Since we are asked to show that **f**, **g**, and **h** are linearly independent over the entire real line, we also suppose that

$$af(x) + bg(x) + ch(x) \equiv 0$$

Thus

$$af'(x) + bg'(x) + ch'(x) \equiv 0$$

and

$$af''(x) + bg''(x) + ch''(x) \equiv 0$$

For any given value of x, the three equations above are a homogeneous system in the three unknowns a, b, and c. The determinant of the coefficients matrix of this system is $w(x)$. If $w(x) \not\equiv 0$, then $w(x_0) \neq 0$ for some x_0. Thus for $x = x_0$ the above system has only the trivial solution $a = b = c = 0$. Hence $a = b = c = 0$ is the only choice of constants which will guarantee that $a\mathbf{f} + b\mathbf{g} + c\mathbf{h} = \mathbf{0}$; that is, **f**,**g**,and **h** are linearly independent.

Incidentally, the converse of this result does not hold. Can you see why it might be possible to have **f**, **g**, and **h** linearly independent while $w(x) \equiv 0$?

20. Any nonzero vector forms a linearly independent set. The only scalar multiple of a nonzero vector which can equal the zero vector is the zero scalar times the vector.

23. Suppose that $\mathbf{v}_i$ is the zero vector. Then

$$0\mathbf{v}_1 + \cdots + 0\mathbf{v}_{i-1} + 1\mathbf{v}_i + 0\mathbf{v}_{i+1} + \cdots + \mathbf{v}_n = \mathbf{0}$$

Thus the set is linearly dependent.

24. Use Theorem 4.4.1, Part (a).

EXERCISE SET 4.5

2. **(a)** This set is a basis. It has the correct number of vectors and neither is a multiple of the other.

(c) This set is not a basis since one vector is a multiple of the other.

3. **(a)** This set has the correct number of vectors and they are linearly independent because

$$\begin{vmatrix} 1 & 2 & 3 \\ 0 & 2 & 3 \\ 0 & 0 & 3 \end{vmatrix} = 6 \neq 0$$

Hence, the set is a basis.

(c) The vectors in this set are linearly dependent because

$$\begin{vmatrix} 2 & 4 & 0 \\ -3 & 1 & -7 \\ 1 & 1 & 1 \end{vmatrix} = 0$$

Hence, the set is not a basis.

4. **(a)** The vectors in this set are linearly dependent because

$$\begin{vmatrix} 1 & 1 & 1 \\ -3 & 1 & -7 \\ 2 & 4 & 0 \end{vmatrix} = 0$$

Thus, the set is not a basis. (Compare with Problem 3(c), above.)

4. **(c)** This set has the correct number of vectors and

$$\begin{vmatrix} 1 & 0 & 0 \\ 1 & 1 & 0 \\ 1 & 1 & 1 \end{vmatrix} = 1 \neq 0$$

Hence, the vectors are linearly independent and therefore are a basis.

5. The set has the correct number of vectors. To show that they are linearly independent, we consider the equation

$$a\begin{bmatrix} 3 & 6 \\ 3 & -6 \end{bmatrix} + b\begin{bmatrix} 0 & -1 \\ -1 & 0 \end{bmatrix} + c\begin{bmatrix} 0 & -8 \\ -12 & -4 \end{bmatrix} + d\begin{bmatrix} 1 & 0 \\ -1 & 2 \end{bmatrix} = \begin{bmatrix} 0 & 0 \\ 0 & 0 \end{bmatrix}$$

If we add matrices and equate corresponding entries, we obtain the following system of equations:

$$\begin{aligned} 3a \qquad\qquad\qquad + \ d &= 0 \\ 6a - b - \ 8c \qquad\quad &= 0 \\ 3a - b - 12c - \ d &= 0 \\ -6a \qquad - \ 4c + 2d &= 0 \end{aligned}$$

Since the determinant of the coefficient matrix is nonzero, the system of equations has only the trivial solution; hence, the vectors are linearly independent.

6. **(a)** Recall that $\cos 2x = \cos^2 x - \sin^2 x$; that is,

$$1\mathbf{v}_1 + (-1)\mathbf{v}_2 + (-1)\mathbf{v}_3 = \mathbf{0}$$

Hence, S is not a linearly independent set of vectors.

(b) We can use the above identity to write any one of the vectors $\mathbf{v}_i$ as a linear combination of the other two. Since no one of these vectors is a multiple of any other, they are pairwise linearly independent. Thus any two form a basis for V.

7. **(a)** Clearly $\mathbf{w} = 3\mathbf{u}_1 - 7\mathbf{u}_2$, so the coordinate vector relative to $\{\mathbf{u}_1, \mathbf{u}_2\}$ is $(3,-7)$

(b) If $\mathbf{w} = a\mathbf{u}_1 + b\mathbf{u}_2$, then equating coordinates yields the system of equations

$$\begin{aligned} 2a + 3b &= 1 \\ -4a + 8b &= 1 \end{aligned}$$

This system has the solution $a = 5/28$, $b = 3/14$. Thus the desired coordinate vector is $(5/28, 3/14)$.

8. **(a)** If $\mathbf{v} = a\mathbf{v}_1 + b\mathbf{v}_2 + c\mathbf{v}_3$, then

$$\begin{aligned} a + 2b + 3c &= 2 \\ 2b + 3c &= -1 \\ 3c &= 3 \end{aligned}$$

From the third equation, $c = 1$. Plugging this value into the second equation yields $b = -2$, and finally, the first equation yields $a = 3$. Thus the desired coordinate vector is $(3,-2,1)$.

9. **(b)** If $\mathbf{p} = a\mathbf{p}_1 + b\mathbf{p}_2 + c\mathbf{p}_3$, then equating coefficients yields the system of equations

$$\begin{aligned} a + b &= 2 \\ a + c &= -1 \\ b + c &= 1 \end{aligned}$$

This system has the solution $a = 0$, $b = 2$, and $c = -1$. Thus the desired coordinate vector is $(0,2,-1)$.

11. The augmented matrix of the system reduces to

$$\begin{bmatrix} 1 & 0 & -1 & 0 \\ 0 & 1 & 0 & 0 \\ 0 & 0 & 0 & 0 \end{bmatrix}$$

$x_1 = s$
$x_2 = 0$
$x_3 = s$

$$\begin{bmatrix} x_1 \\ x_2 \\ x_3 \end{bmatrix} = s \begin{bmatrix} 1 \\ 0 \\ 1 \end{bmatrix}$$

Hence, $x_1 = s$, $x_2 = 0$, and $x_3 = s$. Thus the solution space is spanned by $(1,0,1)$ and has dimension 1.

14. If we reduce the augmented matrix to row-echelon form, we obtain

$$
\begin{array}{ccc} x_1 & x_2 & x_3 \end{array}
$$

$$
\begin{bmatrix} 1 & -3 & 1 & 0 \\ 0 & 0 & 0 & 0 \\ 0 & 0 & 0 & 0 \end{bmatrix} \quad \begin{array}{l} x_1 - 3x_2 - x_3 \\ x_2 = r \\ x_3 = s \\ x_1 = 3r - s \end{array}
$$

Thus $x_1 = 3r - s$, $x_2 = r$, and $x_3 = s$, and the solution vector is

$$
\begin{bmatrix} x_1 \\ x_2 \\ x_3 \end{bmatrix} = \begin{bmatrix} 3r-s \\ r \\ s \end{bmatrix} = \begin{bmatrix} 3 \\ 1 \\ 0 \end{bmatrix} r + \begin{bmatrix} -1 \\ 0 \\ 1 \end{bmatrix} s
$$

Since $(3,1,0)$ and $(-1,0,1)$ are linearly independent, they form a basis for the solution space and the dimension of the solution space is 2.

15. Since the determinant of the system is not zero, the only solution is $x_1 = x_2 = x_3 = 0$. Hence there is no basis for the solution space and its dimension is zero.

17. **(a)** Any two linearly independent vectors in the plane form a basis. For instance, $(1,-1,-1)$ and $(0,5,2)$ are a basis because they satisfy the plane equation and neither is a multiple of the other.

(c) Any nonzero vector which lies on the line forms a basis. For instance, $(2,-1,4)$ will work, as will any nonzero multiple of this vector.

(d) The vectors $(1,1,0)$ and $(0,1,1)$ form a basis because they are linearly independent and

$$
a(1,1,0) + c(0,1,1) = (a,\ a + c,\ c)
$$

19. This space is spanned by the vectors $(0,0,0,0)$, $(0,x,0,0)$, $(0,0,x^2,0)$ and $(0,0,0,x^3)$. Only the last three vectors form a linearly independent triple. Thus the space has dimension 3.

20. Since $\{\mathbf{u}_1,\mathbf{u}_2,\mathbf{u}_3\}$ has the correct number of elements, we need only show that they are linearly independent. Let

$$a\mathbf{u}_1 + b\mathbf{u}_2 + c\mathbf{u}_3 = \mathbf{0}$$

Thus

$$a\mathbf{v}_1 + b(\mathbf{v}_1 + \mathbf{v}_2) + c(\mathbf{v}_1 + \mathbf{v}_2 + \mathbf{v}_3) = \mathbf{0}$$

or

$$(a + b + c)\mathbf{v}_1 + (b + c)\mathbf{v}_2 + c\mathbf{v}_3 = \mathbf{0}$$

Since $\{\mathbf{v}_1,\mathbf{v}_2,\mathbf{v}_3\}$ is a linearly independent set, the above equation implies that $a + b + c = b + c = c = 0$. Thus, $a = b = c = 0$ and $\{u_1,u_2,u_3\}$ is also linearly independent.

21. Suppose that the space of real-valued functions has finite dimensions – say n. Since the functions $1, x, x^2, \ldots ,x^n$ are a set of $n + 1$ linearly independent vectors (see Example 5) in an n-dimensional space, we have a contradiction to Theorem 4.5.2. Thus the assumption that the vector space has finite dimension n leads to a contradiction for every n; therefore, the space has infinite dimension.

22. Let n be the dimension of the vector space V and let U be a subspace of V. We shall show that the dimension of U is finite by constructing a basis for U which has no more than n elements.

If $U = \{0\}$, then the dimension of U is zero. If $U \neq \{0\}$, then there is at least one nonzero vector $\mathbf{u}_1 \in U$. If $U = \text{lin}\{\mathbf{u}_1\}$, then $\{\mathbf{u}_1\}$ is a basis for U and U has dimension one. If $U \neq \text{lin}\{\mathbf{u}_1\}$, then there exists at least one nonzero vector $\mathbf{u}_2 \in U$ which is not contained in $\text{lin}\{\mathbf{u}_1\}$.

Thus $\mathbf{u}_1$ and $\mathbf{u}_2$ are linearly independent. If $U = \text{lin}\{\mathbf{u}_1,\mathbf{u}_2\}$, then $\{\mathbf{u}_1,\mathbf{u}_2\}$ is a basis for U and U has dimension two. If $U \neq \text{lin}\{\mathbf{u}_1,\mathbf{u}_2\}$, then there is at least one nonzero vector $\mathbf{u}_3 \in U$ which is not in $\text{lin}\{\mathbf{u}_1,\mathbf{u}_2\}$, and so on.

At each stage of the construction described above, we produce a set of vectors $C_k = \{\mathbf{u}_1,\mathbf{u}_2,\ldots,\mathbf{u}_k\}$. We now verify that C_k is linearly independent. We know that $C_2 = \{\mathbf{u}_1,\mathbf{u}_2\}$ is linearly independent since neither vector is a multiple of the other. Now suppose that i is the smallest index such that C_i is linearly dependent. Then there are constants $c_1,c_2,\ldots,c_i$ not all zero such that

$$c_1\mathbf{u}_1 + c_2\mathbf{u}_2 + \cdots + c_i\mathbf{u}_i = \mathbf{0}$$

If $c_i = 0$, then C_{i-1} must be linearly dependent, contrary to the assumption that i is the smallest index for which C_i is linearly dependent. On the other hand, if $c_i \neq 0$, then $\mathbf{u}_i$ must belong to $\text{lin}\{\mathbf{u}_1,\ldots,\mathbf{u}_{i-1}\}$. In either case, we have a contradiction and hence, there cannot be a smallest i for which C_i is linearly dependent. Thus, the construction described above will always produce a linearly independent set of vectors C_k.

At each stage of the construction, we check to see whether C_k forms a basis for U. If so, U must have finite dimension k. If not, then we consider C_{k+1}. This process must terminate because, by Theorem 4.5.2, k cannot exceed n. Thus, U has dimension at most n.

25. First notice that if $\mathbf{v}$ and $\mathbf{w}$ are vectors in V and a and b are scalars, then $(a\mathbf{v} + b\mathbf{w})_S = a(\mathbf{v})_S + b(\mathbf{w})_S$. This follows from the definition of coordinate vectors. Clearly, this result applies to any finite sum of vectors. Also notice that if $(\mathbf{v})_S = (\mathbf{0})_S$, then $\mathbf{v} = \mathbf{0}$. Why?

Now suppose that $k_1\mathbf{v}_1 + \cdots + k_r\mathbf{v}_r = \mathbf{0}$. Then

$$\begin{aligned}(k_1\mathbf{v}_1 + \cdots + k_r\mathbf{v}_r)_S &= k_1(\mathbf{v}_1)_S + \cdots + k_r(\mathbf{v}_r)_S \\ &= (\mathbf{0})_S\end{aligned}$$

Conversely, if $k_1(\mathbf{v}_1)_S + \cdots + k_r(\mathbf{v}_r)_S = (\mathbf{0})_S$, then

$$(k_1\mathbf{v}_1 + \cdots + k_r\mathbf{v}_r)_S = (\mathbf{0})_S, \quad \text{or} \quad k_1\mathbf{v}_1 + \cdots + k_r\mathbf{v}_r = \mathbf{0}$$

Thus the vectors $\mathbf{v}_1,\ldots,\mathbf{v}_r$ are linearly independent in V if and only if the coordinate vectors $(\mathbf{v}_1)_S,\ldots,(\mathbf{v}_r)_S$ are linearly independent in R^n.

26. If every vector $\mathbf{v}$ in V can be written as a linear combination $\mathbf{v} = a_1\mathbf{v}_1 + \cdots + a_r\mathbf{v}_r$ of $\mathbf{v}_1,\ldots,\mathbf{v}_r$, then, as in Exercise 25, we have $(\mathbf{v})_S = a_1(\mathbf{v}_1)_S + \cdots + a_r(\mathbf{v}_r)_S$. Hence, the vectors $(\mathbf{v}_1)_S,\ldots,(\mathbf{v}_r)_S$ span a subspace of R^n. But since V is an n-dimensional space with, say, the basis $S = \{\mathbf{u}_1,\ldots,\mathbf{u}_n\}$, then if $\mathbf{u} = b_1\mathbf{u}_1 + \cdots + b_n\mathbf{u}_n$, we have $(\mathbf{u})_S = (b_1,\ldots,b_n)$; that is, every vector in R^n represents a vector in V. Hence $\{(\mathbf{v}_1)_S,\ldots,(\mathbf{v}_r)_S\}$ spans R^n.

Conversely, if $\{(\mathbf{v}_1)_S,\ldots,(\mathbf{v}_r)_S\}$ spans R^n, then for every vector $(b_1,\ldots,b_n)$ in R^n, there is an r-tuple $(a_1,\ldots,a_r)$ such that

$$\begin{aligned}(b_1,\ldots,b_n) &= a_1(\mathbf{v}_1)_S + \cdots + a_r(\mathbf{v}_r)_S \\ &= (a_1\mathbf{v}_1 + \cdots + a_r\mathbf{v}_r)_S\end{aligned}$$

Thus $a_1\mathbf{v}_1 + \cdots + a_r\mathbf{v}_r = b_1\mathbf{u}_1 + \cdots + b_n\mathbf{u}_n$, so that every vector in V can be represented as a linear combination of $\mathbf{v}_1,\ldots,\mathbf{v}_r$.

27. **(a)** Let $\mathbf{v}_1$, $\mathbf{v}_2$, and $\mathbf{v}_3$ denote the vectors. Since $S = \{1,x,x^2\}$ is the standard basis for P_2, we have $(\mathbf{v}_1)_S = (-1,1,-2)$, $(\mathbf{v}_2)_S = (3,3,6)$, and $(\mathbf{v}_3)_S = (9,0,0)$. Since $\{(-1,1,-2), (3,3,6), (9,0,0)\}$ is a linearly independent set of three vectors in R^3, then it spans R^3. Thus, by Exercises 25 and 26, $\{\mathbf{v}_1,\mathbf{v}_2,\mathbf{v}_3\}$ is linearly independent and spans P_2. Hence it is a basis for P_2.

28. **(a)** It is clear from the picture that the x'-y' coordinates of (1,1) are $(0,\sqrt{2})$.

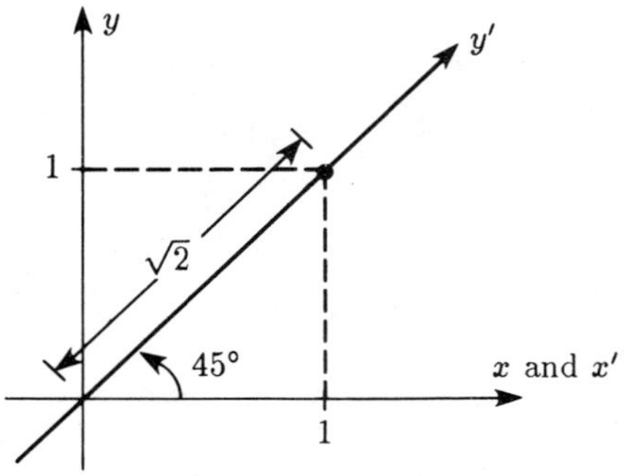

(d) Let (a,b) and (a',b') denote the coordinates of a point with respect to the x-y and x'-y' coordinate systems, respectively. If (a,b) is positioned as in Figure (1), then we have

$$a' = a - b \qquad \text{and} \qquad b' = \sqrt{2}b$$

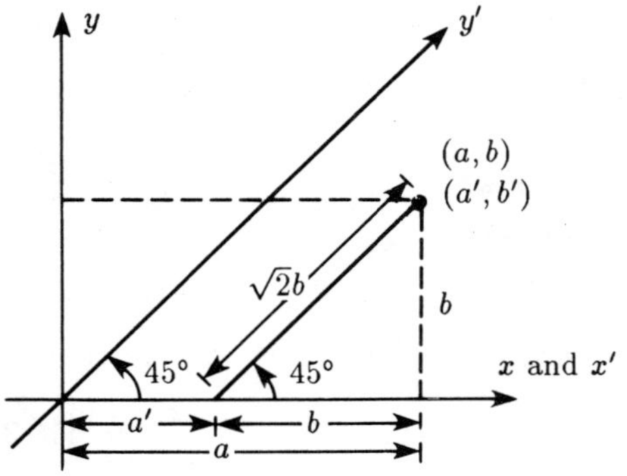

Figure 1

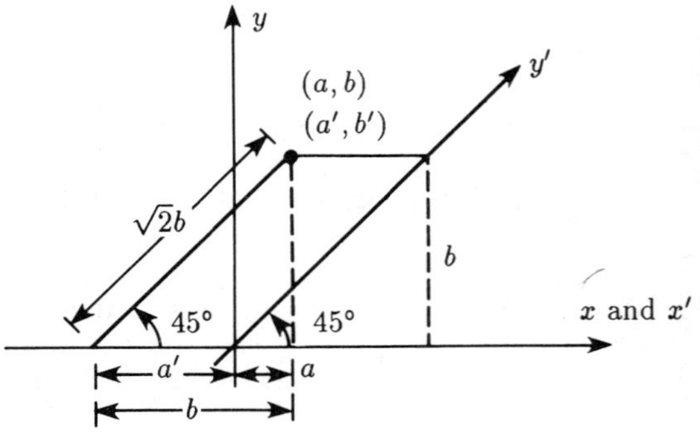

Figure 2

These formulas hold no matter where (a,b) lies in relation to the coordinate axes. Figure (2) shows another configuration, and you should draw similar pictures for all of the remaining cases.

30. If S fails to be a basis for V, then it must fail to span V. That is, there must be a vector $\mathbf{v}$ in V which is not in lin(S). But then $\mathbf{v}$ and the vectors in S would form a linearly independent set of $n + 1$ vectors in the n-dimensional space V, contrary to Theorem 4.5.2.

31. If the spanning set S is not a basis, then S must be a linearly dependent set. In that case, we can express at least one of the vectors in S as a linear combination of the remaining $n-1$ vectors. This yields a spanning set with $n-1$ vectors, contrary to the assumption that V has dimension n.

32. Since $r < n$, there must be at least one vector, call it $\mathbf{v}_{r+1}$, which is in V but not in lin(S). We claim that the set $\{\mathbf{v}_1,\ldots,\mathbf{v}_r,\mathbf{v}_{r+1}\}$ is linearly independent. (To verify this, suppose the contrary. Then there exist constants $c_1,\ldots,c_{r+1}$ not all zero such that

$$c_1\mathbf{v}_1 + \cdots + c_r\mathbf{v}_r + c_{r+1}\mathbf{v}_{r+1} = \mathbf{0}$$

Clearly either $c_{r+1} = 0$ or $c_{r+1} \neq 0$. In either case we obtain a contradiction. Why?) Now if $r + 1 = n$, then the set $\{\mathbf{v}_1,\ldots,\mathbf{v}_{r+1}\}$ contains n linearly independent vectors, which must, by Exercise 30, form a basis for V. If $r + 1 < n$, then we repeat the process, adding another vector, $\mathbf{v}_{r+2}$. We continue until we have produced n linearly independent vectors, which must, as above, form a basis for V.

33. We have seen that there cannot be more than n linearly independent vectors in an n-dimensional vector space V. Thus any basis for W consists of at most n vectors, or $\dim(W) \leq \dim(V)$. If $\dim(W) = \dim(V)$, then each vector space has, by Exercise 30, a basis which spans the other. Hence $V = W$. Conversely, if $V = W$, then $\dim(V) = \dim(W)$.

EXERCISE SET 4.6

3. (a) Since the reduced row-echelon form of A is

$$\begin{bmatrix} 1 & 0 & -16 \\ 0 & 1 & -19 \\ 0 & 0 & 0 \end{bmatrix}$$

the solution to the equation $AX = 0$ is $x_1 = 16t$, $x_2 = 19t$, $x_3 = t$. Thus

$$\begin{bmatrix} 16 \\ 19 \\ 1 \end{bmatrix}$$

is a basis for the nullspace.

(c) Since the reduced row-echelon form of A is

$$\begin{bmatrix} 1 & 0 & 1 & -2/7 \\ 0 & 1 & 1 & 4/7 \\ 0 & 0 & 0 & 0 \end{bmatrix}$$

one solution to the equation $AX = 0$ is $x_1 = -s + 2t$, $x_2 = -s - 4t$, $x_3 = s$, $x_4 = 7t$. Thus the set of vectors

$$\begin{bmatrix} -1 \\ -1 \\ 1 \\ 0 \end{bmatrix} \quad \text{and} \quad \begin{bmatrix} 2 \\ -4 \\ 0 \\ 7 \end{bmatrix}$$

is a basis for the nullspace.

3. **(e)** The reduced row-echelon form of A is

$$\begin{bmatrix} 1 & 0 & 0 & 2 & 4/3 \\ 0 & 1 & 0 & 0 & -1/6 \\ 0 & 0 & 1 & 0 & -5/12 \\ 0 & 0 & 0 & 0 & 0 \\ 0 & 0 & 0 & 0 & 0 \end{bmatrix}$$

One solution to the equation $AX = 0$ is $x_1 = -2s - 16t$, $x_2 = 2t$, $x_3 = 5t$, $x_4 = s$, and $x_5 = 12t$. Hence the set of vectors

$$\begin{bmatrix} -2 \\ 0 \\ 0 \\ 1 \\ 0 \end{bmatrix} \quad \text{and} \quad \begin{bmatrix} -16 \\ 2 \\ 5 \\ 0 \\ 12 \end{bmatrix}$$

is a basis for the nullspace of A.

4. **(a)** From a row-echelon form of A, we have that the vectors (1,-1,3) and (0,1,-19) are a basis for the row space of A.

(c) From a row-echelon form of A, we have that the vectors (1,4,5,2) and (0,1,1,4/7) are a basis for the row space of A.

5. **(a)** A row-echelon form of A^t is

$$\begin{bmatrix} 1 & 5 & 7 \\ 0 & 1 & 1 \\ 0 & 0 & 0 \end{bmatrix}$$

Thus a basis for the column space of A is

$$\begin{bmatrix} 1 \\ 5 \\ 7 \end{bmatrix} \quad \text{and} \quad \begin{bmatrix} 0 \\ 1 \\ 1 \end{bmatrix}$$

(c) A row-echelon form of A^t is

$$\begin{bmatrix} 1 & 2 & -1 \\ 0 & 1 & -1 \\ 0 & 0 & 0 \\ 0 & 0 & 0 \end{bmatrix}$$

Thus a basis for the column space of A is

$$\begin{bmatrix} 1 \\ 2 \\ -1 \end{bmatrix} \quad \text{and} \quad \begin{bmatrix} 0 \\ 1 \\ -1 \end{bmatrix}$$

6. **(a)** Since, by 4(a), the row space of A has dimension 2, any two linearly independent row vectors of A will form a basis for the row space. Because no row of A is a multiple of another, any two rows will do. In particular, the first two rows form a basis and Row 3 = Row 2 + 2(Row 1).

(c) Refer to 4(c) and use the solution to 6(a), above. In particular, the first two rows form a basis and Row 3 = Row 1 − Row 2.

(e) Let $\mathbf{r}_1$, $\mathbf{r}_2$, $\cdots$, $\mathbf{r}_5$ denote the rows of A. If we observe, for instance, that $\mathbf{r}_1 = -\mathbf{r}_3 + \mathbf{r}_4$ and that $\mathbf{r}_2 = 2\mathbf{r}_1 + \mathbf{r}_5$, then we see that $\{\mathbf{r}_3, \mathbf{r}_4, \mathbf{r}_5\}$ spans the row space. Since the dimension of this space is 3 (see the solution to Exercise 3(e)), the set forms a basis.

For those who don't wish to rely on insight, set $a\mathbf{r}_1 + b\mathbf{r}_2 + c\mathbf{r}_3 + d\mathbf{r}_4 + e\mathbf{r}_5 = \mathbf{0}$ and solve the resulting homogeneous system of equations by finding the reduced row-echelon form for A^t. This yields

$$\begin{aligned} a &= -s + 2t \\ b &= -t \\ c &= -s \\ d &= s \\ e &= t \end{aligned}$$

so that $(-s + 2t)\mathbf{r}_1 - t\mathbf{r}_2 - s\mathbf{r}_3 + s\mathbf{r}_4 + t\mathbf{r}_5 = \mathbf{0}$, or

$$s(-\mathbf{r}_1 - \mathbf{r}_3 + \mathbf{r}_4) + t(2\mathbf{r}_1 - \mathbf{r}_2 + \mathbf{r}_5) = \mathbf{0}$$

Since this equation must hold for all values of s and t, we have

$$\mathbf{r}_1 = -\mathbf{r}_3 + \mathbf{r}_4 \quad \text{and} \quad \mathbf{r}_2 = 2\mathbf{r}_1 + \mathbf{r}_5$$

which is the result obtained above.

7. **(a)** The space spanned by these vectors is the row space of the matrix

$$\begin{bmatrix} 1 & 1 & -4 & -3 \\ 2 & 0 & 2 & -2 \\ 2 & -1 & 3 & 2 \end{bmatrix}$$

The row-echelon form of the above matrix is

$$\begin{bmatrix} 1 & 1 & -4 & -3 \\ 0 & 1 & -5 & -2 \\ 0 & 0 & 1 & -1/2 \end{bmatrix}$$

and the reduced row-echelon form is

$$\begin{bmatrix} 1 & 0 & 0 & -1/2 \\ 0 & 1 & 0 & -9/2 \\ 0 & 0 & 1 & -1/2 \end{bmatrix}$$

Thus {(1,1,-4,-3), (0,1,-5,-2), (0,0,1,-1/2)} is one basis. Another basis is {(1,0,0,-1/2), (0,1,0,-9/2), (0,0,1,-1/2)}.

8. **(a)** If we solve the vector equation

(*) $$a\mathbf{v}_1 + b\mathbf{v}_2 + c\mathbf{v}_3 + d\mathbf{v}_4 = \mathbf{0}$$

we obtain the homogeneous system

$$\begin{aligned} a - 3b - c - 5d &= 0 \\ 3b + 3c + 3d &= 0 \\ a + 7b + 9c + 5d &= 0 \\ a + b + 3c - d &= 0 \end{aligned}$$

The reduced row-echelon form of the augmented matrix is

$$\begin{bmatrix} 1 & 0 & 2 & -2 & 0 \\ 0 & 1 & 1 & 1 & 0 \\ 0 & 0 & 0 & 0 & 0 \\ 0 & 0 & 0 & 0 & 0 \end{bmatrix}$$

Thus $\{\mathbf{v}_1, \mathbf{v}_2\}$ forms a basis for the space. The solution is $a = -2s + 2t$, $b = -s - t$, $c = s$, $d = t$. This yields

$$(-2s + 2t)\mathbf{v}_1 + (-s - t)\mathbf{v}_2 + s\mathbf{v}_3 + t\mathbf{v}_4 = \mathbf{0}$$

or

$$s(-2\mathbf{v}_1 - \mathbf{v}_2 + \mathbf{v}_3) + t(2\mathbf{v}_1 - \mathbf{v}_2 + \mathbf{v}_4) = \mathbf{0}$$

Since s and t are arbitrary, set $s = 1$, $t = 0$ and then $s = 0$, $t = 1$ to obtain the dependency equations

$$-2\mathbf{v}_1 - \mathbf{v}_2 + \mathbf{v}_3 = \mathbf{0}$$

$$2\mathbf{v}_1 - \mathbf{v}_2 + \mathbf{v}_4 = \mathbf{0}$$

Thus

$$\mathbf{v}_3 = 2\mathbf{v}_1 + \mathbf{v}_2$$

and

$$\mathbf{v}_4 = -2\mathbf{v}_1 + \mathbf{v}_2$$

(c) If we solve the vector equation

$$(*) \qquad a\mathbf{v}_1 + b\mathbf{v}_2 + c\mathbf{v}_3 + d\mathbf{v}_4 + e\mathbf{v}_5 = \mathbf{0}$$

we obtain the homogeneous system

$$\begin{aligned} a - 2b + 4c \phantom{{}+4d} - 7e &= 0 \\ -a + 3b - 5c + 4d + 18e &= 0 \\ 5a + b + 9c + 2d + 2e &= 0 \\ 2a \phantom{{}+3b} + 4c - 3d - 8e &= 0 \end{aligned}$$

The reduced row-echelon form of the augmented matrix is

$$\begin{bmatrix} 1 & 0 & 2 & 0 & -1 & 0 \\ 0 & 1 & -1 & 0 & 3 & 0 \\ 0 & 0 & 0 & 1 & 2 & 0 \\ 0 & 0 & 0 & 0 & 0 & 0 \end{bmatrix}$$

This tells us that $\{\mathbf{v}_1, \mathbf{v}_2, \mathbf{v}_4\}$ is the desired basis. The solution is $a = -2s + t$, $b = s - 3t$, $c = s$, $d = -2t$, and $e = t$. This yields

$$(-2s + t)\mathbf{v}_1 + (s - 3t)\mathbf{v}_2 + s\mathbf{v}_3 - 2t\mathbf{v}_4 + t\mathbf{v}_5 = \mathbf{0}$$

or

$$s(-2\mathbf{v}_1 + \mathbf{v}_2 + \mathbf{v}_3) + t(\mathbf{v}_1 - 3\mathbf{v}_2 - 2\mathbf{v}_4 + \mathbf{v}_5) = \mathbf{0}$$

Since s and t are arbitrary, set $s = 1$, $t = 0$ and then $s = 0$, $t = 1$ to obtain the dependency equations

$$-2\mathbf{v}_1 + \mathbf{v}_2 + \mathbf{v}_3 = \mathbf{0}$$

$$\mathbf{v}_1 - 3\mathbf{v}_2 - 2\mathbf{v}_4 + \mathbf{v}_5 = \mathbf{0}$$

Thus

$$\mathbf{v}_3 = 2\mathbf{v}_1 - \mathbf{v}_2$$

and

$$\mathbf{v}_5 = -\mathbf{v}_1 + 3\mathbf{v}_2 + 2\mathbf{v}_4$$

10. Call the nonzero row vectors $\mathbf{r}_1, \cdots, \mathbf{r}_s$. Label the rows in the usual top to bottom fashion and consider the equation

$$a_1\mathbf{r}_1 + a_2\mathbf{r}_2 + \cdots + a_s\mathbf{r}_s = \mathbf{0}$$

Since $\mathbf{r}_1$ has its leading 1 in, say, the tth column, then the tth coordinate of the left-hand side is a_1. Since this must equal zero, we have $a_1 = 0$. But then the same argument applies to $\mathbf{r}_2$ and hence $a_2 = 0$.

In fact, all of the numbers $a_1 = a_2 = \cdots = a_s = 0$, and hence the row vectors must be linearly independent.

Alternatively, we prove this result by contradiction. Suppose that the rows are linearly dependent. Then one of them can be expressed as a linear combination of the others; say $\mathbf{r}_1 = c_2\mathbf{r}_2 + \cdots + c_s\mathbf{r}_s$. But then the succession of elementary row operations which adds $-c_i$ times Row i to Row 1 for $i = 2,\ldots,s$ will reduce Row 1 to the zero vector. Consequently, the matrix could not have been in row-echelon form. Thus if it is in row-echelon form, the nonzero row vectors must be linearly independent.

11. Call the original matrix A, the given row-echelon form B, and its reduced row-echelon form C. We know from Exercise 10 that the rows of B (or C) which contain leading entries 1 are linearly independent. An argument similar to the first one given in Exercise 10 shows that the columns of B (or C) which contain leading entries 1 are also linearly independent. By Theorem 4.6.3, these columns will form a basis for the column space of A provided they form a basis for the column space of B (or C). Thus if we can show that the appropriate columns of C span its column space, then the corresponding columns of B will span the column space of A. But the columns in question in C each have a single 1 and all the other entries are zero. The remaining columns of C only have nonzero entries in rows with a 1 in one of the "basis" columns. Thus the "basis" columns can easily be used to form linear combinations equal to any other column. Hence they span the column space of C and therefore are, indeed, a basis for it. Thus the corresponding column vectors of B are a basis for the column space of A.

EXERCISE SET 4.7

1. **(a)** The reduced row-echelon form for A is

$$\begin{bmatrix} 1 & 0 & -16 \\ 0 & 1 & -19 \\ 0 & 0 & 0 \end{bmatrix}$$

Thus rank $(A) = 2$. The solution to $A\mathbf{x} = 0$ is $x = 16t$, $y = 19t$, $z = t$, so that the nullity is one. There are three columns, so we have $2 + 1 = 3$.

(c) The reduced row-echelon form for A is

$$\begin{bmatrix} 1 & 0 & 1 & -2/7 \\ 0 & 1 & 1 & 4/7 \\ 0 & 0 & 0 & 0 \end{bmatrix}$$

Thus rank $(A) = 2$. The null space will have dimension two since the solution to $A\mathbf{x} = 0$ has two parameters. There are four columns, so we have $2 + 2 = 4$.

2. **(a)** As in the proof of Theorem 4.7.4, we consider the equations $A\mathbf{x} = \mathbf{b}$. Since $x_1 = 1$, $x_2 = -1$ is a solution to this system of equations, it follows that $\mathbf{b}$ is in the column space of A and

$$\mathbf{b} = (1)\begin{bmatrix} 1 \\ 4 \end{bmatrix} + (-1)\begin{bmatrix} 3 \\ -6 \end{bmatrix}$$

2. (b) The rank of the augmented matrix $[A:\mathbf{b}]$ is 3, but rank $(A) = 2$. Thus, by Theorem 4.7.5, the system is inconsistent, and therefore **b** is not in the column space of A.

4. (a) The general solution is $x_1 = 1 + 3t$, $x_2 = t$. Its vector form is

$$\begin{bmatrix} 1 \\ 0 \end{bmatrix} + t\begin{bmatrix} 3 \\ 1 \end{bmatrix}$$

Thus the vector form of the general solution to $A\mathbf{x} = 0$ is

$$t\begin{bmatrix} 3 \\ 1 \end{bmatrix}$$

(c) The general solution is $x_1 = -1 + 2s - t$, $x_2 = s$, $x_3 = t$, $x_4 = 0$. Its vector form is

$$\begin{bmatrix} -1 \\ 0 \\ 0 \\ 0 \end{bmatrix} + s\begin{bmatrix} 2 \\ 1 \\ 0 \\ 0 \end{bmatrix} + t\begin{bmatrix} -1 \\ 0 \\ 1 \\ 0 \end{bmatrix}$$

Thus the vector form of the general solution to $A\mathbf{x} = 0$ is

$$s\begin{bmatrix} 2 \\ 1 \\ 0 \\ 0 \end{bmatrix} + t\begin{bmatrix} -1 \\ 0 \\ 1 \\ 0 \end{bmatrix}$$

7. Use Theorems 4.7.5 and 4.7.7.

(a) The system is consistent because the two ranks are equal. Since $n = r = 3$, $n - r = 0$ and therefore the number of parameters is 0.

(b) The system is inconsistent because the two ranks are not equal.

(d) The system is consistent because the two ranks are equal. Here $n = 9$ and $r = 2$, so that $n - r = 7$ parameters will appear in the solution.

(f) Since the ranks are equal, the system in consistent. However A must be the zero matrix, so the system gives no information at all about its solution. This is reflected in the fact that $n - r = 4 - 0 = 4$, so that there will be 4 parameters in the solution for the 4 variables.

9. Suppose that A has rank 2. Then two of its column vectors are linearly independent. Thus, by Theorem 4.7.3, at least one of the 2×2 submatrices has nonzero determinant.

Conversely, if at least one of the determinants of the 2×2 submatrices is nonzero, then, by Theorem 4.7.3, at least two of the column vectors must be linearly independent. Thus the rank of A must be at least 2. But since the dimension of the row space of A is at most 2, A has rank at most 2. Thus, the rank of A is exactly 2.

10. If the nullspace of A is a line through the origin, then it has the form $x = at$, $y = bt$, $z = ct$ where t is the only parameter. Thus nullity(A) = $3 - \text{rank}(A) = 1$. That is, the row and column spaces of A have dimension 2, so neither space can be a line. Why?

12. **(a)** If A is a 3×5 matrix, then it has 5 columns, each of which is a vector in R^3. Thus, by Theorem 4.5.2, they form a linearly dependent set.

14. If A is an $n \times n$ invertible matrix, then, by Theorem 4.7.3, its row vectors are linearly independent. Since there are n of these vectors, each with n components, they form a basis for R^n.

15. Suppose that A is an $m \times n$ matrix. If $A\mathbf{x} = 0$ has only the trivial solution, then $n\text{-rank}(A) = 0$ or $\text{rank}(A) = n$. Thus, by Theorem 4.7.7, there are no parameters in any solution to $A\mathbf{x} = \mathbf{b}$. That is, if the system is consistent, it can have at most one solution.

16. Suppose that the system $A\mathbf{x} = \mathbf{b}$ is consistent. Then, by Theorem 4.7.4, $\mathbf{b}$ is in the column space of A; that is, $\mathbf{b}$ can be written as a linear combination of the columns of A. Thus the rank of A and the rank of the augmented matrix (A with $\mathbf{b}$ added as an additional column) are equal.

Conversely, if the two ranks are equal, then $\mathbf{b}$ can be written as a linear combination of the columns of A, so that $\mathbf{b}$ is in the column space of A. Hence, by Theorem 4.7.4, $A\mathbf{x} = \mathbf{b}$ is consistent.

SUPPLEMENTARY EXERCISES 4

1. **(b)** The augmented matrix of this system reduces to

$$\begin{bmatrix} 2 & -3 & 1 & 0 \\ 0 & 0 & 0 & 0 \\ 0 & 0 & 0 & 0 \end{bmatrix}$$

Therefore, the solution space is a plane with equation $2x - 3y + z = 0$

(c) The solution is $x = 2t$, $y = t$, $z = 0$, which is a line.

3. **(a)** The identities

$$\sin(x + \theta) = \cos\theta \sin x + \sin\theta \cos x$$
$$\cos(x + \theta) = \cos\theta \cos x - \sin\theta \sin x$$

hold for all values of x and θ. Hence

$$(*) \qquad \begin{aligned} \mathbf{f}_1 &= (\cos\theta)\mathbf{f} + (\sin\theta)\mathbf{g} \\ \mathbf{g}_1 &= (-\sin\theta)\mathbf{f} + (\cos\theta)\mathbf{g} \end{aligned}$$

That is, $\mathbf{f}_1$ and $\mathbf{g}_1$ are linear combinations of $\mathbf{f}$ and $\mathbf{g}$ and therefore belong to W.

(b) If we solve the system $(*)$ for $\mathbf{f}$ and $\mathbf{g}$, we obtain

$$\mathbf{f} = (\cos\theta)\mathbf{f}_1 + (-\sin\theta)\mathbf{g}_1$$
$$\mathbf{g} = (\sin\theta)\mathbf{f}_1 + (\cos\theta)\mathbf{g}_1$$

Hence, any linear combination of $\mathbf{f}$ and $\mathbf{g}$ is also a linear combination of $\mathbf{f}_1$ and $\mathbf{g}_1$ and thus $\mathbf{f}_1$ and $\mathbf{g}_1$ span W. Since the dimension of W is 2 (it is spanned by 2 linearly independent vectors), Theorem 4.5.4(b) implies that $\mathbf{f}_1$ and $\mathbf{g}_1$ form a basis for W.

4. **(a)** We look for constants a, b, and c such that

$$\mathbf{v} = a\mathbf{v}_1 + b\mathbf{v}_2 + c\mathbf{v}_3, \quad \text{or} \quad \begin{aligned} a + 3b + 2c &= 1 \\ -a \qquad + c &= 1 \end{aligned}$$

This system has the solution

$$a = t - 1 \qquad b = \frac{2}{3} - t \qquad c = t$$

where t is arbitrary. If we set $t = 0$ and $t = 1$, we obtain $\mathbf{v} = (-1)\mathbf{v}_1 + \frac{2}{3}\mathbf{v}_2$ and $\mathbf{v} = \left(-\frac{1}{3}\right)\mathbf{v}_2 + \mathbf{v}_3$, respectively. There are infinitely many other possibilities.

(b) Since $\mathbf{v}_1$, $\mathbf{v}_2$, and $\mathbf{v}_3$ all belong to R^2 and $\dim\left(R^2\right) = 2$, it follows from Theorem 4.5.2 that these three vectors do not form a basis for R^2. Hence, Theorem 4.5.1 does not apply.

5. Suppose that there are constants $c_1, \ldots, c_n$, not all zero, such that $c_1A\mathbf{v}_1 + \cdots + c_nA\mathbf{v}_n = 0$. Then

$$A\left(c_1\mathbf{v}_1 + \cdots + c_n\mathbf{v}_n\right) = 0$$

Since the vectors $\mathbf{v}_1, \ldots, \mathbf{v}_n$ are linearly independent, the $n \times 1$ matrix $c_1\mathbf{v}_1 + \cdots + c_n\mathbf{v}_n$ cannot equal 0. By Theorem 1.6.3, this is true if and only if A is invertible.

6. Consider the polynomials x and $x + 1$ in P_1. Verify that these polynomials form a basis for P_1.

7. (c) Since the odd numbered rows are all repeats of Row 1 and the even numbered rows are all repeats of Row 2, while Rows 1 and 2 are linearly independent, an $n \times n$ checker board matrix has rank 2 whenever $n \geq 2$. Since the nullity is n minus the rank, we have nullity $= n - 2$.

9. (a) If p belongs to the set, then it contains only even powers of x. Since this set is closed under polynomial addition and scalar multiplication (Why?), it is a subspace of P_n. One basis is the set $\{1, x^2, x^4, \ldots, x^{2m}\}$ where $2m = n$ if n is even and $2m = n - 1$ if n is odd.

(b) If p belongs to this set, then its constant term must be zero. Since this set is closed under polynomial addition and scalar multiplication (Why?), it is a subspace of P_n. One basis is the set $\{x, x^2, \ldots, x^n\}$.

11. (a) Since $\begin{vmatrix} 1 & 0 \\ 2 & -1 \end{vmatrix} = -1 \neq 0$, the rank is 2.

(b) Since all three 2×2 subdeterminants are zero, the rank is 1.

(c) Since the determinant of the matrix is zero, its rank is less than 3. Since $\begin{vmatrix} 1 & 0 \\ 2 & -1 \end{vmatrix} = -1 \neq 0$, the rank is 2.

(d) Since the determinant of the 3×3 submatrix obtained by deleting the last column is $30 \neq 0$, the rank of the matrix is 3.

12. Call the matrix A. Since the determinant of every 5×5, 4×4, and 3×3 submatrix is zero, $\text{rank}(A) \leq 2$. Since

$$\det\left(\begin{bmatrix} 0 & a_{i6} \\ a_{5j} & a_{56} \end{bmatrix}\right) = -a_{5j}a_{i6}$$

for $i = 1,2,\ldots,4$ and $j = 1,\ldots,5$, then $\text{rank}(A) = 2$ if any of these determinants is nonzero. Otherwise, if any of the numbers $a_{ij} \neq 0$, then $\text{rank}(A) = 1$ and if $a_{ij} = 0$ for all i and j, then $\text{rank}(A) = 0$.

13. **(b)** Let $S = \{\mathbf{v}_1,\ldots,\mathbf{v}_n\}$ and let $\mathbf{u} = u_1\mathbf{v}_1 + \cdots + u_n\mathbf{v}_n$. Thus $(\mathbf{u})_S = (u_1,\cdots,u_n)$. We have

$$k\mathbf{u} = ku_1\mathbf{v}_1 + \cdots + ku_n\mathbf{v}_n$$

so that $(k\mathbf{u})_S = (ku_1,\cdots,ku_n) = k(u_1,\cdots,u_n)$. Therefore $(k\mathbf{u})_S = k(\mathbf{u})_S$.

EXERCISE SET 5.1

1. (c) Since $\mathbf{v} + \mathbf{w} = (3,11)$, we have

$$\langle \mathbf{u}, \mathbf{v} + \mathbf{w} \rangle = 3(3) + (-2)(11) = -13$$

On the other hand,

$$\langle \mathbf{u}, \mathbf{v} \rangle = 3(4) + (-2)(5) = 2$$

and

$$\langle \mathbf{u}, \mathbf{w} \rangle = 3(-1) + (-2)(6) = -15$$

(d) Since $k\mathbf{u} = (-12,8)$ and $k\mathbf{v} = (-16,-20)$, we have

$$\langle k\mathbf{u}, \mathbf{v} \rangle = (-12)(4) + (8)(5) = -8$$

and

$$\langle \mathbf{u}, k\mathbf{v} \rangle = 3(-16) + (-2)(-20) = -8$$

Since $\langle \mathbf{u}, \mathbf{v} \rangle = 2$, $k\langle \mathbf{u}, \mathbf{v} \rangle = -8$.

2. (c) Since $\mathbf{v} + \mathbf{w} = (3,11)$, we have

$$\langle \mathbf{u}, \mathbf{v} + \mathbf{w} \rangle = 4(3)(3) + 5(-2)(11) = -74$$

On the other hand,

$$\langle \mathbf{u}, \mathbf{v} \rangle = 4(3)(4) + 5(-2)(5) = -2$$

and

$$\langle \mathbf{u}, \mathbf{w} \rangle = 4(3)(-1) + 5(-2)(6) = -72$$

3. **(a)** $\langle \mathbf{u},\mathbf{v}\rangle = 3(-1) - 2(3) + 4(1) + 8(1) = 3$

4. **(a)** $\langle \mathbf{p},\mathbf{q}\rangle = (-2)(4) + 1(0) + 3(-7) = -29$

5. **(a)** By Formula (2),

$$\langle \mathbf{u},\mathbf{v}\rangle = \begin{bmatrix} v_1 & v_2 \end{bmatrix} \begin{bmatrix} 3 & 0 \\ 0 & 2 \end{bmatrix} \begin{bmatrix} 3 & 0 \\ 0 & 2 \end{bmatrix} \begin{bmatrix} u_1 \\ u_2 \end{bmatrix}$$

$$= \begin{bmatrix} v_1 & v_2 \end{bmatrix} \begin{bmatrix} 9 & 0 \\ 0 & 4 \end{bmatrix} \begin{bmatrix} u_1 \\ u_2 \end{bmatrix}$$

$$= \begin{bmatrix} 9v_1 & 4v_2 \end{bmatrix} \begin{bmatrix} u_1 \\ u_2 \end{bmatrix}$$

$$= 9u_1v_1 + 4u_2v_2$$

(b) We have $\langle \mathbf{u},\mathbf{v}\rangle = 9(-3)(1) + 4(2)(7) = 29$.

6. **(a)** Since

$$A^tA = \begin{bmatrix} 5 & -1 \\ -1 & 10 \end{bmatrix}$$

then

$$\begin{bmatrix} v_1 & v_2 \end{bmatrix} \begin{bmatrix} 5 & -1 \\ -1 & 10 \end{bmatrix} \begin{bmatrix} u_1 \\ u_2 \end{bmatrix} = \begin{bmatrix} 5v_1 - v_2 & -v_1 + 10v_2 \end{bmatrix} \begin{bmatrix} u_1 \\ u_2 \end{bmatrix}$$

$$= 5u_1v_1 - u_1v_2 - u_2v_1 + 10u_2v_2$$

7. **(a)** By Formula (3), we have $\langle \mathbf{u},\mathbf{v}\rangle = \mathbf{v}^tA^tA\mathbf{u}$ where

$$A = \begin{bmatrix} \sqrt{3} & 0 \\ 0 & \sqrt{5} \end{bmatrix}$$

8. **(a)** (1) $\langle \mathbf{u},\mathbf{v}\rangle = 3u_1v_1 + 5u_2v_2 = 3v_1u_1 + 5v_2u_2 = \langle \mathbf{v},\mathbf{u}\rangle$

(2) If $\mathbf{w} = (w_1,w_2)$, then

$$\begin{aligned}\langle \mathbf{u}+\mathbf{v},\mathbf{w}\rangle &= 3(u_1 + v_1)w_1 + 5(u_2 + v_2)w_2\\ &= (3u_1w_1 + 5u_2w_2) + (3v_1w_1 + 5v_2w_2)\\ &= \langle \mathbf{u},\mathbf{w}\rangle + \langle \mathbf{v},\mathbf{w}\rangle\end{aligned}$$

(3) $$\begin{aligned}\langle k\mathbf{u},\mathbf{v}\rangle &= 3(ku_1)v_1 + 5(ku_2)v_2\\ &= k(3u_1v_1 + 5u_2v_2)\\ &= k\langle \mathbf{u},\mathbf{v}\rangle\end{aligned}$$

(4) $\langle \mathbf{v},\mathbf{v}\rangle = 3v_1^2 + 5v_2^2 \geq 0$

Moreover, $\langle \mathbf{v},\mathbf{v}\rangle = 0$ if and only if $v_1 = v_2 = 0$, or $\mathbf{v} = \mathbf{0}$.

9. **(b)** Axioms 1 and 4 are easily checked. However, if $\mathbf{w} = (w_1,w_2,w_3)$, then

$$\begin{aligned}\langle \mathbf{u}+\mathbf{v},\mathbf{w}\rangle &= (u_1+v_1)^2w_1^2 + (u_2+v_2)^2w_2^2 + (u_3+v_3)^2w_3^2\\ &= \langle \mathbf{u},\mathbf{w}\rangle + \langle \mathbf{v},\mathbf{w}\rangle + 2u_1v_1w_1^2 + 2u_2v_2w_2^2 + 2u_3v_3w_3^2\end{aligned}$$

If, for instance, $\mathbf{u} = \mathbf{v} = \mathbf{w} = (1,0,0)$, then Axiom 2 fails.

To check Axiom 3, we note that $\langle k\mathbf{u},\mathbf{v}\rangle = k^2\langle \mathbf{u},\mathbf{v}\rangle$. Thus $\langle k\mathbf{u},\mathbf{v}\rangle \neq k\langle \mathbf{u},\mathbf{v}\rangle$ unless $k = 0$ or $k = 1$, so Axiom 3 fails.

(c) (1) Axiom 1 follows from the commutativity of multiplication in R.

(2) If $\mathbf{w} = (w_1,w_2,w_3)$, then

$$\begin{aligned}\langle \mathbf{u}+\mathbf{v},\mathbf{w}\rangle &= 2(u_1+v_1)w_1 + (u_2+v_2)w_2 + 4(u_3+v_3)w_3\\ &= 2u_1w_1 + u_2w_2 + 4u_3w_3 + 2v_1w_1 + v_2w_2 + 4v_3w_3\\ &= \langle \mathbf{u},\mathbf{w}\rangle + \langle \mathbf{v},\mathbf{w}\rangle\end{aligned}$$

(3) $\langle k\mathbf{u},\mathbf{v}\rangle = 2(ku_1)v_1 + (ku_2)v_2 + 4(ku_3)v_3 = k\langle \mathbf{u},\mathbf{v}\rangle$

(4) $\langle \mathbf{v},\mathbf{v}\rangle = 2v_1^2 + v_2^2 + 4v_3^2 \geq 0$

$= 0$ if and only if $v_1 = v_2 = v_3 = 0$, or $\mathbf{v} = \mathbf{0}$

Thus this is an inner product for R^3.

10. (c) Since $A^tA = \begin{bmatrix} 2 & -1 \\ -1 & 13 \end{bmatrix}$, we have

$$\|\mathbf{w}\|^2 = \left(\begin{bmatrix} -1 & 3 \end{bmatrix}\begin{bmatrix} 2 & -1 \\ -1 & 13 \end{bmatrix}\right)\begin{bmatrix} -1 \\ 3 \end{bmatrix}$$

$$= \begin{bmatrix} -5 & 40 \end{bmatrix}\begin{bmatrix} -1 \\ 3 \end{bmatrix} = 125$$

Thus $\|\mathbf{w}\| = \sqrt{125} = 5\sqrt{5}$.

11. We have $\mathbf{u} - \mathbf{v} = (-3,-3)$.

(b) $d(\mathbf{u},\mathbf{v}) = \|(-3,-3)\| = [3(9) + 2(9)]^{1/2} = \sqrt{45} = 3\sqrt{5}$

(c) From Problem 10(c), we have

$$[d(\mathbf{u},\mathbf{v})]^2 = \begin{bmatrix} -3 & -3 \end{bmatrix}\begin{bmatrix} 2 & -1 \\ -1 & 13 \end{bmatrix}\begin{bmatrix} -3 \\ -3 \end{bmatrix} = 117$$

Thus

$$d(\mathbf{u},\mathbf{v}) = \sqrt{117} = 3\sqrt{13}$$

12. (a) $\|\mathbf{p}\| = \left[(-2)^2 + (3)^2 + (2)^2\right]^{1/2} = \sqrt{17}$

13. (a) $\|A\| = \left[(-2)^2 + (5)^2 + (3)^2 + (6)^2\right]^{1/2} = \sqrt{74}$

14. Since $\mathbf{p} - \mathbf{q} = 1 - x - 4x^2$, we have

$$d(\mathbf{p},\mathbf{q}) = \langle \mathbf{p}-\mathbf{q},\ \mathbf{p}-\mathbf{q}\rangle^{1/2} = \left[1^2 + (-1)^2 + (-4)^2\right]^{1/2} = 3\sqrt{2}$$

15. (a) Since $A - B = \begin{bmatrix} 6 & -1 \\ 8 & -2 \end{bmatrix}$, we have

$$d(A,B) = \langle A-B,\ A-B\rangle^{1/2} = \left[6^2 + (-1)^2 + 8^2 + (-2)^2\right]^{1/2} = \sqrt{105}$$

16. (a) We have $\langle \mathbf{u} + \mathbf{v},\ \mathbf{v} + \mathbf{w}\rangle = \langle \mathbf{u},\mathbf{v}\rangle + \langle \mathbf{u},\mathbf{w}\rangle + \langle \mathbf{v},\mathbf{v}\rangle + \langle \mathbf{v},\mathbf{w}\rangle = 2 + 5 + (2)^2 - 3 = 8$.

(d) We have $\|\mathbf{u} + \mathbf{v}\|^2 = \langle \mathbf{u} + \mathbf{v},\ \mathbf{u} + \mathbf{v}\rangle = \langle \mathbf{u},\mathbf{u}\rangle + 2\langle \mathbf{u},\mathbf{v}\rangle + \langle \mathbf{v},\mathbf{v}\rangle = (1)^2 + 2(2) + (2)^2 = 9$. Hence $\|\mathbf{u} + \mathbf{v}\| = 3$.

17. (a) For instance, $\|x\| = \left[\int_{-1}^{1} x^2\,dx\right]^{1/2} = \left[\left.\frac{x^3}{3}\right]_{-1}^{1}\right]^{1/2} = \left[\frac{2}{3}\right]^{1/2}$.

(b) We have

$$\begin{aligned} d(\mathbf{p},\mathbf{q}) &= \|\mathbf{p} - \mathbf{q}\| \\ &= \|1 - x\| \\ &= \left[\int_{-1}^{1} (1 - x)^2\,dx\right]^{1/2} \\ &= \left[\int_{-1}^{1} \left[1 - 2x + x^2\right] dx\right]^{1/2} \\ &= \left[\left[x - x^2 + \frac{x^3}{3}\right]_{-1}^{1}\right]^{1/2} \\ &= 2\left[\frac{2}{3}\right]^{1/2} \end{aligned}$$

18. **(a)** Since we are looking for points (x,y) with $\|x,y\| = 1$, we have $\left[\frac{1}{4}x^2 + \frac{1}{16}y^2\right]^{1/2} = 1$ or $\frac{1}{4}x^2 + \frac{1}{16}y^2 = 1$. This is the ellipse shown in the figure.

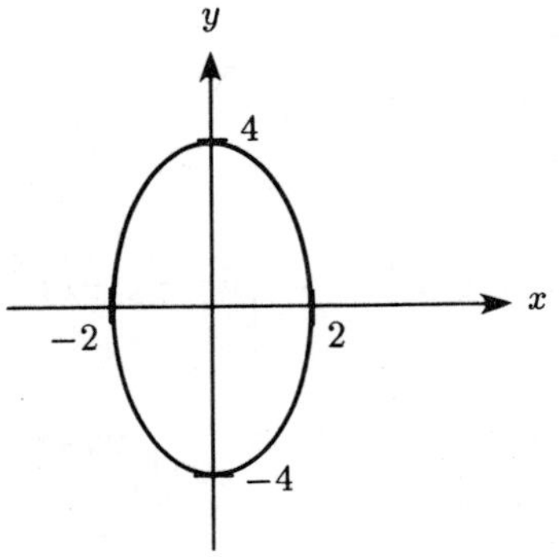

20. Observe that

(*) $$\|\mathbf{u} + \mathbf{v}\|^2 = \langle \mathbf{u} + \mathbf{v},\ \mathbf{u} + \mathbf{v}\rangle = \langle \mathbf{u},\mathbf{u}\rangle + 2\langle \mathbf{u},\mathbf{v}\rangle + \langle \mathbf{v},\mathbf{v}\rangle$$

and

(**) $$\|\mathbf{u} - \mathbf{v}\|^2 = \langle \mathbf{u} - \mathbf{v},\ \mathbf{u} - \mathbf{v}\rangle = \langle \mathbf{u},\mathbf{u}\rangle - 2\langle \mathbf{u},\mathbf{v}\rangle + \langle \mathbf{v},\mathbf{v}\rangle$$

Thus if we add (*) and (**), we obtain the equation

$$\|\mathbf{u} + \mathbf{v}\|^2 + \|\mathbf{u} - \mathbf{v}\|^2 = 2\|\mathbf{u}\|^2 + 2\|\mathbf{v}\|^2$$

21. If, in the solution to Exercise 20, we subtract (**) from (*) and divide by 4, we obtain the desired result.

22. If, for instance, $V = \begin{bmatrix} 0 & 1 \\ -1 & 0 \end{bmatrix}$, then $\langle V, V\rangle = -2 < 0$. Thus Axiom 4 fails.

23. Axioms 1 and 3 are easily verified. So is Axiom 2, as shown: Let $\mathbf{r} = r(x)$ be a polynomial in P_2. Then

$$\begin{aligned}\langle \mathbf{p}+\mathbf{q},\mathbf{r}\rangle &= (p+q)(0)r(0) + (p+q)(1/2)r(1/2) + (p+q)(1)r(1) \\ &= p(0)r(0) + p(1/2)r(1/2) + p(1)r(1) + q(0)r(0) \\ &\qquad + q(1/2)r(1/2) + q(1)r(1) \\ &= \langle \mathbf{p},\mathbf{r}\rangle + \langle \mathbf{q},\mathbf{r}\rangle\end{aligned}$$

It remains to verify Axiom 4:

$$\begin{aligned}\langle \mathbf{p},\mathbf{p}\rangle &= [p(0)]^2 + [p(1/2)]^2 + [p(1)]^2 \geq 0 \\ &= 0 \text{ if and only if } p(0) = p(1/2) = p(1) = 0\end{aligned}$$

But a quadratic polynomial can have at most two zeros unless it is identically zero. Thus $\langle \mathbf{p},\mathbf{p}\rangle = 0$ if and only if $\mathbf{p}$ is identically zero, or $\mathbf{p} = \mathbf{0}$.

24. Using the hint and properties of the transpose, we have

$$\langle \mathbf{u},A\mathbf{v}\rangle = (A\mathbf{v})^t\mathbf{u} = (\mathbf{v}^tA^t)\mathbf{u} = \mathbf{v}^t(A^t\mathbf{u}) = \langle A^t\mathbf{u},\mathbf{v}\rangle$$

26. Axiom 4 fails to hold for nonnegative constants c_1, c_2, c_3, although it holds for positive values of these constants. Consider, for example, $c_1 = 0$, $c_2 = c_3 = 1$, and $\mathbf{u} = (1,0,0)$.

27. **(b)**
$$\begin{aligned}\langle \mathbf{p},\mathbf{q}\rangle &= \int_{-1}^{1}(x - 5x^3)\,(2 + 8x^2)\,dx \\ &= \int_{-1}^{1}(2x - 2x^3 - 40x^5)\,dx \\ &= x^2 - x^4/2 - 20x^6/3\Big]_{-1}^{1} \\ &= 0\end{aligned}$$

28. **(a)**

$$\langle \mathbf{f},\mathbf{g}\rangle = \int_0^1 \cos 2\pi x \sin 2\pi x \, dx$$

$$= (1/4\pi)\sin^2 2\pi x \Big]_0^1$$

$$= 0$$

29. We have $\langle U, V\rangle = u_1v_1 + u_2v_2 + u_3v_3 + u_4v_4$ and

$$\mathrm{tr}(U^tV) = \mathbf{tr}\left[\begin{bmatrix} u_1 & u_3 \\ u_2 & u_4 \end{bmatrix}\begin{bmatrix} v_1 & v_2 \\ v_3 & v_4 \end{bmatrix}\right]$$

$$= \mathbf{tr}\left[\begin{bmatrix} u_1v_1 + u_3v_3 & u_1v_2 + u_3v_4 \\ u_2v_1 + u_4v_3 & u_2v_2 + u_4v_4 \end{bmatrix}\right]$$

$$= u_1v_1 + u_3v_3 + u_2v_2 + u_4v_4$$

which does, indeed, equal $\langle U, V\rangle$.

30. We use the formula $\langle \mathbf{u},\mathbf{v}\rangle = \mathbf{v}^tA^tA\mathbf{u}$ as suggested in the hint.

(1) We must show that $\langle \mathbf{u},\mathbf{v}\rangle = \langle \mathbf{v},\mathbf{u}\rangle$. Since $\langle \mathbf{v},\mathbf{u}\rangle$ is a real number which can be thought of as a 1×1 matrix, it will suffice to show that $\langle \mathbf{u},\mathbf{v}\rangle = \langle \mathbf{v},\mathbf{u}\rangle^t$. But

$$\langle \mathbf{v},\mathbf{u}\rangle^t = (\mathbf{u}^tA^tA\mathbf{v})^t = \mathbf{v}^tA^tA\mathbf{u} = \langle \mathbf{u},\mathbf{v}\rangle$$

(2) We have

$$\begin{aligned}\langle \mathbf{u} + \mathbf{v},\mathbf{w}\rangle &= \mathbf{w}^tA^tA(\mathbf{u} + \mathbf{v}) \\ &= \mathbf{w}^tA^tA\mathbf{u} + \mathbf{w}^tA^tA\mathbf{v} \\ &= \langle \mathbf{u},\mathbf{w}\rangle + \langle \mathbf{v},\mathbf{w}\rangle\end{aligned}$$

(3) Here $\langle k\mathbf{u},\mathbf{v}\rangle = \mathbf{v}^tA^tA(k\mathbf{u}) = k(\mathbf{v}^tA^tA\mathbf{u}) = k\langle \mathbf{u},\mathbf{v}\rangle$.

(4) Let R_i be the ith row of A and let $\mathbf{v} = [v_1 \cdots v_n]^t$. Then

$$\begin{aligned}
\langle \mathbf{v},\mathbf{v}\rangle &= \mathbf{v}^t A^t A\mathbf{v} \\
&= \begin{bmatrix} v_1 & \cdots & v_n \end{bmatrix} \begin{bmatrix} R_1^t & \cdots & R_n^t \end{bmatrix} \begin{bmatrix} R_1 \\ \vdots \\ R_n \end{bmatrix} \begin{bmatrix} v_1 \\ \vdots \\ v_n \end{bmatrix} \\
&= \begin{bmatrix} \mathbf{v}\cdot R_1 & \cdots & \mathbf{v}\cdot R_n \end{bmatrix} \begin{bmatrix} R_1\cdot\mathbf{v} \\ \vdots \\ R_n\cdot\mathbf{v} \end{bmatrix} \\
&= \left(\mathbf{v}\cdot R_1\right)^2 + \cdots + \left(\mathbf{v}\cdot R_n\right)^2 \\
&\geq 0
\end{aligned}$$

Therefore $\langle \mathbf{v},\mathbf{v}\rangle \geq 0$ for every vector $\mathbf{v}$. It is clear that if $\mathbf{v} = \mathbf{0}$, then $\langle \mathbf{v},\mathbf{v}\rangle = 0$. It remains to show that if $\mathbf{v} \neq \mathbf{0}$, then $\langle \mathbf{v},\mathbf{v}\rangle \neq 0$.

To see this, we notice that the only way that $\langle \mathbf{v},\mathbf{v}\rangle$ can equal zero is for $\mathbf{v} \cdot R_i$ to equal zero for every row R_i of A. But this implies that a linear combination of the columns of A is the $n \times 1$ zero matrix. That is, if we let C_i be the ith column of A, then

$$v_1C_1 + v_2C_2 + \cdots + v_nC_n = 0$$

Therefore the column vectors of A are linearly dependent, and hence A is not invertible. This is why the matrix A has to be invertible to generate an inner product. If A is invertible, then $\langle \mathbf{v},\mathbf{v}\rangle$ can only $= 0$ when $\mathbf{v} = \mathbf{0}$.

31. (1) Axiom 1 follows immediately from the commutativity of multiplication in the real numbers.

(2) If $\mathbf{w} = (w_1, \cdots, w_n)$, then

$$\begin{aligned}\langle \mathbf{u} + \mathbf{v}, \mathbf{w}\rangle &= c_1(u_1 + v_1)w_1 + \cdots + c_n(u_n + v_n)w_n \\ &= (c_1u_1w_1 + \cdots + c_nu_nw_n) + (c_1v_1w_1 + \cdots + c_nv_nw_n) \\ &= \langle \mathbf{u}, \mathbf{w}\rangle + \langle \mathbf{v}, \mathbf{w}\rangle\end{aligned}$$

(3) If k is any constant, then

$$\begin{aligned}\langle k\mathbf{u}, \mathbf{v}\rangle &= c_1(ku_1)v_1 + \cdots + c_n(ku_n)v_n \\ &= k(c_1u_1v_1 + \cdots + c_nu_nv_n) \\ &= k\langle \mathbf{u}, \mathbf{v}\rangle\end{aligned}$$

(4) Since $\langle \mathbf{v}, \mathbf{v}\rangle = c_1v_1^2 + \cdots + c_nv_n^2$ where $c_1, \cdots, c_n$ are positive reals, then $\langle \mathbf{v}, \mathbf{v}\rangle \geq 0$, and $\langle \mathbf{v}, \mathbf{v}\rangle = 0$ if and only if $v_1 = \cdots = v_n = 0$ or $\mathbf{v} = \mathbf{0}$. If one of the constants, c_i, were negative or zero, then Axiom 4 would fail to hold.

32. The fact that $\langle \mathbf{0}, \mathbf{v}\rangle = \langle \mathbf{v}, \mathbf{0}\rangle$ follows from the symmetry axiom. By Theorem 4.2.1, $\langle \mathbf{0}, \mathbf{v}\rangle = \langle 0\mathbf{0}, \mathbf{v}\rangle = 0\langle \mathbf{0}, \mathbf{v}\rangle = 0$. Alternatively, $\langle \mathbf{0} + \mathbf{0}, \mathbf{v}\rangle = \langle \mathbf{0}, \mathbf{v}\rangle + \langle \mathbf{0}, \mathbf{v}\rangle = \langle \mathbf{0}, \mathbf{v}\rangle$, which can only be true if $\langle \mathbf{0}, \mathbf{v}\rangle = 0$.

33. We have

$$\begin{aligned}\langle \mathbf{u}, k\mathbf{v}\rangle &= \langle k\mathbf{v}, \mathbf{u}\rangle \quad \text{(by symmetry)} \\ &= k\langle \mathbf{v}, \mathbf{u}\rangle \quad \text{(by homogeneity)} \\ &= k\langle \mathbf{u}, \mathbf{v}\rangle \quad \text{(by symmetry)}\end{aligned}$$

EXERCISE SET 5.2

1. (e) Since $\mathbf{u} \cdot \mathbf{v} = 0 + 6 + 2 + 0 = 8$, the vectors are not orthogonal.

2. We have $\langle \mathbf{u},\mathbf{w}_1\rangle = \langle \mathbf{u},\mathbf{w}_3\rangle = 0$ but $\langle \mathbf{u},\mathbf{w}_2\rangle = -2 \neq 0$. Hence, $\mathbf{u}$ is not orthogonal to W.

3. (a) $\cos\theta = \dfrac{\langle(1,-3),\ (2,4)\rangle}{\|(1,-3)\|\ \|(2,4)\|} = \dfrac{2 - 12}{\sqrt{10}\ \sqrt{20}} = \dfrac{-1}{\sqrt{2}}$

   (c) $\cos\theta = \dfrac{\langle(-1,5,2),\ (2,4,-9)\rangle}{\|(-1,5,2)\|\ \|(2,4,-9)\|} = \dfrac{-2 + 20 - 18}{\sqrt{30}\ \sqrt{101}} = 0$

   (e) $\cos\theta = \dfrac{\langle(1,0,1,0),\ (-3,-3,-3,-3)\rangle}{\|(1,0,1,0,)\|\ \|(-3,-3,-3,-3)\|} = \dfrac{-3 - 3}{\sqrt{2}\ \sqrt{36}} = \dfrac{-1}{\sqrt{2}}$

4. (a) $\cos\theta = \dfrac{\langle -1 + 5x + 2x^2,\ 2 + 4x - 9x^2\rangle}{\|-1 + 5x + 2x^2\|\ \|2 + 4x - 9x^2\|} = \dfrac{-2 + 20 - 18}{\sqrt{30}\ \sqrt{101}} = 0$

   (Compare with 3.(c), above.)

5. $\langle \mathbf{p},\mathbf{q}\rangle = (1)(0) + (-1)(2) + (2)(1) = 0$

6. (a) $\cos\theta = \dfrac{\langle A,B\rangle}{\|A\|\ \|B\|} = \dfrac{6 + 12 + 1 + 0}{\sqrt{50}\ \sqrt{14}} = \dfrac{19}{10\sqrt{7}}$

7. **(b)** $\left\langle \begin{bmatrix} 2 & 1 \\ -1 & 3 \end{bmatrix}, \begin{bmatrix} 1 & 1 \\ 0 & -1 \end{bmatrix} \right\rangle = (2)(1) + (1)(1) + (-1)(0) + (3)(-1) = 0$

Thus the matrices are orthogonal.

(d) $\left\langle \begin{bmatrix} 2 & 1 \\ -1 & 3 \end{bmatrix}, \begin{bmatrix} 2 & 1 \\ 5 & 2 \end{bmatrix} \right\rangle = 4 + 1 - 5 + 6 = 6 \neq 0$

Thus the matrices are not orthogonal.

8. **(a)** We look for values of k such that

$$\langle \mathbf{u}, \mathbf{v} \rangle = 2 + 7 + 3k = 0$$

Clearly $k = -3$ is the only possible value.

9. We must find two vectors $\mathbf{x} = \left[x_1, x_2, x_3, x_4\right]$ such that $\langle \mathbf{x}, \mathbf{x} \rangle = 1$ and $\langle \mathbf{x}, \mathbf{u} \rangle = \langle \mathbf{x}, \mathbf{v} \rangle = \langle \mathbf{x}, \mathbf{w} \rangle = 0$. Thus x_1, x_2, x_3, and x_4 must satisfy the equations

$$\begin{aligned} x_1^2 + x_2^2 + x_3^2 + x_4^2 &= 1 \\ 2x_1 + x_2 - 4x_3 &= 0 \\ -x_1 - x_2 + 2x_3 + 2x_4 &= 0 \\ 3x_1 + 2x_2 + 5x_3 + 4x_4 &= 0 \end{aligned}$$

The solution to the three linear equations is $x_1 = -34t$, $x_2 = 44t$, $x_3 = -6t$, and $x_4 = 11t$. If we substitute these values into the quadratic equation, we get

$$\left[(-34)^2 + (44)^2 + (-6)^2 + (11)^2\right] t^2 = 1$$

or

$$t = \pm \frac{1}{\sqrt{3249}}$$

Therefore, the two vectors are

$$\pm \frac{1}{\sqrt{3249}} (-34,44,-6,11)$$

10. **(a)** Here $\langle \mathbf{u},\mathbf{v}\rangle^2 = \left[3(4) + 2(-1)\right]^2 = 100$, while

$$\langle \mathbf{u},\mathbf{u}\rangle\langle \mathbf{v},\mathbf{v}\rangle = \left[3^2 + 2^2\right]\left[4^2 + (-1)^2\right] = 221.$$

11. **(a)** Here $\langle \mathbf{u},\mathbf{v}\rangle^2 = \left[3(-2)(1) + 2(1)(0)\right]^2 = 36$, while

$$\langle \mathbf{u},\mathbf{u}\rangle\langle \mathbf{v},\mathbf{v}\rangle = \left[3(-2)^2 + 2(1)^2\right]\left[3(1)^2 + 2(0)^2\right] = 42.$$

12. If **u** and **v** are orthogonal vectors with norm 1, then

$$\begin{aligned} \|\mathbf{u} - \mathbf{v}\| &= \langle \mathbf{u}-\mathbf{v}, \mathbf{u}-\mathbf{v}\rangle^{1/2} \\ &= [\langle \mathbf{u},\mathbf{u}\rangle - 2\langle \mathbf{u},\mathbf{v}\rangle + \langle \mathbf{v},\mathbf{v}\rangle]^{1/2} \\ &= [1 - 2(0) + 1]^{1/2} \\ &= \sqrt{2} \end{aligned}$$

13. If a vector **z** is orthogonal to all of the basis vectors, then, by the result of Exercise 19, it is orthogonal to every linear combination of the basis vectors, and hence to every vector in V. Thus, in particular, $\langle \mathbf{z},\mathbf{z}\rangle = 0$, so, by Axiom 4, $\mathbf{z} = \mathbf{0}$.

14. **(a)** Let S denote the set of all vectors in V orthogonal to **v**. We must show (see Theorem 4.3.1) that S is closed under vector addition and scalar multiplication.

If **x** and **y** are in S, then both are orthogonal to **v**. We must show that $\mathbf{x} + \mathbf{y}$ is also orthogonal to **v**, or that $\langle \mathbf{v}, \mathbf{x}+\mathbf{y}\rangle = 0$. But

$$\langle \mathbf{v}, \mathbf{x}+\mathbf{y}\rangle = \langle \mathbf{v},\mathbf{x}\rangle + \langle \mathbf{v},\mathbf{y}\rangle = 0 + 0 = 0$$

We must also show that if $\mathbf{x}$ is in S, then $k\mathbf{x}$ is in S. But if $\langle\mathbf{v},\mathbf{x}\rangle = 0$, then $\langle\mathbf{v},k\mathbf{x}\rangle = k\langle\mathbf{v},\mathbf{x}\rangle = 0$. Thus we have verified both closure properties.

(b) We assume that each vector in R^2 or R^3 has its initial point at the origin. In R^2, S is the set of all vectors lying on the line which is perpendicular to $\mathbf{v}$ and which passes through the origin. In R^3, S is the set of all vectors lying in the plane which is perpendicular to $\mathbf{v}$ and which passes through the origin.

15. Note that

$$\begin{aligned}\|\mathbf{v}_1 + \cdots + \mathbf{v}_r\|^2 &= \langle\mathbf{v}_1 + \cdots + \mathbf{v}_r,\ \mathbf{v}_1 + \cdots + \mathbf{v}_r\rangle \\ &= \sum_{i=1}^{r} \langle\mathbf{v}_i,\mathbf{v}_i\rangle + \sum_{i\neq j} \langle\mathbf{v}_i,\mathbf{v}_j\rangle \\ &= \|\mathbf{v}_1\|^2 + \cdots + \|\mathbf{v}_r\|^2 + \sum_{i\neq j} \langle\mathbf{v}_i,\mathbf{v}_j\rangle\end{aligned}$$

If $\langle\mathbf{v}_i,\mathbf{v}_j\rangle = 0$ whenever $i \neq j$, then the result follows.

17. We wish to show that $\angle ABC$ is a right angle, or that $\overrightarrow{AB}$ and $\overrightarrow{BC}$ are orthogonal. Observe that $\overrightarrow{AB} = \mathbf{u} - (-\mathbf{v})$ and $\overrightarrow{BC} = \mathbf{v} - \mathbf{u}$ where $\mathbf{u}$ and $\mathbf{v}$ are radii of the circle, as shown in the figure. Thus $\|\mathbf{u}\| = \|\mathbf{v}\|$. Hence

$$\begin{aligned}\langle\overrightarrow{AB}, \overrightarrow{BC}\rangle &= \langle\mathbf{u} + \mathbf{v},\ \mathbf{v} - \mathbf{u}\rangle \\ &= \langle\mathbf{u},\mathbf{v}\rangle + \langle\mathbf{v},\mathbf{v}\rangle + \langle\mathbf{u},-\mathbf{u}\rangle + \langle\mathbf{v},-\mathbf{u}\rangle \\ &= \langle\mathbf{v},\mathbf{u}\rangle + \langle\mathbf{v},\mathbf{v}\rangle - \langle\mathbf{u},\mathbf{u}\rangle - \langle\mathbf{v},\mathbf{u}\rangle \\ &= \|\mathbf{v}\|^2 - \|\mathbf{u}\|^2 \\ &= 0\end{aligned}$$

18. If $\langle \mathbf{w},\mathbf{u}_1\rangle = \langle \mathbf{w},\mathbf{u}_2\rangle = 0$, then

$$\begin{aligned}\langle \mathbf{w}, k_1\mathbf{u}_1 + k_2\mathbf{u}_2\rangle &= \langle \mathbf{w}, k_1\mathbf{u}_1\rangle + \langle \mathbf{w}, k_2\mathbf{u}_2\rangle \\ &= k_1\langle \mathbf{w},\mathbf{u}_1\rangle + k_2\langle \mathbf{w},\mathbf{u}_2\rangle \\ &= 0\end{aligned}$$

Thus $\mathbf{w}$ is orthogonal to $k_1\mathbf{u}_1 + k_2\mathbf{u}_2$.

Now consider R^3 with the Euclidean inner product. If $\mathbf{w}$ is perpendicular to both $\mathbf{u}_1$ and $\mathbf{u}_2$ and they determine a plane, then $\mathbf{w}$ is perpendicular to every vector in that plane. If $\mathbf{u}_1$ and $\mathbf{u}_2$ determine a line, then $\mathbf{w}$ must be perpendicular to every vector in that line. If $\mathbf{u}_1$ and $\mathbf{u}_2$ determine neither a plane nor a line, then $\mathbf{u}_1 = \mathbf{u}_2 = \mathbf{0}$ and the result is not of interest.

19. By definition, $\mathbf{u} \in \text{lin}\{\mathbf{u}_1,\mathbf{u}_2,\ldots,\mathbf{u}_r\}$ if and only if there exist constants $c_1, c_2, \ldots, c_r$ such that

$$\mathbf{u} = c_1\mathbf{u}_1 + c_2\mathbf{u}_2 + \cdots + c_r\mathbf{u}_r$$

But if $\langle \mathbf{w},\mathbf{u}_1\rangle = \langle \mathbf{w},\mathbf{u}_2\rangle = \cdots = \langle \mathbf{w},\mathbf{u}_r\rangle = 0$, then $\langle \mathbf{w},\mathbf{u}\rangle = 0$.

20. This is just the Cauchy-Schwarz Inequality using the inner product on R^n generated by A (see Formula (2)).

21. If we let R^2 have the Euclidean inner product and apply the Cauchy-Schwarz Inequality to the vectors $\mathbf{u} = (a,b)$ and $\mathbf{v} = (\cos\theta, \sin\theta)$, the result follows directly.

23. Suppose that

$$(*) \qquad \langle \mathbf{u},\mathbf{v}\rangle^2 = \langle \mathbf{u},\mathbf{u}\rangle\langle \mathbf{v},\mathbf{v}\rangle$$

Following the proof of Theorem 5.2.1, we see that (*) can hold if and only if either $\mathbf{u} = \mathbf{0}$ or the quadratic equation $\langle t\mathbf{u} + \mathbf{v}, t\mathbf{u} + \mathbf{v}\rangle = 0$ has just one real root. If $\mathbf{u} = \mathbf{0}$, then $\mathbf{u}$ and $\mathbf{v}$ are linearly dependent. If the equation has one real root, t_0, then $\langle t_0\mathbf{u} + \mathbf{v},\ t_0\mathbf{u} + \mathbf{v}\rangle = 0$. Thus, by Axiom 4, $t_0\mathbf{u} + \mathbf{v} = \mathbf{0}$, and again $\mathbf{u}$ and $\mathbf{v}$ are linearly dependent.

Conversely, suppose that $\mathbf{u}$ and $\mathbf{v}$ are linearly dependent. Then either $\mathbf{u} = \mathbf{0}$ or $\mathbf{v}$ can be uniquely expressed as a multiple, $-t_0$, of $\mathbf{u}$; that is, $t_0\mathbf{u} + \mathbf{v} = \mathbf{0}$. Thus if we work backward through the above argument, we can verify that $\langle \mathbf{u},\mathbf{v}\rangle^2 = \langle \mathbf{u},\mathbf{u}\rangle\langle \mathbf{v},\mathbf{v}\rangle$.

24. Recall that $\cos(\alpha + \beta) = \cos\alpha\cos\beta - \sin\alpha\sin\beta$. Thus

$$\begin{aligned}\cos(\alpha - \beta) &= \cos\alpha\cos(-\beta) - \sin\alpha\sin(-\beta)\\ &= \cos\alpha\cos\beta + \sin\alpha\sin\beta\end{aligned}$$

Adding these two equations gives the identity

$$\cos(\alpha + \beta) + \cos(\alpha - \beta) = 2\cos\alpha\cos\beta$$

Thus if $k \neq \ell$, then $k + \ell$ and $k - \ell$ are nonzero and

$$\begin{aligned}\langle \mathbf{f}_k,\mathbf{f}_\ell\rangle &= \int_0^\pi \cos(kx)\cos(\ell x)\,dx\\ &= \int_0^\pi \frac{1}{2}\,[\cos(k + \ell)x + \cos(k - \ell)x]\,dx\\ &= \frac{1}{2}\left[\frac{\sin(k + \ell)x}{k + \ell} + \frac{\sin(k - \ell)x}{k - \ell}\right]_0^\pi\\ &= 0\end{aligned}$$

25. **(a)** As noted in Example 9, $\int_0^1 f(x)g(x)\,dx$ is an inner product on $C[0,1]$. Thus the Cauchy-Schwarz inequality must hold, and that is exactly what we're asked to prove.

(b) In the notation of Exercise 24, we must show that

$$\langle \mathbf{f} + \mathbf{g},\ \mathbf{f} + \mathbf{g}\rangle^{1/2} \le \langle \mathbf{f},\mathbf{f}\rangle^{1/2} + \langle \mathbf{g},\mathbf{g}\rangle^{1/2}$$

or, squaring both sides, that

$$\langle \mathbf{f}+\mathbf{g},\ \mathbf{f}+\mathbf{g}\rangle \le \langle \mathbf{f},\mathbf{f}\rangle + 2\langle \mathbf{f},\mathbf{f}\rangle^{1/2}\langle \mathbf{g},\mathbf{g}\rangle^{1/2} + \langle \mathbf{g},\mathbf{g}\rangle$$

For any inner product, we know that

$$\langle \mathbf{f}+\mathbf{g},\ \mathbf{f}+\mathbf{g}\rangle = \langle \mathbf{f},\mathbf{f}\rangle + 2\langle \mathbf{f},\mathbf{g}\rangle + \langle \mathbf{g},\mathbf{g}\rangle$$

By the Cauchy-Schwarz inequality

$$\langle \mathbf{f},\mathbf{g}\rangle^2 \le \langle \mathbf{f},\mathbf{f}\rangle\langle \mathbf{g},\mathbf{g}\rangle$$

or

$$\langle \mathbf{f},\mathbf{g}\rangle \le \langle \mathbf{f},\mathbf{f}\rangle^{1/2}\langle \mathbf{g},\mathbf{g}\rangle^{1/2}$$

If we substitute the above inequality into the equation for $\langle \mathbf{f}+\mathbf{g},\ \mathbf{f}+\mathbf{g}\rangle$, we obtain

$$\langle \mathbf{f}+\mathbf{g},\ \mathbf{f}+\mathbf{g}\rangle \le \langle \mathbf{f},\mathbf{f}\rangle + 2\langle \mathbf{f},\mathbf{f}\rangle^{1/2}\langle \mathbf{g},\mathbf{g}\rangle^{1/2} + \langle \mathbf{g},\mathbf{g}\rangle$$

as required.

26. Suppose that $\langle \mathbf{u},\mathbf{v}\rangle = au_1v_1 + bu_2v_2$ is a weighted Euclidean inner product on R^2. For **u** and **v** to be orthogonal unit vectors with respect to this inner product, we must have

$$\begin{aligned}\langle \mathbf{u},\mathbf{u}\rangle &= a + 3b = 1\\ \langle \mathbf{v},\mathbf{v}\rangle &= a + 3b = 1\\ \langle \mathbf{u},\mathbf{v}\rangle &= -a + 3b = 0\end{aligned}$$

which yields $a = 1/2$ and $b = 1/6$.

EXERCISE SET 5.3

2. **(a)** The vectors are orthogonal, but since $\|(2,0)\| \neq 1$, they are not orthonormal.

   (c) The vectors are not orthogonal and therefore not orthonormal.

4. **(a)** The set is not orthonormal because the last two vectors are not orthogonal.

   (c) The set is not orthonormal because the last two vectors are not orthogonal.

5. See Exercise 3, Parts (b) and (c).

6. **(a)** Denote the vectors by $\mathbf{v}_1$, $\mathbf{v}_2$, $\mathbf{v}_3$, and $\mathbf{v}_4$. Then

$$\langle \mathbf{v}_1, \mathbf{v}_i \rangle = 0 \text{ for } i = 2,3,4$$

$$\langle \mathbf{v}_2, \mathbf{v}_3 \rangle = \frac{4}{9} - \frac{2}{9} - \frac{2}{9} = 0$$

$$\langle \mathbf{v}_2, \mathbf{v}_4 \rangle = \frac{2}{9} + \frac{2}{9} - \frac{4}{9} = 0$$

$$\langle \mathbf{v}_3, \mathbf{v}_4 \rangle = \frac{2}{9} - \frac{4}{9} + \frac{2}{9} = 0$$

Thus, the vectors are orthogonal. Moreover, $\|\mathbf{v}_1\| = 1$ and $\|\mathbf{v}_2\| = \|\mathbf{v}_3\| = \|\mathbf{v}_4\| = \frac{4}{9} + \frac{1}{9} + \frac{4}{9} = 1$. The vectors are therefore orthonormal.

7. **(b)** Call the vectors $\mathbf{u}_1, \mathbf{u}_2$ and $\mathbf{u}_3$. Then $\langle \mathbf{u}_1, \mathbf{u}_2 \rangle = 2 - 2 = 0$, $\langle \mathbf{u}_1, \mathbf{u}_3 \rangle = 0$, and $\langle \mathbf{u}_2, \mathbf{u}_3 \rangle = 0$. The set is therefore orthogonal. Moreover, $\|\mathbf{u}_1\| = \sqrt{2}$, $\|\mathbf{u}_2\| = \sqrt{8} = 2\sqrt{2}$, and $\|\mathbf{u}_3\| = \sqrt{25} = 5$. Thus $\left\{\frac{1}{\sqrt{2}}\mathbf{u}_1, \frac{1}{2\sqrt{2}}\mathbf{u}_2, \frac{1}{5}\mathbf{u}_3\right\}$ is an orthonormal set.

9. It is easy to verify that $\mathbf{v}_1 \cdot \mathbf{v}_2 = \mathbf{v}_1 \cdot \mathbf{v}_3 = \mathbf{v}_2 \cdot \mathbf{v}_3 = 0$ and that $\|\mathbf{v}_3\| = 1$. Moreover, $\|\mathbf{v}_1\|^2 = (-3/5)^2 + (4/5)^2 = 1$ and $\|\mathbf{v}_2\| = (4/5)^2 + (3/5)^2 = 1$. Thus $\{\mathbf{v}_1, \mathbf{v}_2, \mathbf{v}_3\}$ is an orthonormal set in R^3. It will be an orthonormal basis provided that the three vectors are linearly independent, which is guaranteed by Theorem 5.3.3.

(b) By Theorem 5.3.1, we have

$$(3,-7,4) = \left[-\frac{9}{5} - \frac{28}{5} + 0\right]\mathbf{v}_1 + \left[\frac{12}{5} - \frac{21}{5} + 0\right]\mathbf{v}_2 + 4\mathbf{v}_3$$

$$= (-37/5)\mathbf{v}_1 + (-9/5)\mathbf{v}_2 + 4\mathbf{v}_3$$

10. It is easy to check that $\mathbf{v}_1 \cdot \mathbf{v}_2 = \mathbf{v}_1 \cdot \mathbf{v}_3 = \mathbf{v}_1 \cdot \mathbf{v}_4 = \mathbf{v}_2 \cdot \mathbf{v}_3 = \mathbf{v}_2 \cdot \mathbf{v}_4 = \mathbf{v}_3 \cdot \mathbf{v}_4 = 0$, so that we have 4 orthogonal vectors in R^4. Thus, by Theorem 5.3.3, they form an orthogonal basis.

(b) By Formula (1), we have $(\sqrt{2}, -3\sqrt{2}, 5\sqrt{2}, -\sqrt{2}) =$

$$\frac{\sqrt{2} + 3\sqrt{2} + 10\sqrt{2} + \sqrt{2}}{1 + 1 + 4 + 1}\mathbf{v}_1 + \frac{-2\sqrt{2} - 6\sqrt{2} + 15\sqrt{2} - 2\sqrt{2}}{4 + 4 + 9 + 4}\mathbf{v}_2$$

$$+ \frac{\sqrt{2} - 6\sqrt{2} + 0 + \sqrt{2}}{1 + 4 + 0 + 1}\mathbf{v}_3 + \frac{\sqrt{2} + 0 + 0 - \sqrt{2}}{1 + 0 + 0 + 1}\mathbf{v}_4$$

$$= \frac{15\sqrt{2}}{7}\mathbf{v}_1 + \frac{5\sqrt{2}}{21}\mathbf{v}_2 - \frac{2\sqrt{2}}{3}\mathbf{v}_3 + 0\,\mathbf{v}_4$$

11. **(a)** Since $(\mathbf{w})_S = (\langle \mathbf{w},\mathbf{u}_1\rangle, \langle \mathbf{w},\mathbf{u}_2\rangle) = \left(\frac{-4}{\sqrt{2}}, \frac{10}{\sqrt{2}}\right)$

$= (-2\sqrt{2}, 5\sqrt{2}),$

$$[\mathbf{w}]_S = \begin{bmatrix} -2\sqrt{2} \\ 5\sqrt{2} \end{bmatrix}$$

12. **(a)** We have $\mathbf{u} = \mathbf{w}_1 + \mathbf{w}_2 = \left(\frac{7}{5}, -\frac{1}{5}\right)$ and $\mathbf{v} = -\mathbf{w}_1 + 4\mathbf{w}_2 = \left(\frac{13}{5}, \frac{16}{5}\right)$.

(b) By Theorem 5.3.2, these numbers can be computed directly using the Euclidean inner product without reference to the basis vectors. We find that $\|\mathbf{u}\| = \sqrt{2}$, $d(\mathbf{u},\mathbf{v}) = \sqrt{13}$, and $\langle \mathbf{u},\mathbf{v}\rangle = 3$. Similar computations with the versions of $\mathbf{u}$ and $\mathbf{v}$ obtained in Part (a) fortunately yield the same results.

14. **(a)** $\|\mathbf{u}\| = (1 + 4 + 1 + 9)^{1/2} = \sqrt{15}$

$\|\mathbf{v} - \mathbf{w}\| = \|(2,1,-2,4)\| = (4 + 1 + 4 + 16)^{1/2} = 5$

$\|\mathbf{v} + \mathbf{w}\| = \|(-2,-7,4,6)\| = (4 + 49 + 16 + 36)^{1/2} = \sqrt{105}$

$\langle \mathbf{v},\mathbf{w}\rangle = 0 + 12 + 3 + 5 = 20$

16. **(a)** Let

$$\mathbf{v}_1 = \frac{\mathbf{u}_1}{\|\mathbf{u}_1\|} = \left(\frac{1}{\sqrt{10}}, -\frac{3}{\sqrt{10}}\right)$$

Since $\langle \mathbf{u}_2,\mathbf{v}_1\rangle = \frac{2}{\sqrt{10}} - \frac{6}{\sqrt{10}} = -\frac{4}{\sqrt{10}}$, we have

$$\mathbf{u}_2 - \langle \mathbf{u}_2,\mathbf{v}_1\rangle\mathbf{v}_1 = (2,2) + \frac{4}{\sqrt{10}}\left(\frac{1}{\sqrt{10}}, -\frac{3}{\sqrt{10}}\right)$$

$$= \left(\frac{12}{5}, \frac{4}{5}\right)$$

This vector has norm $\left\|\left(\frac{12}{5}, \frac{4}{5}\right)\right\| = \frac{4\sqrt{10}}{5}$. Thus

$$\mathbf{v}_2 = \frac{5}{4\sqrt{10}}\left(\frac{12}{5}, \frac{4}{5}\right) = \left(\frac{3}{\sqrt{10}}, \frac{1}{\sqrt{10}}\right)$$

and $\{\mathbf{v}_1, \mathbf{v}_2\}$ is the desired orthonormal basis.

17. (a) Let

$$\mathbf{v}_1 = \frac{\mathbf{u}_1}{\|\mathbf{u}_1\|} = \left(\frac{1}{\sqrt{3}}, \frac{1}{\sqrt{3}}, \frac{1}{\sqrt{3}}\right)$$

Since $<\mathbf{u}_2, \mathbf{v}_1> = 0$, we have

$$\mathbf{v}_2 = \frac{\mathbf{u}_2}{\|\mathbf{u}_2\|} = \left(-\frac{1}{\sqrt{2}}, \frac{1}{\sqrt{2}}, 0\right)$$

Since $<\mathbf{u}_3, \mathbf{v}_1> = \frac{4}{\sqrt{3}}$ and $<\mathbf{u}_3, \mathbf{v}_2> = \frac{1}{\sqrt{2}}$, we have

$$\begin{aligned}\mathbf{u}_3 - <\mathbf{u}_3, \mathbf{v}_1>\mathbf{v}_1 - <\mathbf{u}_3, \mathbf{v}_2>\mathbf{v}_2 \\ &= (1,2,1) - \frac{4}{\sqrt{3}}\left(\frac{1}{\sqrt{3}}, \frac{1}{\sqrt{3}}, \frac{1}{\sqrt{3}}\right) - \frac{1}{\sqrt{2}}\left(-\frac{1}{\sqrt{2}}, \frac{1}{\sqrt{2}}, 0\right) \\ &= \left(\frac{1}{6}, \frac{1}{6}, -\frac{1}{3}\right)\end{aligned}$$

This vector has norm $\left\|\left(\frac{1}{6}, \frac{1}{6}, -\frac{1}{3}\right)\right\| = \frac{1}{\sqrt{6}}$. Thus

$$\mathbf{v}_3 = \left(\frac{1}{\sqrt{6}}, \frac{1}{\sqrt{6}}, -\frac{2}{\sqrt{6}}\right)$$

and $\{\mathbf{v}_1, \mathbf{v}_2, \mathbf{v}_3\}$ is the desired orthonormal basis.

19. Let $\mathbf{u}_1 = (0,1,2)$ and $\mathbf{u}_2 = (-1,0,1)$. Then

$$\mathbf{v}_1 = \frac{\mathbf{u}_1}{\|\mathbf{u}_1\|} = \left(0, \frac{1}{\sqrt{5}}, \frac{2}{\sqrt{5}}\right)$$

Since $<\mathbf{u}_2,\mathbf{v}_1> = \frac{2}{\sqrt{5}}$, then

$$\mathbf{u}_2 - <\mathbf{u}_2,\mathbf{v}_1>\mathbf{v}_1 = \left(-1, -\frac{2}{5}, \frac{1}{5}\right)$$

where $\left\|\left(-1, -\frac{2}{5}, \frac{1}{5}\right)\right\| = \frac{\sqrt{30}}{5}$. Hence

$$\mathbf{v}_2 = \left(-\frac{5}{\sqrt{30}}, \frac{-2}{\sqrt{30}}, \frac{1}{\sqrt{30}}\right)$$

Thus $\{\mathbf{v}_1,\mathbf{v}_2\}$ is an orthonormal basis.

21. Note that $\mathbf{u}_1$ and $\mathbf{u}_2$ are orthonormal. Thus we apply Theorem 5.3.5 to obtain

$$\begin{aligned}\mathbf{w}_1 &= <\mathbf{w},\mathbf{u}_1>\mathbf{u}_1 + <\mathbf{w},\mathbf{u}_2>\mathbf{u}_2 \\ &= -\left(\frac{4}{5}, 0, -\frac{3}{5}\right) + 2(0,1,0) \\ &= \left(-\frac{4}{5}, 2, \frac{3}{5}\right)\end{aligned}$$

and

$$\begin{aligned}\mathbf{w}_2 &= \mathbf{w} - \mathbf{w}_1 \\ &= \left(\frac{9}{5}, 0, \frac{12}{5}\right)\end{aligned}$$

22. Since $\mathbf{u}_1$ and $\mathbf{u}_2$ are not orthogonal, we must apply the Gram-Schmidt process before we can invoke Theorem 5.3.5. Thus we let

$$\mathbf{v}_1 = \frac{\mathbf{u}_1}{\|\mathbf{u}_1\|} = \left(\frac{1}{\sqrt{3}}, \frac{1}{\sqrt{3}}, \frac{1}{\sqrt{3}}\right)$$

Since $<\mathbf{u}_2,\mathbf{v}_1> = \dfrac{1}{\sqrt{3}}$, we have

$$\mathbf{u}_2 - <\mathbf{u}_2,\mathbf{v}_1>\mathbf{v}_1 = \left(\frac{5}{3}, -\frac{1}{3}, -\frac{4}{3}\right)$$

where $\left\|\left(\dfrac{5}{3}, -\dfrac{1}{3}, -\dfrac{4}{3}\right)\right\| = \dfrac{\sqrt{42}}{3}$. Thus

$$\mathbf{v}_2 = \left(\frac{5}{\sqrt{42}}, -\frac{1}{\sqrt{42}}, -\frac{4}{\sqrt{42}}\right)$$

Hence $\{\mathbf{v}_1,\mathbf{v}_2\}$ is the desired orthonormal set. Next, we compute $<\mathbf{w},\mathbf{v}_1> = 2\sqrt{3}$ and $<\mathbf{w},\mathbf{v}_2> = -\dfrac{7\sqrt{42}}{6}$. Theorem 5.3.5 then gives

$$\mathbf{w}_1 = <\mathbf{w},\mathbf{v}_1>\mathbf{v}_1 + <\mathbf{w},\mathbf{v}_2>\mathbf{v}_2 = \left(\frac{13}{14}, \frac{31}{14}, \frac{40}{14}\right)$$

and

$$\mathbf{w}_2 = \mathbf{w} - \mathbf{w}_1 = \left(\frac{1}{14}, -\frac{3}{14}, \frac{2}{14}\right)$$

24. By Theorem 5.3.1, we know that

$$\mathbf{w} = a_1\mathbf{v}_1 + a_2\mathbf{v}_2 + a_3\mathbf{v}_3$$

where $a_i = <\mathbf{w},\mathbf{v}_i>$. Thus

$$\|\mathbf{w}\|^2 = <\mathbf{w},\mathbf{w}>$$

$$= \sum_{i=1}^{3} a_i^2<\mathbf{v}_i,\mathbf{v}_i> + \sum_{i\neq j} a_ia_j<\mathbf{v}_i,\mathbf{v}_j>$$

But $\langle \mathbf{v}_i, \mathbf{v}_j \rangle = 0$ if $i \neq j$ and $\langle \mathbf{v}_i, \mathbf{v}_i \rangle = 1$ because the set $\{\mathbf{v}_1, \mathbf{v}_2, \mathbf{v}_3\}$ is orthonormal. Hence

$$\begin{aligned} \|\mathbf{w}\|^2 &= a_1^{\ 2} + a_2^{\ 2} + a_3^{\ 2} \\ &= \langle \mathbf{w}, \mathbf{v}_1 \rangle^2 + \langle \mathbf{w}, \mathbf{v}_2 \rangle^2 + \langle \mathbf{w}, \mathbf{v}_3 \rangle^2 \end{aligned}$$

25. Generalize the solution of Exercise 24.

26. Suppose the contrary; that is, suppose that

(*) $$\mathbf{u}_3 - \langle \mathbf{u}_3, \mathbf{v}_1 \rangle \mathbf{v}_1 - \langle \mathbf{u}_3, \mathbf{v}_2 \rangle \mathbf{v}_2 = \mathbf{0}$$

Then (*) implies that $\mathbf{u}_3$ is a linear combination of $\mathbf{v}_1$ and $\mathbf{v}_2$. But $\mathbf{v}_1$ is a multiple of $\mathbf{u}_1$ while $\mathbf{v}_2$ is a linear combination of $\mathbf{u}_1$ and $\mathbf{u}_2$. Hence, (*) implies that $\mathbf{u}_3$ is a linear combination of $\mathbf{u}_1$ and $\mathbf{u}_2$ and therefore that $\{\mathbf{u}_1, \mathbf{u}_2, \mathbf{u}_3\}$ is linearly dependent, contrary to the hypothesis that $\{\mathbf{u}_1, \ldots, \mathbf{u}_n\}$ is linearly independent. Thus, the assumption that (*) holds leads to a contradiction.

27. We have $\mathbf{u}_1 = 1$, $\mathbf{u}_2 = x$, and $\mathbf{u}_3 = x^2$. Since

$$\|\mathbf{u}_1\|^2 = \langle \mathbf{u}_1, \mathbf{u}_1 \rangle = \int_{-1}^{1} 1 \, dx = 2$$

we let

$$\mathbf{v}_1 = \frac{1}{\sqrt{2}}$$

Then

$$\langle \mathbf{u}_2, \mathbf{v}_1 \rangle = \frac{1}{\sqrt{2}} \int_{-1}^{1} x \, dx = 0$$

and thus $\mathbf{v}_2 = \mathbf{u}_2/\|\mathbf{u}_2\|$ where

$$\|\mathbf{u}_2\|^2 = \int_{-1}^{1} x^2\,dx = \frac{2}{3}$$

Hence

$$\mathbf{v}_2 = \sqrt{\frac{3}{2}}\,x$$

In order to compute $\mathbf{v}_3$, we note that

$$\langle \mathbf{u}_3, \mathbf{v}_1 \rangle = \frac{1}{\sqrt{2}} \int_{-1}^{1} x^2\,dx = \frac{\sqrt{2}}{3}$$

and

$$\langle \mathbf{u}_3, \mathbf{v}_2 \rangle = \sqrt{\frac{3}{2}} \int_{-1}^{1} x^3\,dx = 0$$

Thus

$$\mathbf{u}_3 - \langle \mathbf{u}_3, \mathbf{v}_1 \rangle \mathbf{v}_1 - \langle \mathbf{u}_3, \mathbf{v}_2 \rangle \mathbf{v}_2 = x^2 - \frac{1}{3}$$

and

$$\left\| x^2 - \frac{1}{3} \right\|^2 = \int_{-1}^{1} \left[x^2 - \frac{1}{3} \right]^2 dx = \frac{8}{45}$$

Hence,

$$\mathbf{v}_3 = \sqrt{\frac{45}{8}} \left[x^2 - \frac{1}{3} \right] \quad \text{or} \quad \mathbf{v}_3 = \frac{\sqrt{5}}{2\sqrt{2}} \left[3x^2 - 1 \right]$$

28. (b) If we call the three polynomials $\mathbf{v}_1$, $\mathbf{v}_2$, and $\mathbf{v}_3$, and let $\mathbf{u} = 2 - 7x^2$, then, by Theorem 5.3.1,

$$\mathbf{u} = \langle \mathbf{u}, \mathbf{v}_1 \rangle \mathbf{v}_1 + \langle \mathbf{u}, \mathbf{v}_2 \rangle \mathbf{v}_2 + \langle \mathbf{u}, \mathbf{v}_3 \rangle \mathbf{v}_3$$

Here

$$\langle \mathbf{u},\mathbf{v}_1\rangle = \frac{1}{\sqrt{2}}\int_{-1}^{1}\left[2 - 7x^2\right]dx = -\frac{\sqrt{2}}{3}$$

$$\langle \mathbf{u},\mathbf{v}_2\rangle = \sqrt{\frac{3}{2}}\int_{-1}^{1}\left[2x - 7x^3\right]dx = 0$$

$$\langle \mathbf{u},\mathbf{v}_3\rangle = \frac{1}{2}\sqrt{\frac{5}{2}}\int_{-1}^{1}\left[2 - 7x^2\right]\left[3x^2 - 1\right]dx = -\frac{28}{15}\sqrt{\frac{5}{2}}$$

It is easy to check that $\mathbf{u} = -\frac{\sqrt{2}}{3}\mathbf{v}_1 - \frac{28}{15}\sqrt{\frac{5}{2}}\mathbf{v}_3$.

29. This is similar to Exercise 27 except that the lower limit of integration is changed from −1 to 0. If we again set $\mathbf{u}_1 = 1$, $\mathbf{u}_2 = x$, and $\mathbf{u}_3 = x^2$, then $\|\mathbf{u}_1\| = 1$ and thus

$$\mathbf{v}_1 = 1$$

Then $\langle \mathbf{u}_2,\mathbf{v}_1\rangle = \int_0^1 x\,dx = \frac{1}{2}$ and thus

$$\mathbf{v}_2 = \frac{x - 1/2}{\|x - 1/2\|} = \sqrt{12}(x - 1/2)$$

or

$$\mathbf{v}_2 = \sqrt{3}(2x - 1)$$

Finally,

$$\langle \mathbf{u}_3,\mathbf{v}_1\rangle = \int_0^1 x^2dx = \frac{1}{3}$$

and

$$\langle \mathbf{u}_3, \mathbf{v}_2 \rangle = \sqrt{3}\int_0^1 \left[2x^3 - x^2\right] dx = \frac{\sqrt{3}}{6}$$

Thus

$$\mathbf{v}_3 = \frac{x^2 - \frac{1}{3} - \frac{1}{2}(2x - 1)}{\left\| x^2 - \frac{1}{3} - \frac{1}{2}(2x - 1) \right\|} = 6\sqrt{5}\left[x^2 - x + \frac{1}{6}\right]$$

or

$$\mathbf{v}_3 = \sqrt{5}\left[6x^2 - 6x + 1\right]$$

30. First, we find an orthonormal basis for the plane. Since the point (0,1,3), for instance, lies in the plane, we choose

$$\mathbf{v}_1 = \frac{(0,1,3)}{\|(0,1,3)\|} = \frac{1}{\sqrt{10}}(0,1,3)$$

Obviously, $\|\mathbf{v}_1\| = 1$. We now look for a vector which is orthogonal to $\mathbf{v}_1$. Because the vector (5,-3,1) is orthogonal to every vector in the plane (refer to Section 3.5), then the vector

$$(0,1,3) \times (5,-3,1) = (10,15,-5)$$

will lie in the plane and will also be orthogonal to $\mathbf{v}_1$. Hence, we choose

$$\mathbf{v}_2 = \frac{(2,3,-1)}{\|(2,3,-1)\|} = \frac{1}{\sqrt{14}}(2,3,-1)$$

Obviously, $\|\mathbf{v}_2\| = 1$. Thus, $\{\mathbf{v}_1, \mathbf{v}_2\}$ is an orthonormal basis for the plane.

Theorem 5.3.7 now tells us that the point Q is given by the equation

$$\begin{aligned} \overrightarrow{OQ} &= \text{proj}_{\{\mathbf{v}_1, \mathbf{v}_2\}} \overrightarrow{OP} \\ &= \langle \overrightarrow{OP}, \mathbf{v}_1 \rangle \mathbf{v}_1 + \langle \overrightarrow{OP}, \mathbf{v}_2 \rangle \mathbf{v}_2 \end{aligned}$$

But $\langle \overrightarrow{OP}, \mathbf{v}_1 \rangle = \sqrt{10}$ and $\langle \overrightarrow{OP}, \mathbf{v}_2 \rangle = -8/\sqrt{14}$. Hence

$$\overrightarrow{OQ} = \left(-\frac{8}{7}, -\frac{5}{7}, \frac{25}{7}\right) \quad \text{or} \quad Q = \left(-\frac{8}{7}, -\frac{5}{7}, \frac{25}{7}\right)$$

The distance between Q and the plane is

$$\|\overrightarrow{OP} - \overrightarrow{OQ}\| = \frac{3\sqrt{35}}{7}$$

31. First, we find an orthonormal basis for the line. If we put $t = 1$, we see that the point $P = (2,-1,4)$ lies on the line and hence $\overrightarrow{OP}$ is a basis vector for the line. Thus, we normalize to obtain

$$\mathbf{v} = \frac{1}{\sqrt{21}}(2,-1,4)$$

which is an orthonormal basis for the line.

Theorem 5.3.7 now tells us that the point Q is given by

$$\begin{aligned}\overrightarrow{OQ} &= \text{proj}_{\{\mathbf{v}\}}\overrightarrow{OP} \\ &= \langle \overrightarrow{OP}, \mathbf{v} \rangle \mathbf{v} = \left(-\frac{8}{7}, \frac{4}{7}, -\frac{16}{7}\right)\end{aligned}$$

Thus

$$Q = \left(-\frac{8}{7}, \frac{4}{7}, -\frac{16}{7}\right)$$

32. Theorem 5.3.5 provides us with two vectors $\mathbf{w}_1$ and $\mathbf{w}_2$ where $\mathbf{w}_1 = \langle \mathbf{u}, \mathbf{v}_1 \rangle \mathbf{v}_1 + \cdots + \langle \mathbf{u}, \mathbf{v}_r \rangle \mathbf{v}_r$ and $\mathbf{w}_2 = \mathbf{u} - \mathbf{w}_1$. We must verify that: (i) $\mathbf{u} = \mathbf{w}_1 + \mathbf{w}_2$, (ii) $\mathbf{w}_1$ lies in the space W spanned by $\{\mathbf{v}_1, \mathbf{v}_2, \ldots, \mathbf{v}_r\}$, and (iii) $\mathbf{w}_2$ is orthogonal to W.

Part (i) is obvious from the definition of $\mathbf{w}_2$.

Part (ii) follows immediately because $\mathbf{w}_1$ is, by definition, a linear combination of the vectors $\mathbf{v}_1, \mathbf{v}_2, \ldots, \mathbf{v}_r$.

To prove Part (iii), we use the facts that $<\mathbf{v}_i,\mathbf{v}_j> = 0$ if $i \neq j$ and $<\mathbf{v}_i,\mathbf{v}_i> = 1$, together with the definition of $\mathbf{w}_1$, to show that for $i = 1,\ldots,r$

$$\begin{aligned}<\mathbf{w}_2,\mathbf{v}_i> &= <\mathbf{u}-\mathbf{w}_1,\mathbf{v}_i>\\ &= <\mathbf{u},\mathbf{v}_i> - <\mathbf{w}_1,\mathbf{v}_i>\\ &= <\mathbf{u},\mathbf{v}_i> - <\mathbf{u},\mathbf{v}_i><\mathbf{v}_i,\mathbf{v}_i>\\ &= <\mathbf{u},\mathbf{v}_i> - <\mathbf{u},\mathbf{v}_i>\\ &= 0\end{aligned}$$

Thus, $\mathbf{w}_2$ is orthogonal to each of the vectors $\mathbf{v}_1,\mathbf{v}_2,\ldots,\mathbf{v}_r$ and hence $\mathbf{w}_2$ is orthogonal to W.

33. Let $S = \{\mathbf{w}_1,\mathbf{w}_2,\ldots,\mathbf{w}_n\}$ be the orthonormal basis. Then $(\mathbf{u})_S = (u_1,u_2,\ldots,u_n)$ means that $\mathbf{u} = u_1\mathbf{w}_1 + u_2\mathbf{w}_2 + \cdots + u_n\mathbf{w}_n$. Thus

$$\begin{aligned}\|\mathbf{u}\|^2 &= <u_1\mathbf{w}_1 + \cdots + u_n\mathbf{w}_n,\ u_1\mathbf{w}_1 + \cdots\ u_n\mathbf{w}_n>\\ &= \sum_{i=1}^{n} u_iu_i<\mathbf{w}_i,\mathbf{w}_i> + \sum_{i\neq j} u_iu_j<\mathbf{w}_i,\mathbf{w}_j>\\ &= \sum_{i=1}^{n} u_i^{\,2} \qquad \text{(because } S \text{ is orthonormal)}\end{aligned}$$

Taking positive square roots yields Theorem 5.3.2(*a*).

Note that this result also follows from Theorem 5.3.1 and Exercise 25.

34. Since $d(\mathbf{u},\mathbf{v}) = \|\mathbf{u} - \mathbf{v}\|$, we can apply Theorem 5.3.2(*a*) to $\mathbf{u} - \mathbf{v}$ to obtain Theorem 5.3.2(*b*).

35. Let $S = \{\mathbf{w}_1, \mathbf{w}_2, \ldots, \mathbf{w}_n\}$ be the orthonormal basis. Then

$$\mathbf{u} = \sum_{i=1}^{n} u_i \mathbf{w}_i \qquad \text{and} \qquad \mathbf{v} = \sum_{i=1}^{n} v_i \mathbf{w}_i$$

Hence

$$\begin{aligned} \langle \mathbf{u}, \mathbf{v} \rangle &= \langle u_1 \mathbf{w}_1 + \cdots + u_n \mathbf{w}_n, v_1 \mathbf{w}_1 + \cdots + v_n \mathbf{w}_n \rangle \\ &= \sum_{i=1}^{n} u_i v_i \langle \mathbf{w}_i, \mathbf{w}_i \rangle + \sum_{i \neq j} u_i v_j \langle \mathbf{w}_i, \mathbf{w}_j \rangle \\ &= \sum_{i=1}^{n} u_i v_i \qquad \text{(because } S \text{ is orthonormal)} \end{aligned}$$

EXERCISE SET 5.4

1. **(b)** We have $(\mathbf{w})_S = (a,b)$ where $\mathbf{w} = a\mathbf{u}_1 + b\mathbf{u}_2$. Thus

$$\begin{aligned} 2a + 3b &= 1 \\ -4a + 8b &= 1 \end{aligned}$$

or $a = \frac{5}{28}$ and $b = \frac{3}{14}$. Hence $(\mathbf{w})_S = \left(\frac{5}{28}, \frac{3}{14}\right)$ and

$$[\mathbf{w}]_S = \begin{bmatrix} \frac{5}{28} \\ \frac{3}{14} \end{bmatrix}$$

2. **(a)** Let $\mathbf{v} = a\mathbf{v}_1 + b\mathbf{v}_2 + c\mathbf{v}_3$. Then

$$\begin{aligned} a + 2b + 3c &= 2 \\ 2b + 3c &= -1 \\ 3c &= 3 \end{aligned}$$

so that $a = 3$, $b = -2$, and $c = 1$. Thus $(\mathbf{v})_S = (3,-2,1)$ and

$$[\mathbf{v}]_S = \begin{bmatrix} 3 \\ -2 \\ 1 \end{bmatrix}$$

3. **(b)** Let $\mathbf{p} = a\mathbf{p}_1 + b\mathbf{p}_2 + c\mathbf{p}_3$. Then

$$\begin{aligned} a + b \quad &= 2 \\ a \quad + c &= -1 \\ b + c &= 1 \end{aligned}$$

or $a = 0$, $b = 2$, and $c = -1$. Thus $(\mathbf{v})_S = (0,2,-1)$ and

$$[\mathbf{v}]_S = \begin{bmatrix} 0 \\ 2 \\ -1 \end{bmatrix}$$

4. Let $A = aA_1 + bA_2 + cA_3 + dA_4$. Then

$$\begin{aligned} -a + b \quad &= 2 \\ a + b \quad &= 0 \\ c &= -1 \\ d &= 3 \end{aligned}$$

so that $a = -1$, $b = 1$, $c = -1$, and $d = 3$. Thus $(A)_S = (-1,1,-1,3)$ and

$$[A]_S = \begin{bmatrix} -1 \\ 1 \\ -1 \\ 3 \end{bmatrix}$$

5. **(a)** We have $\mathbf{w} = 6\mathbf{v}_1 - \mathbf{v}_2 + 4\mathbf{v}_3 = (16,10,12)$.

(c) We have $B = -8A_1 + 7A_2 + 6A_3 + 3A_4 = \begin{bmatrix} 15 & -1 \\ 6 & 3 \end{bmatrix}$.

7. **(a)** Since $\mathbf{v}_1 = \frac{13}{10}\mathbf{u}_1 - \frac{2}{5}\mathbf{u}_2$ and $\mathbf{v}_2 = -\frac{1}{2}\mathbf{u}_1 + 0\mathbf{u}_2$, the transition matrix is

$$Q = \begin{bmatrix} \frac{13}{10} & -\frac{1}{2} \\ -\frac{2}{5} & 0 \end{bmatrix}$$

(b) Since $\mathbf{u}_1 = 0\mathbf{v}_1 - 2\mathbf{v}_2$ and $\mathbf{u}_2 = -\frac{5}{2}\mathbf{v}_1 - \frac{13}{2}\mathbf{v}_2$, the transition matrix is

$$P = \begin{bmatrix} 0 & -\frac{5}{2} \\ -2 & -\frac{13}{2} \end{bmatrix}$$

Note that $P = Q^{-1}$.

(c) We find that $\mathbf{w} = -\frac{17}{10}\mathbf{u}_1 + \frac{8}{5}\mathbf{u}_2$; that is

$$[\mathbf{w}]_B = \begin{bmatrix} -\frac{17}{10} \\ \frac{8}{5} \end{bmatrix}$$

and hence

$$[\mathbf{w}]_{B'} = \begin{bmatrix} 0 & -\frac{5}{2} \\ -2 & -\frac{13}{2} \end{bmatrix} \begin{bmatrix} -\frac{17}{10} \\ \frac{8}{5} \end{bmatrix} = \begin{bmatrix} -4 \\ -7 \end{bmatrix}$$

(d) Verify that $\mathbf{w} = (-4)\mathbf{v}_1 + (-7)\mathbf{v}_2$.

8. **(a)** We must find constants c_{ij} such that

$$\mathbf{u}_j = c_{1j}\mathbf{v}_1 + c_{2j}\mathbf{v}_2 + c_{3j}\mathbf{v}_3$$

for $j = 1,2,3$. For $j = 1$, this reduces to the system

$$\begin{aligned} -6c_{11} - 2c_{21} - 2c_{31} &= -3 \\ -6c_{11} - 6c_{21} - 3c_{31} &= 0 \\ 4c_{21} + 7c_{31} &= -3 \end{aligned}$$

which has the solution $c_{11} = 3/4$, $c_{21} = -3/4$, $c_{31} = 0$. For $j = 2$ and $j = 3$, we have systems with the same coefficient matrix but with different right-hand sides. These have solutions $c_{12} = 3/4$, $c_{22} = -17/12$, $c_{32} = 2/3$, $c_{13} = 1/12$, $c_{23} = -17/12$, and $c_{33} = 2/3$. Thus the transition matrix is

$$P = \begin{bmatrix} 3/4 & 3/4 & 1/12 \\ -3/4 & -17/12 & -17/12 \\ 0 & 2/3 & 2/3 \end{bmatrix}$$

8. **(b)** We must find constants a, b, c such that $\mathbf{w} = a\mathbf{u}_1 + b\mathbf{u}_2 + c\mathbf{u}_3$. This means solving the system of equations

$$\begin{aligned} -3a - 3b + c &= -5 \\ 2b + 6c &= 8 \\ -3a - b - c &= -5 \end{aligned}$$

The solution is $a = 1$, $b = 1$, $c = 1$. Thus

$$[\mathbf{w}]_B = \begin{bmatrix} 1 \\ 1 \\ 1 \end{bmatrix}$$

and hence

$$[\mathbf{w}]_{B'} = \begin{bmatrix} 3/4 & 3/4 & 1/12 \\ -3/4 & -17/12 & -17/12 \\ 0 & 2/3 & 2/3 \end{bmatrix} \begin{bmatrix} 1 \\ 1 \\ 1 \end{bmatrix} = \begin{bmatrix} 19/12 \\ -43/12 \\ 4/3 \end{bmatrix}$$

(c) Let $\mathbf{w} = a\mathbf{v}_1 + b\mathbf{v}_2 + c\mathbf{v}_3$ and solve the equations

$$\begin{aligned} -6a - 2b - 2c &= -5 \\ -6a - 6b - 3c &= 8 \\ 4b + 7c &= -5 \end{aligned}$$

for a, b, and c.

10. **(a)** The transition matrix from B' to B is

$$\begin{bmatrix} 3/4 & 7/2 \\ 3/2 & 1 \end{bmatrix}^{-1} = \begin{bmatrix} -2/9 & 7/9 \\ 1/3 & -1/6 \end{bmatrix}$$

(Use the result of Example 7, Section 1.5, or any other method to compute this inverse.)

(b) Since $\mathbf{p}_1 = (3/4)\mathbf{q}_1 + (3/2)\mathbf{q}_2$ and $\mathbf{p}_2 = (7/2)\mathbf{q}_1 + \mathbf{q}_2$, the transition matrix is

$$\begin{bmatrix} 3/4 & 7/2 \\ 3/2 & 1 \end{bmatrix}$$

(c) Let $\mathbf{p} = a\mathbf{p}_1 + b\mathbf{p}_2$. Thus

$$\begin{aligned} 6a + 10b &= -4 \\ 3a + 2b &= 1 \end{aligned}$$

so that $a = 1$ and $b = -1$. Therefore $[\mathbf{p}]_B = \begin{bmatrix} 1 \\ -1 \end{bmatrix}$ and hence

$$[\mathbf{p}]_{B'} = \begin{bmatrix} 3/4 & 7/2 \\ 3/2 & 1 \end{bmatrix}\begin{bmatrix} 1 \\ -1 \end{bmatrix} = \begin{bmatrix} -11/4 \\ 1/2 \end{bmatrix}$$

(d) Let $\mathbf{p} = a\mathbf{q}_1 + b\mathbf{q}_2$ and solve for a and b.

11. **(a)** By hypothesis, $\mathbf{f}_1$ and $\mathbf{f}_2$ span V. Since neither is a multiple of the other, then $\{\mathbf{f}_1,\mathbf{f}_2\}$ is linearly independent and hence is a basis for V. Now by inspection, $\mathbf{f}_1 = \frac{1}{2}\,\mathbf{g}_1 + \left(-\frac{1}{6}\right)\mathbf{g}_2$ and $\mathbf{f}_2 = \frac{1}{3}\,\mathbf{g}_2$. Therefore, $\{\mathbf{g}_1,\mathbf{g}_2\}$ must also be a basis for V because it is a spanning set which contains the correct number of vectors.

(b) The transition matrix is

$$\begin{bmatrix} \frac{1}{2} & 0 \\ -\frac{1}{6} & \frac{1}{3} \end{bmatrix}^{-1} = \begin{bmatrix} 2 & 0 \\ 1 & 3 \end{bmatrix}$$

(c) From the observations in Part (a), we have

$$P = \begin{bmatrix} \frac{1}{2} & 0 \\ -\frac{1}{6} & \frac{1}{3} \end{bmatrix}$$

(d) Since $\mathbf{h} = 2\mathbf{f}_1 + (-5)\mathbf{f}_2$, we have $[\mathbf{h}]_B = \begin{bmatrix} 2 \\ -5 \end{bmatrix}$; thus

$$[\mathbf{h}]_{B'} = \begin{bmatrix} \frac{1}{2} & 0 \\ -\frac{1}{6} & \frac{1}{3} \end{bmatrix}\begin{bmatrix} 2 \\ -5 \end{bmatrix} = \begin{bmatrix} 1 \\ -2 \end{bmatrix}$$

12. **(a)** Equation (9) yields

$$\begin{bmatrix} x' \\ y' \end{bmatrix} = \begin{bmatrix} \cos(3\pi/4) & \sin(3\pi/4) \\ -\sin(3\pi/4) & \cos(3\pi/4) \end{bmatrix}\begin{bmatrix} x \\ y \end{bmatrix}$$

$$= \begin{bmatrix} -1/\sqrt{2} & 1/\sqrt{2} \\ -1/\sqrt{2} & -1/\sqrt{2} \end{bmatrix}\begin{bmatrix} -2 \\ 6 \end{bmatrix} = \begin{bmatrix} 4\sqrt{2} \\ -2\sqrt{2} \end{bmatrix}$$

(b) If we compute the inverse of the transition matrix in (a), we have

$$\begin{bmatrix} x \\ y \end{bmatrix} \begin{bmatrix} -1/\sqrt{2} & -1/\sqrt{2} \\ 1/\sqrt{2} & -1/\sqrt{2} \end{bmatrix} \begin{bmatrix} 5 \\ 2 \end{bmatrix} = \begin{bmatrix} -7/\sqrt{2} \\ 3/\sqrt{2} \end{bmatrix}$$

14. **(a)** $$\begin{bmatrix} x' \\ y' \\ z' \end{bmatrix} = \begin{bmatrix} \cos(\pi/4) & \sin(\pi/4) & 0 \\ -\sin(\pi/4) & \cos(\pi/4) & 0 \\ 0 & 0 & 1 \end{bmatrix} \begin{bmatrix} -1 \\ 2 \\ 5 \end{bmatrix} = \begin{bmatrix} 1/\sqrt{2} \\ 3/\sqrt{2} \\ 5 \end{bmatrix}$$

(b) $$\begin{bmatrix} x \\ y \\ z \end{bmatrix} = \begin{bmatrix} 1/\sqrt{2} & -1/\sqrt{2} & 0 \\ 1/\sqrt{2} & 1/\sqrt{2} & 0 \\ 0 & 0 & 1 \end{bmatrix} \begin{bmatrix} 1 \\ 6 \\ -3 \end{bmatrix} = \begin{bmatrix} -5/\sqrt{2} \\ 7/\sqrt{2} \\ -3 \end{bmatrix}$$

15. The general transition matrix will be

$$\begin{bmatrix} \cos\theta & 0 & -\sin\theta \\ 0 & 1 & 0 \\ \sin\theta & 0 & \cos\theta \end{bmatrix}$$

In particular, if we rotate through $\theta = \frac{\pi}{3}$, then the transition matrix is

$$\begin{bmatrix} \frac{1}{2} & 0 & -\frac{\sqrt{3}}{2} \\ 0 & 1 & 0 \\ \frac{\sqrt{3}}{2} & 0 & \frac{1}{2} \end{bmatrix}$$

19. **(a)** Call the row vectors $\mathbf{r}_1$ and $\mathbf{r}_2$. Then $\|\mathbf{r}_1\|^2 = \cos^2\theta + \sin^2\theta = 1$, $\|\mathbf{r}_2\|^2 = \sin^2\theta + \cos^2\theta = 1$, and $\mathbf{r}_1 \cdot \mathbf{r}_2 = \cos\theta\sin\theta - \sin\theta\cos\theta = 0$. Thus the row vectors form an orthonormal set and we can apply Theorem 5.4.3.

21. **(a)** Since $P^{-1} = P^t$, we have

$$\begin{bmatrix} x' \\ y' \end{bmatrix} = \begin{bmatrix} -3/5 & 4/5 \\ -4/5 & -3/5 \end{bmatrix}\begin{bmatrix} 2 \\ -1 \end{bmatrix} = \begin{bmatrix} -2 \\ -1 \end{bmatrix}$$

22. From the transition matrix, we see that (1,0) and (0,1) in the $x'y'$-system have coordinates (-3/5,4/5) and (-4/5,-3/5), respectively, in the xy-system. Hence we have the following sketch:

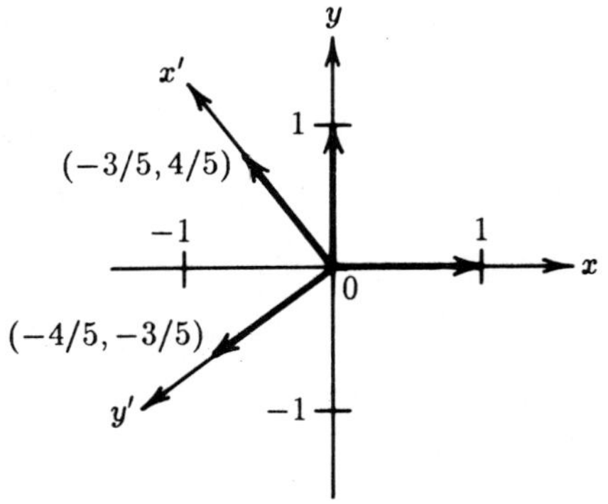

23. The matrices in Parts (a) and (c) are orthogonal and have determinant 1. Hence they represent rotations. The remaining matrices are orthogonal and have determinant -1. Therefore, they represent rotations combined with reflections.

24. **(a)** The vectors (1,0) and (0,1) in the $x'y'$-system have coordinates $(1/\sqrt{2}, -1/\sqrt{2})$ and $(1/\sqrt{2}, 1/\sqrt{2})$, respectively, in the xy-system. The transformation corresponds to a clockwise rotation through an angle of $\pi/4$ radians or 45 degrees.

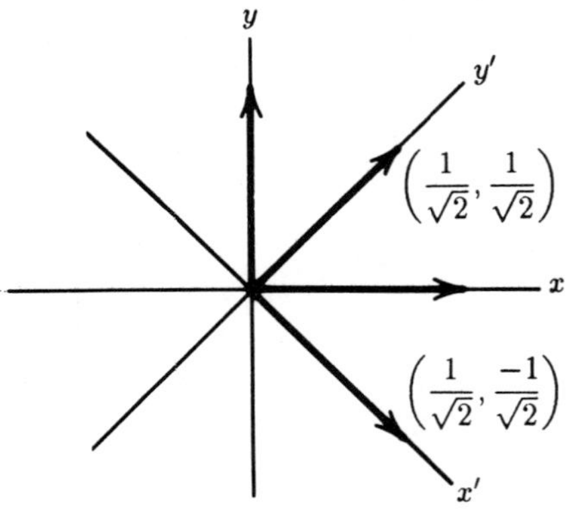

(b) The vectors (1,0) and (0,1) in the $x'y'$-system have coordinates $(-1/\sqrt{2},\ 1/\sqrt{2})$ and $(1/\sqrt{2},\ 1/\sqrt{2})$, respectively, in the xy-system. The transformation corresponds to a counterclockwise rotation through an angle of $3\pi/4$ radians or 135 degrees followed by a reflection about the x'-axis.

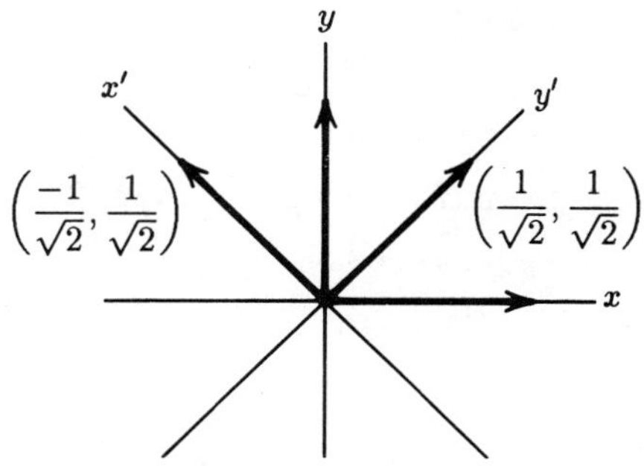

25. **(a)** $$\begin{bmatrix} x' \\ y' \\ z' \end{bmatrix} = \begin{bmatrix} 4/5 & 3/5 & 0 \\ -3/5 & 4/5 & 0 \\ 0 & 0 & 1 \end{bmatrix} \begin{bmatrix} 3 \\ 0 \\ -7 \end{bmatrix} = \begin{bmatrix} 12/5 \\ -9/5 \\ -7 \end{bmatrix}$$

26. Since (1,0,0), (0,1,0) and (0,0,1) in the $x'y'z'$-system have coordinates (4/5,3/5,0), (-3/5,4/5,0) and (0,0,1), respectively, in the xyz-system, the transformation represents a clockwise rotation about the z-axis through an angle arcsin(0.6) radians (approximately 37°).

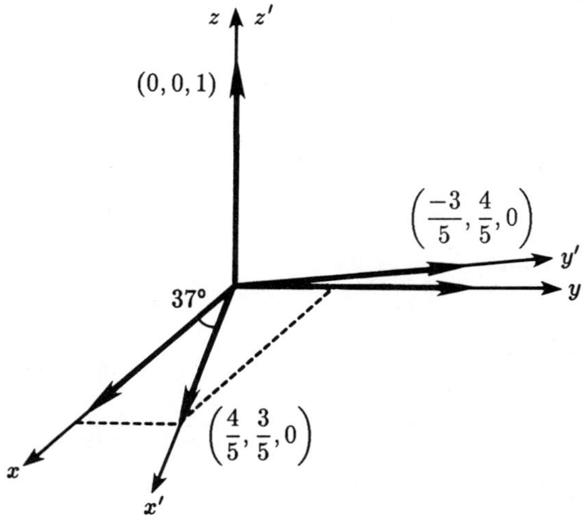

27. **(a)** The matrix P is orthonormal, but $\det(P) = -1$. Thus P does not represent a rotation.

28. **(a)** We find that (1,0,0), (0,1,0), and (0,0,1) in the $x'y'z'$-system are transformed into (4/5,3/5,0), (0,0,1), and (-3/5,4/5,0) in the xyz-system.

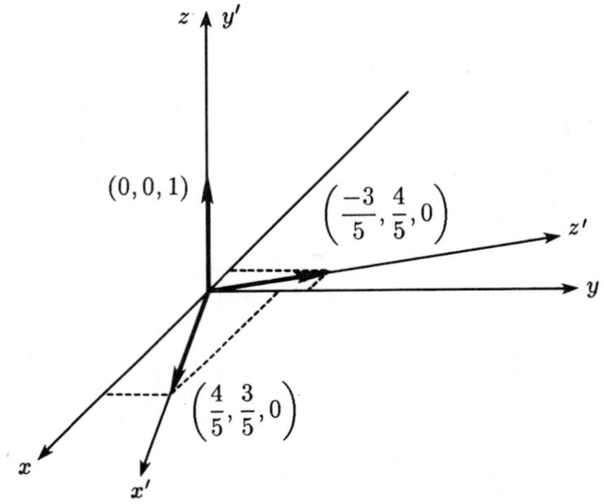

29. **(a)** See Exercise 15, above.

30. Use the result of Example 5 to find P_1 where $\mathbf{x}' = P_1\mathbf{x}$. Now use the result of Exercise 29(a), above, to find P_2 where $\mathbf{x}'' = P_2\mathbf{x}'$. Then $\mathbf{x}'' = P_2P_1\mathbf{x}$, and hence $A = P_2P_1$.

31. <u>First proof</u>: We are given that $A^{-1} = A^t$; we want to show that $\left[A^t\right]^{-1} = \left[A^t\right]^t$. But

$$\left[A^t\right]^{-1} = \left[A^{-1}\right]^{-1} \qquad \text{(Since } A \text{ is orthogonal)}$$

$$= A \qquad \text{(by Theorem 1.5.7)}$$

$$= \left[A^t\right]^t \qquad \text{(property of the transpose)}$$

Second proof: If A is orthogonal, then by Theorem 5.4.3, its rows form an orthonormal set. Hence the columns of A^t form an orthonormal set, and so A^t is orthogonal.

Note that the first proof is independent of Theorem 5.4.3. This is desirable since we use the result of Exercise 31 to prove part of Theorem 5.4.3 in Exercise 33.

Remark: We have just proved that if A is orthogonal, then so is A^t. Thus if A^t is orthogonal, so is $\left[A^t\right]^t = A$. That is, A is orthogonal if and only if A^t is orthogonal.

32. This exercise asks us to prove part of Theorem 5.4.3. By definition, A is orthogonal if and only if $A^{-1} = A^t$; moreover $A^{-1} = A^t$ if and only if $AA^t = I$. Thus A is orthogonal if and only if $AA^t = I$.

Now let $\mathbf{r}_1, \mathbf{r}_2, \ldots, \mathbf{r}_n$ denote the row vectors of A. The ij^{th} entry in AA^t is $\mathbf{r}_i \cdot \mathbf{r}_j$. But $AA^t = I$ if and only if

$$\mathbf{r}_i \cdot \mathbf{r}_j = \begin{cases} 0 & i \neq j \\ 1 & i = j \end{cases}$$

Therefore, $\{\mathbf{r}_i\}$ is an orthonormal set of vectors with respect to the Euclidean inner product in R^n if and only if A is orthogonal.

33. We know from Exercise 31 that A is orthogonal if and only if A^t is orthogonal. We also know from Exercise 32 that A^t is orthogonal if and only if its rows form an orthonormal set. But this is true if and only if the columns of A form an orthonormal set, which completes the proof.

34. Suppose that P is orthogonal. Then $P^{-1} = P^t$ and hence $PP^t = I$. Thus $\det\left[PP^t\right] = \det(I) = 1$. But by Theorems 2.3.1 and 2.3.3, $\det\left[P^t\right] = \det(P)$ and $\det\left[PP^t\right] = \det(P)\det\left[P^t\right]$. Thus $[\det(P)]^2 = 1$, or $\det(P) = \pm 1$.

SUPPLEMENTARY EXERCISES 5

1. **(a)** We must find a vector $\mathbf{x} = (x_1, x_2, x_3, x_4)$ such that

$$\mathbf{x} \cdot \mathbf{u}_1 = 0, \quad \mathbf{x} \cdot \mathbf{u}_4 = 0, \quad \text{and} \quad \frac{\mathbf{x} \cdot \mathbf{u}_2}{\|\mathbf{x}\| \, \|\mathbf{u}_2\|} = \frac{\mathbf{x} \cdot \mathbf{u}_3}{\|\mathbf{x}\| \, \|\mathbf{u}_3\|}$$

The first two conditions guarantee that $x_1 = x_4 = 0$. The third condition implies that $x_2 = x_3$. Thus any vector of the form $(0, a, a, 0)$ will satisfy the given conditions provided $a \neq 0$.

(b) We must find a vector $\mathbf{x} = (x_1, x_2, x_3, x_4)$ such that $\mathbf{x} \cdot \mathbf{u}_1 = \mathbf{x} \cdot \mathbf{u}_4 = 0$. This implies that $x_1 = x_4 = 0$. Moreover, since $\|\mathbf{x}\| = \|\mathbf{u}_2\| = \|\mathbf{u}_3\| = 1$, the cosine of the angle between $\mathbf{x}$ and $\mathbf{u}_2$ is $\mathbf{x} \cdot \mathbf{u}_2$ and the cosine of the angle between $\mathbf{x}$ and $\mathbf{u}_3$ is $\mathbf{x} \cdot \mathbf{u}_3$. Thus we are looking for a vector $\mathbf{x}$ such that $\mathbf{x} \cdot \mathbf{u}_2 = 2\mathbf{x} \cdot \mathbf{u}_3$, or $x_2 = 2x_3$. Since $\|\mathbf{x}\| = 1$, we have $\mathbf{x} = (0, 2x_3, x_3, 0)$ where $4x_3^2 + x_3^2 = 1$ or $x_3 = \pm \frac{1}{\sqrt{5}}$. Therefore

$$\mathbf{x} = \pm \left(0, \frac{2}{\sqrt{5}}, \frac{1}{\sqrt{5}}, 0\right)$$

2. **(a)** Let $\mathbf{r}_i = (a_{i1}, a_{i2}, \ldots, a_{in})$ be the ith row of A for $i = 1, \ldots, m$ and let $\mathbf{c}_j = (b_{1j}, b_{2j}, \ldots, b_{nj})^t$ be the jth column of B for $j = 1, \ldots, p$.

If $C = AB$, then the entry in row i and column j of C is given by

$$c_{ij} = \sum_{k=1}^{n} a_{ik}b_{kj} = \mathbf{r}_i \cdot \mathbf{c}_j$$

2. **(b)** Note that the columns of A^t are the rows of A. Thus by (a), if $AA^t = I$ and $\mathbf{r}_i$ is the ith row of A, then

$$\mathbf{r}_i \cdot \mathbf{r}_j = \begin{cases} 0 & \text{if } i \neq j \\ 1 & \text{if } i = j \end{cases}$$

That is, $\{\mathbf{r}_1,\ldots,\mathbf{r}_n\}$ is orthonormal in R^n with respect to the Euclidean inner product.

(c) If $AA^t = I$, then A and A^t are inverses of one another. Thus $A^tA = I$. Therefore, since $\left[A^t\right]^t = A$, we can apply Part (b) to A^t to deduce that the row vectors of A^t form an orthonormal set in R^n. But since the row vectors of A^t are the column vectors of A, the desired result follows.

4. Let $\mathbf{u} = (\sqrt{a_1}, \ldots, \sqrt{a_n})$ and $\mathbf{v} = (1/\sqrt{a_1}, \ldots, 1/\sqrt{a_n})$. By the Cauchy-Schwarz Inequality,

$$(\mathbf{u} \cdot \mathbf{v})^2 = \underbrace{(1 + \cdots + 1)^2}_{n \text{ terms}} \leq \|\mathbf{u}\|^2\|\mathbf{v}\|^2$$

or

$$n^2 \leq \left[a_1 + \cdots + a_n\right]\left[\frac{1}{a_1} + \cdots + \frac{1}{a_n}\right]$$

6. If $\mathbf{v} = (a,b,c)$ where $\|\mathbf{v}\| = 1$ and $\mathbf{v} \cdot \mathbf{u}_1 = \mathbf{v} \cdot \mathbf{u}_2 = \mathbf{v} \cdot \mathbf{u}_3 = 0$, then

$$\begin{aligned} a + b - c &= 0 \\ -2a - b + 2c &= 0 \\ -a \quad + c &= 0 \end{aligned}$$

where $a^2 + b^2 + c^2 = 1^2$. The above system has the solution $a = t$, $b = 0$, and $c = t$ where t is arbitrary. But $a^2 + b^2 + c^2 = 1^2$ implies that $2t^2 = 1^2$ or $t = \pm 1/\sqrt{2}$. The two vectors are therefore $\mathbf{v} = \pm(1/\sqrt{2},\ 0,\ 1/\sqrt{2})$.

7. Let

$$(*) \qquad \langle \mathbf{u},\mathbf{v} \rangle = w_1u_1v_1 + w_2u_2v_2 + \cdots + w_nu_nv_n$$

be the weighted Euclidean inner product. Since $\langle \mathbf{v}_i,\mathbf{v}_j \rangle = 0$ whenever $i \neq j$, the vectors $\{\mathbf{v}_1,\mathbf{v}_2,\ldots,\mathbf{v}_n\}$ form an orthogonal set with respect to (*) for <u>any</u> choice of the constants $w_1,w_2,\ldots,w_n$. We must now choose the positive constants $w_1,w_2,\ldots,w_n$ so that $\|\mathbf{v}_k\| = 1$ for all k. But $\|\mathbf{v}_k\|^2 = kw_k$. If we let $w_k = 1/k$ for $k = 1,2,\ldots,n$, the given vectors will then form an orthonormal set with respect to (*).

8. Suppose that $\langle \mathbf{u},\mathbf{v} \rangle = w_1u_1v_1 + w_2u_2v_2$ were such an inner product. Then

$$\begin{aligned} \langle (1,2),(3,-1) \rangle = 0 &\Rightarrow 3w_1 - 2w_2 = 0 \\ \|(1,2)\| = 1 &\Rightarrow w_1 + 4w_2 = 1 \\ \|(3,-1)\| = 1 &\Rightarrow 9w_1 + w_2 = 1 \end{aligned}$$

But since the three equations for w_1 and w_2 above are inconsistent (why?), there is no such inner product.

9. **(a)** Let $\mathbf{w}$ be an arbitrary vector in $\text{lin}(\mathbf{u}_1,\mathbf{u}_2)$. Then there exist constants a and b such that $\mathbf{w} = a\mathbf{u}_1 + b\mathbf{u}_2$. Hence $\langle \mathbf{v},\mathbf{w} \rangle = a\langle \mathbf{v},\mathbf{u}_1 \rangle + b\langle \mathbf{v},\mathbf{u}_2 \rangle = 0$ since $\langle \mathbf{v},\mathbf{u}_1 \rangle = \langle \mathbf{v},\mathbf{u}_2 \rangle = 0$ by hypothesis.

9. (b) If $\mathbf{u}_1$ and $\mathbf{u}_2$ are linearly independent, then lin($\mathbf{u}_1,\mathbf{u}_2$) is a plane in R^3 and the result says that $\mathbf{v}$ is orthogonal to every vector in that plane.

10. From the definition of the norm, we have

$$\begin{aligned}\|\mathbf{u} - \mathbf{v}\|^2 &= \langle\mathbf{u} - \mathbf{v},\ \mathbf{u} - \mathbf{v}\rangle \\ &= \langle\mathbf{u},\mathbf{u}\rangle + \langle\mathbf{v},\mathbf{v}\rangle - 2\langle\mathbf{u},\mathbf{v}\rangle \\ &= \|\mathbf{u}\|^2 + \|\mathbf{v}\|^2 - 2\|\mathbf{u}\|\ \|\mathbf{v}\| \cos\theta\end{aligned}$$

11. (a) The length of each "side" of this "cube" is $|k|$. The length of the "diagonal" is $\sqrt{n}|k|$. The inner product of any "side" with the "diagonal" is k^2. Therefore,

$$\cos\theta = \frac{k^2}{|k|\sqrt{n}|k|} = \frac{1}{\sqrt{n}}$$

(b) As $n \longrightarrow +\infty$, $\cos\theta \longrightarrow 0$, so that $\theta \longrightarrow \pi/2$.

12. (a) We have that $\mathbf{u} + \mathbf{v}$ and $\mathbf{u} - \mathbf{v}$ are orthogonal if and only if

$$\langle\mathbf{u} + \mathbf{v},\ \mathbf{u} - \mathbf{v}\rangle = 0$$

But

$$\langle\mathbf{u} + \mathbf{v},\ \mathbf{u} - \mathbf{v}\rangle = \|\mathbf{u}\| - \|\mathbf{v}\|$$

by the properties of an inner product (why?). Thus $\mathbf{u} + \mathbf{v}$ and $\mathbf{u} - \mathbf{v}$ are orthogonal if and only if $\|\mathbf{u}\| = \|\mathbf{v}\|$.

(b) Since $\mathbf{u} + \mathbf{v}$ and $\mathbf{u} - \mathbf{v}$ can be thought of as diagonals of a parallelogram with sides $\mathbf{u}$ and $\mathbf{v}$, the above result says that the two diagonals are perpendicular if and only if the parallelogram has equal sides.

13. Recall that **u** can be expressed as the linear combination

$$\mathbf{u} = a_1\mathbf{v}_1 + \cdots + a_n\mathbf{v}_n$$

where $a_i = \langle \mathbf{u}, \mathbf{v}_i \rangle$ for $i = 1, \ldots, n$. Thus

$$\cos^2 \alpha_i = \left[\frac{\langle \mathbf{u}, \mathbf{v}_i \rangle}{\|\mathbf{u}\|\ \|\mathbf{v}_i\|}\right]^2$$

$$= \left[\frac{a_i}{\|\mathbf{u}\|}\right]^2 \qquad (\|\mathbf{v}_i\| = 1)$$

$$= \frac{a_i^2}{a_1^2 + a_2^2 + \cdots + a_n^2} \qquad \text{(why?)}$$

Therefore

$$\cos^2 \alpha_1 + \cdots + \cos^2 \alpha_n = \frac{a_1^2 + a_2^2 + \cdots + a_n^2}{a_1^2 + a_2^2 + \cdots + a_n^2} = 1$$

14. We must show that $\langle \mathbf{u}, \mathbf{v} \rangle$ satisfies the four axioms which define an inner product. We show (2) and (4), and leave the others to you.

(2) Using Axiom 3 for $\langle \mathbf{u}, \mathbf{v} \rangle_1$ and $\langle \mathbf{u}, \mathbf{v} \rangle_2$, we have

$$\begin{aligned} \langle \mathbf{u}+\mathbf{v}, \mathbf{w} \rangle &= \langle \mathbf{u}+\mathbf{v}, \mathbf{w} \rangle_1 + \langle \mathbf{u}+\mathbf{v}, \mathbf{w} \rangle_2 \\ &= \langle \mathbf{u}, \mathbf{w} \rangle_1 + \langle \mathbf{v}, \mathbf{w} \rangle_1 + \langle \mathbf{u}, \mathbf{w} \rangle_2 + \langle \mathbf{v}, \mathbf{w} \rangle_2 \\ &= \langle \mathbf{u}, \mathbf{w} \rangle_1 + \langle \mathbf{u}, \mathbf{w} \rangle_2 + \langle \mathbf{v}, \mathbf{w} \rangle_1 + \langle \mathbf{v}, \mathbf{w} \rangle_2 \\ &= \langle \mathbf{u}, \mathbf{w} \rangle + \langle \mathbf{v}, \mathbf{w} \rangle \end{aligned}$$

(4) Using Axiom 4 for $\langle \mathbf{u}, \mathbf{v} \rangle_1$ and $\langle \mathbf{u}, \mathbf{v} \rangle_2$, we have

$$\langle \mathbf{v}, \mathbf{v} \rangle = \langle \mathbf{v}, \mathbf{v} \rangle_1 + \langle \mathbf{v}, \mathbf{v} \rangle_2 \geq 0$$

since both $\langle \mathbf{v},\mathbf{v}\rangle_1$ and $\langle \mathbf{v},\mathbf{v}\rangle_2$ are nonnegative. Moreover, $\langle \mathbf{v},\mathbf{v}\rangle = 0$ if and only if $\langle \mathbf{v},\mathbf{v}\rangle_1 = \langle \mathbf{v},\mathbf{v}\rangle_2 = 0$, which is true if and only if $\mathbf{v} = \mathbf{0}$.

15. Recall that A is orthogonal provided $A^{-1} = A^t$. Hence

$$\begin{aligned}\langle \mathbf{u},\mathbf{v}\rangle &= \mathbf{v}^t A^t A\mathbf{u} \\ &= \mathbf{v}^t A^{-1} A\mathbf{u} = \mathbf{v}^t\mathbf{u}\end{aligned}$$

which is the Euclidean inner product.

EXERCISE SET 6.1

1. **(a)** Since

$$\det(\lambda I - A) = \det\begin{bmatrix} \lambda - 3 & 0 \\ 8 & \lambda + 1 \end{bmatrix} = (\lambda - 3)(\lambda + 1)$$

the characteristic equation is $\lambda^2 - 2\lambda - 3 = 0$.

(e) Since

$$\det(\lambda I - A) = \det\begin{bmatrix} \lambda & 0 \\ 0 & \lambda \end{bmatrix} = \lambda^2$$

the characteristic equation is $\lambda^2 = 0$.

3. **(a)** The equation $(\lambda I - A)\mathbf{x} = \mathbf{0}$ becomes

$$\begin{bmatrix} \lambda - 3 & 0 \\ 8 & \lambda + 1 \end{bmatrix}\begin{bmatrix} x_1 \\ x_2 \end{bmatrix} = \begin{bmatrix} 0 \\ 0 \end{bmatrix}$$

The eigenvalues are $\lambda = 3$ and $\lambda = -1$. Substituting $\lambda = 3$ into $(\lambda I - A)\mathbf{x} = 0$ yields.

$$\begin{bmatrix} 0 & 0 \\ 8 & 4 \end{bmatrix}\begin{bmatrix} x_1 \\ x_2 \end{bmatrix} = \begin{bmatrix} 0 \\ 0 \end{bmatrix}$$

or

$$8x_1 + 4x_2 = 0$$

Thus $x_1 = \frac{1}{2}s$ and $x_2 = s$ where s is arbitrary, so that a basis for the eigenspace corresponding to $\lambda = 3$ is $\begin{bmatrix} 1/2 \\ 1 \end{bmatrix}$.

Of course, $\begin{bmatrix} 1 \\ 2 \end{bmatrix}$ and $\begin{bmatrix} \pi \\ 2\pi \end{bmatrix}$ are also bases.

Substituting $\lambda = -1$ into $(\lambda I - A)\mathbf{x} = \mathbf{0}$ yields

$$\begin{bmatrix} -4 & 0 \\ 8 & 0 \end{bmatrix}\begin{bmatrix} x_1 \\ x_2 \end{bmatrix} = \begin{bmatrix} 0 \\ 0 \end{bmatrix}$$

or

$$\begin{aligned} -4x_1 &= 0 \\ 8x_1 &= 0 \end{aligned}$$

Hence, $x_1 = 0$ and $x_2 = s$ where s is arbitrary. In particular, if $s = 1$, then a basis for the eigenspace corresponding to $\lambda = -1$ is $\begin{bmatrix} 0 \\ 1 \end{bmatrix}$.

3. **(e)** The equation $(\lambda I - A)\mathbf{x} = \mathbf{0}$ becomes

$$\begin{bmatrix} \lambda & 0 \\ 0 & \lambda \end{bmatrix}\begin{bmatrix} x_1 \\ x_2 \end{bmatrix} = \begin{bmatrix} 0 \\ 0 \end{bmatrix}$$

Clearly, $\lambda = 0$ is the only eigenvalue. Substituting $\lambda = 0$ into the above equation yields $x_1 = s$ and $x_2 = t$ where s and t are arbitrary. In particular, if $s = t = 1$, then we find that $\begin{bmatrix} 1 \\ 0 \end{bmatrix}$ and $\begin{bmatrix} 0 \\ 1 \end{bmatrix}$ form a basis for the eigenspace associated with $\lambda = 0$.

4. (c) Since

$$\det[\lambda I - A] = \det\begin{bmatrix} \lambda + 2 & 0 & -1 \\ 6 & \lambda + 2 & 0 \\ -19 & -5 & \lambda + 4 \end{bmatrix}$$

$$= \lambda^3 + 8\lambda^2 + \lambda + 8$$

the characteristic equation of A is

$$\lambda^3 + 8\lambda^2 + \lambda + 8 = 0$$

5. (c) From the solution to 4.(c), we have

$$\lambda^3 + 8\lambda^2 + \lambda + 8 = (\lambda + 8)(\lambda^2 + 1)$$

Since $\lambda^2 + 1 = 0$ has no real solutions, then $\lambda = -8$ is the only (real) eigenvalue.

6. (c) The only eigenvalue is $\lambda = -8$. If we substitute $\lambda = -8$ into the equation $(\lambda I - A)\mathbf{x} = \mathbf{0}$, we obtain

$$\begin{bmatrix} -6 & 0 & -1 \\ 6 & -6 & 0 \\ -19 & -5 & -4 \end{bmatrix}\begin{bmatrix} x_1 \\ x_2 \\ x_3 \end{bmatrix} = \begin{bmatrix} 0 \\ 0 \\ 0 \end{bmatrix}$$

The augmented matrix of the above system can be reduced to

$$\begin{bmatrix} 1 & 0 & 1/6 & 0 \\ 0 & 1 & 1/6 & 0 \\ 0 & 0 & 0 & 0 \end{bmatrix}$$

Thus, $x_1 = -\frac{1}{6}s$, $x_2 = -\frac{1}{6}s$, $x_3 = s$ is a solution where s is arbitrary. Therefore $\begin{bmatrix} -1/6 \\ -1/6 \\ 1 \end{bmatrix}$ forms a basis for the eigenspace.

6. (f) The eigenvalues are $\lambda = -4$ and $\lambda = 3$. If we substitute $\lambda = -4$ into the equation $(\lambda I - A)\mathbf{x} = \mathbf{0}$, we obtain

$$\begin{bmatrix} -9 & -6 & -2 \\ 0 & -3 & 8 \\ -1 & 0 & -2 \end{bmatrix} \begin{bmatrix} x_1 \\ x_2 \\ x_3 \end{bmatrix} = \begin{bmatrix} 0 \\ 0 \\ 0 \end{bmatrix}$$

If we reduce the augmented matrix to row-echelon form, we obtain

$$\begin{bmatrix} 1 & 0 & 2 & 0 \\ 0 & 1 & -8/3 & 0 \\ 0 & 0 & 0 & 0 \end{bmatrix}$$

This implies that $x_1 = -2s$, $x_2 = \frac{8}{3}s$, $x_3 = s$. If we set $s = 1$, we find that

$$\begin{bmatrix} -2 \\ 8/3 \\ 1 \end{bmatrix}$$

is a basis for the eigenspace associated with $\lambda = -4$.

If we substitue $\lambda = 3$ into the equation $(\lambda I - A)\mathbf{x} = \mathbf{0}$, we obtain

$$\begin{bmatrix} -2 & -6 & -2 \\ 0 & 4 & 8 \\ -1 & 0 & 5 \end{bmatrix} \begin{bmatrix} x_1 \\ x_2 \\ x_3 \end{bmatrix} = \begin{bmatrix} 0 \\ 0 \\ 0 \end{bmatrix}$$

If we reduce the augmented matrix to row-echelon form, we obtain

$$\begin{bmatrix} 1 & 0 & -5 & 0 \\ 0 & 1 & 2 & 0 \\ 0 & 0 & 0 & 0 \end{bmatrix}$$

It then follows that

$$\begin{bmatrix} 5 \\ -2 \\ 1 \end{bmatrix}$$

is a basis for the eigenspace associated with $\lambda = 3$.

7. **(a)** Since

$$\det[\lambda I - A] = \det\begin{bmatrix} \lambda & 0 & -2 & 0 \\ -1 & \lambda & -1 & 0 \\ 0 & -1 & \lambda + 2 & 0 \\ 0 & 0 & 0 & \lambda - 1 \end{bmatrix}$$

$$= \lambda^4 + \lambda^3 - 3\lambda^2 - \lambda + 2$$

$$= (\lambda - 1)^2(\lambda + 2)(\lambda + 1)$$

the characteristic equation is

$$(\lambda - 1)^2(\lambda + 2)(\lambda + 1) = 0$$

9. **(a)** The eigenvalues are $\lambda = 1$, $\lambda = -2$, and $\lambda = -1$. If we set $\lambda = 1$, then $(\lambda I - A)\mathbf{x} = \mathbf{0}$ becomes

$$\begin{bmatrix} 1 & 0 & -2 & 0 \\ -1 & 1 & -1 & 0 \\ 0 & -1 & 3 & 0 \\ 0 & 0 & 0 & 0 \end{bmatrix}\begin{bmatrix} x_1 \\ x_2 \\ x_3 \\ x_4 \end{bmatrix} = \begin{bmatrix} 0 \\ 0 \\ 0 \\ 0 \end{bmatrix}$$

The augmented matrix can be reduced to

$$\begin{bmatrix} 1 & 0 & -2 & 0 & 0 \\ 0 & 1 & -3 & 0 & 0 \\ 0 & 0 & 0 & 0 & 0 \\ 0 & 0 & 0 & 0 & 0 \end{bmatrix}$$

Thus, $x_1 = 2s$, $x_2 = 3s$, $x_3 = s$, and $x_4 = t$ is a solution for all s and t. In particular, if we let $s = t = 1$, we see that

$$\begin{bmatrix} 2 \\ 3 \\ 1 \\ 0 \end{bmatrix} \quad \text{and} \quad \begin{bmatrix} 0 \\ 0 \\ 0 \\ 1 \end{bmatrix}$$

form a basis for the eigenspace associated with $\lambda = 1$.

If we set $\lambda = -2$, then $(\lambda I - A)\mathbf{x} = \mathbf{0}$ becomes

$$\begin{bmatrix} -2 & 0 & -2 & 0 \\ -1 & -2 & -1 & 0 \\ 0 & -1 & 0 & 0 \\ 0 & 0 & 0 & -3 \end{bmatrix} \begin{bmatrix} x_1 \\ x_2 \\ x_3 \\ x_4 \end{bmatrix} = \begin{bmatrix} 0 \\ 0 \\ 0 \\ 0 \end{bmatrix}$$

The augmented matrix can be reduced to

$$\begin{bmatrix} 1 & 0 & 1 & 0 & 0 \\ 0 & 1 & 0 & 0 & 0 \\ 0 & 0 & 0 & 1 & 0 \\ 0 & 0 & 0 & 0 & 0 \end{bmatrix}$$

This implies that $x_1 = -s$, $x_2 = x_4 = 0$, and $x_3 = s$. Therefore the vector

$$\begin{bmatrix} -1 \\ 0 \\ 1 \\ 0 \end{bmatrix}$$

forms a basis for the eigenspace associated with $\lambda = -2$.

Finally, if we set $\lambda = -1$, then $(\lambda I - A)\mathbf{x} = \mathbf{0}$ becomes

$$\begin{bmatrix} -1 & 0 & -2 & 0 \\ -1 & -1 & -1 & 0 \\ 0 & -1 & 1 & 0 \\ 0 & 0 & 0 & -2 \end{bmatrix} \begin{bmatrix} x_1 \\ x_2 \\ x_3 \\ x_4 \end{bmatrix} = \begin{bmatrix} 0 \\ 0 \\ 0 \\ 0 \end{bmatrix}$$

The augmented matrix can be reduced to

$$\begin{bmatrix} 1 & 0 & 2 & 0 & 0 \\ 0 & 1 & -1 & 0 & 0 \\ 0 & 0 & 0 & 1 & 0 \\ 0 & 0 & 0 & 0 & 0 \end{bmatrix}$$

Thus, $x_1 = -2s$, $x_2 = s$, $x_3 = s$, and $x_4 = 0$ is a solution. Therefore the vector

$$\begin{bmatrix} -2 \\ 1 \\ 1 \\ 0 \end{bmatrix}$$

forms a basis for the eigenspace associated with $\lambda = -1$.

11. By Theorem 6.1.2, the eigenvalues of A are 1, 1/2, 0, and 2. Thus by Theorem 6.1.3, the eigenvalues of A^9 are 1^9, $(1/2)^9 = 1/512$, 0, and $2^9 = 512$.

12. Notice that $A^2 = I$. Therefore $A^{25} = A$.

13. The vectors $A\mathbf{x}$ and $\mathbf{x}$ will lie on the same line through the origin if and only if there exists a real number λ such that $A\mathbf{x} = \lambda\mathbf{x}$, that is, if and only if λ is a real eigenvalue for A and $\mathbf{x}$ is the associated eigenvector.

(a) In this case, the eigenvalues are $\lambda = 3$ and $\lambda = 2$, while associated eigenvectors are

$$\begin{bmatrix} 1 \\ 1 \end{bmatrix} \quad \text{and} \quad \begin{bmatrix} 1 \\ 2 \end{bmatrix}$$

respectively. Hence the lines $y = x$ and $y = 2x$ are the only lines which are invariant under A.

(b) In this case, the characteristic equation for A is $\lambda^2 + 1 = 0$. Since A has no real eigenvalues, there are no lines which are invariant under A.

14. Recall (Theorem 4.7.3) that A is invertible if and only if $A\mathbf{x} = \mathbf{0}$ has only the trivial solution. Therefore $A\mathbf{x} = \mathbf{0}$ has a nontrivial solution if and only if A is not invertible.

15. From the hint, we know that the constant term is $\det(0I - A) = \det(-A)$. Now each signed elementary product in $\det(-A)$ is just $(-1)^n$ times the corresponding signed elementary product in $\det(A)$. Thus $\det(-A) = (-1)^n\det(A)$.

16. Let a_{ij} denote the ijth entry of A. Then the characteristic polynomial of A is $\det(\lambda I - A)$ or

$$\det\begin{bmatrix} \lambda - a_{11} & -a_{12} & \cdots & -a_{1n} \\ -a_{21} & \lambda - a_{22} & \cdots & -a_{2n} \\ \vdots & \vdots & & \vdots \\ -a_{n1} & -a_{n2} & \cdots & \lambda - a_{nn} \end{bmatrix}$$

This determinant is a sum each of whose terms is the product of n entries from the given matrix. Each of these entries is either a constant or is of the form $\lambda - a_{ij}$. The only term with a λ in each factor of the product is

$$(\lambda - a_{11})(\lambda - a_{22}) \cdots (\lambda - a_{nn})$$

Therefore, this term must produce the highest power of λ in the characteristic polynomial. This power is clearly n and the coefficient of λ^n is 1.

18. The characteristic equation of A is

$$\lambda^2 - (a + d)\lambda + ad - bc = 0$$

This is a quadratic equation whose discriminant is

$$\begin{aligned}(a + d)^2 - 4\,ad + 4\,bc &= a^2 - 2\,ad + d^2 + 4\,bc \\ &= (a - d)^2 + 4\,bc\end{aligned}$$

The roots are

$$\lambda = \frac{1}{2}\left[\,(a + d) \pm \sqrt{(a - d)^2 + 4\,bc}\,\right]$$

If the discriminant is positive, then the equation has two distinct real roots; if it is zero, then the equation has one real root (repeated); if it is negative, then the equation has no real roots. Since the eigenvalues are assumed to be real numbers, the result follows.

19. We are given the two distinct eigenvalues λ_1 and λ_2. To find the corresponding eigenvectors, we set

$$\begin{bmatrix} \lambda_i - a & -b \\ -c & \lambda_i - d \end{bmatrix} \begin{bmatrix} x_1 \\ x_2 \end{bmatrix} = \begin{bmatrix} 0 \\ 0 \end{bmatrix}$$

This yields the equations

$$\begin{aligned} (\lambda_i - a)x_1 - \qquad bx_2 &= 0 \\ -cx_1 + (\lambda_i - d)x_2 &= 0 \end{aligned}$$

This system of equations is guaranteed to have a nontrivial solution. Since b and $a - \lambda_i$ are not both zero, then any nonzero multiple of $\begin{bmatrix} -b \\ a - \lambda_i \end{bmatrix}$ will be an eigenvector corresponding to λ_i. Therefore

$$\begin{bmatrix} -b \\ a - \lambda_1 \end{bmatrix} \quad \text{and} \quad \begin{bmatrix} -b \\ a - \lambda_2 \end{bmatrix}$$

are eigenvectors corresponding to λ_1 and λ_2, respectively.

20. As in Exercise 18, we have

$$\begin{aligned} \lambda &= \frac{a + d \pm \sqrt{(a - d)^2 + 4bc}}{2} \\ &= \frac{a + d \pm \sqrt{(c - b)^2 + 4bc}}{2} \qquad \text{because } a - d = c - b \\ &= \frac{a + d \pm \sqrt{(c + b)^2}}{2} \end{aligned}$$

$$= \frac{a + b + c + d}{2} \quad \text{or} \quad \frac{a - b - c + d}{2}$$

$$= a + b \quad \text{or} \quad a - c$$

<u>Alternate Solution</u>: Recall that if r_1 and r_2 are roots of the quadratic equation $x^2 + Bx + C = 0$, then $B = -(r_1 + r_2)$ and $C = r_1r_2$. The converse of this result is also true. Thus if we can show that the system of equations

$$\lambda_1 + \lambda_2 = a + d$$

$$\lambda_1\lambda_2 = ad - bc$$

is satisfied by $\lambda_1 = a + b$ and $\lambda_2 = a - c$, we shall be done. This is a straightforward computation and we leave it to you.

EXERCISE SET 6.2

1. Let A denote the matrix. The characteristic equation of A is $(\lambda - 2)^2 = 0$, so $\lambda = 2$ is the only eigenvalue. To find the eigenvectors of A, we look for vectors $\mathbf{x} \neq \mathbf{0}$ such that $(2I - A)\mathbf{x} = \mathbf{0}$. By a routine calculation, all eigenvectors are of the form $\mathbf{x} = \begin{bmatrix} 0 \\ t \end{bmatrix}$. Since the 2 x 2 matrix A does not have 2 linearly independent eigenvectors, it is not diagonalizable.

3. Let A denote the matrix. The eigenvalues of A are $\lambda = 3$ and $\lambda = 2$. Thus the eigenvectors of A are the nontrivial solutions of $(3I - A)\mathbf{x} = \mathbf{0}$ and $(2I - A)\mathbf{x} = \mathbf{0}$. These are of the form

$$\begin{bmatrix} s \\ 0 \\ 0 \end{bmatrix} \quad \text{and} \quad \begin{bmatrix} 0 \\ 0 \\ t \end{bmatrix}$$

respectively. Thus the 3 x 3 matrix A has only 2 linearly independent eigenvectors, so it is not diagonalizable.

5. The characteristic equation is $\lambda^2 - 3\lambda + 2 = 0$, the eigenvalues are $\lambda = 2$ and $\lambda = 1$, and the eigenspaces are spanned by the vectors

$$\begin{bmatrix} 3/4 \\ 1 \end{bmatrix} \quad \text{and} \quad \begin{bmatrix} 4/5 \\ 1 \end{bmatrix}$$

respectively. Thus, if we let

$$P = \begin{bmatrix} 3 & 4 \\ 4 & 5 \end{bmatrix}$$

then

$$P^{-1}AP = \begin{bmatrix} 2 & 0 \\ 0 & 1 \end{bmatrix}$$

Clearly, there are other possibilities for P.

7. The characteristic equation is $\lambda(\lambda - 1)(\lambda - 2) = 0$, the eigenvalues are $\lambda = 0$, $\lambda = 1$, and $\lambda = 2$, and the eigenspaces are spanned by the vectors

$$\begin{bmatrix} 0 \\ -1 \\ 1 \end{bmatrix} \quad \begin{bmatrix} 1 \\ 0 \\ 0 \end{bmatrix} \quad \begin{bmatrix} 0 \\ 1 \\ 1 \end{bmatrix}$$

Thus, if we let

$$P = \begin{bmatrix} 0 & 1 & 0 \\ -1 & 0 & 1 \\ 1 & 0 & 1 \end{bmatrix}$$

then

$$P^{-1}AP = \begin{bmatrix} 0 & 0 & 0 \\ 0 & 1 & 0 \\ 0 & 0 & 2 \end{bmatrix}$$

9. The characteristic equation is $\lambda^3 - 4\lambda^2 + 5\lambda - 2 = 0$, the eigenvalues are $\lambda = 1$ and $\lambda = 2$, and the eigenspaces are spanned by the vectors

$$\begin{bmatrix} 1 \\ 4/3 \\ 1 \end{bmatrix} \quad \text{and} \quad \begin{bmatrix} 3/4 \\ 3/4 \\ 1 \end{bmatrix}$$

respectively. Since A is a 3 x 3 matrix with only 2 linearly independent eigenvectors, it is not diagonalizable.

10. The characteristic equation is $\lambda^3 - 6\lambda^2 + 11\lambda - 6 = 0$, the eigenvalues are $\lambda = 1$, $\lambda = 2$, and $\lambda = 3$, and the eigenspaces are spanned by the vectors

$$\begin{bmatrix} 1 \\ 1 \\ 1 \end{bmatrix} \quad \begin{bmatrix} 2/3 \\ 1 \\ 1 \end{bmatrix} \quad \begin{bmatrix} 1/4 \\ 3/4 \\ 1 \end{bmatrix}$$

Thus, one possibility is

$$P = \begin{bmatrix} 1 & 2 & 1 \\ 1 & 3 & 3 \\ 1 & 3 & 4 \end{bmatrix}$$

and

$$P^{-1}AP = \begin{bmatrix} 1 & 0 & 0 \\ 0 & 2 & 0 \\ 0 & 0 & 3 \end{bmatrix}$$

12. The characteristic equation is $\lambda^2(\lambda - 1) = 0$; thus $\lambda = 0$ and $\lambda = 1$ are the only eigenvalues. The eigenspace associated with $\lambda = 0$ is spanned by the vectors $\begin{bmatrix} 1 \\ 0 \\ -3 \end{bmatrix}$ and $\begin{bmatrix} 0 \\ 1 \\ 0 \end{bmatrix}$; the eigenspace associated with $\lambda = 1$ is spanned by $\begin{bmatrix} 0 \\ 0 \\ 1 \end{bmatrix}$. Thus, one possibility is

$$P = \begin{bmatrix} 1 & 0 & 0 \\ 0 & 1 & 0 \\ -3 & 0 & 1 \end{bmatrix}$$

and hence

$$P^{-1}AP = \begin{bmatrix} 0 & 0 & 0 \\ 0 & 0 & 0 \\ 0 & 0 & 1 \end{bmatrix}$$

15. The eigenvalues of A are $\lambda = 1$ and $\lambda = 2$ and the eigenspaces are spanned by the vectors

$$\begin{bmatrix} 1 \\ 1 \end{bmatrix} \quad \text{and} \quad \begin{bmatrix} 0 \\ 1 \end{bmatrix}$$

Thus if we let

$$P = \begin{bmatrix} 1 & 0 \\ 1 & 1 \end{bmatrix}$$

then

$$P^{-1}AP = \begin{bmatrix} 1 & 0 \\ 0 & 2 \end{bmatrix}$$

Now by Formula (6), we have

$$A^{10} = P\begin{bmatrix} 1 & 0 \\ 0 & 2^{10} \end{bmatrix}P^{-1}$$

$$= \begin{bmatrix} 1 & 0 \\ 1 & 1 \end{bmatrix}\begin{bmatrix} 1 & 0 \\ 0 & 2^{10} \end{bmatrix}\begin{bmatrix} 1 & 0 \\ -1 & 1 \end{bmatrix}$$

$$= \begin{bmatrix} 1 & 0 \\ 1-2^{10} & 2^{10} \end{bmatrix}$$

17. By Theorem 6.1.2, the eigenvalues of A are $\lambda_1 = 1$ and $\lambda_2 = \lambda_3 = -1$. Corresponding eigenvectors are.

$$\begin{bmatrix} 1 \\ 0 \\ 0 \end{bmatrix} \quad \begin{bmatrix} 1 \\ 1 \\ 0 \end{bmatrix} \quad \begin{bmatrix} -4 \\ 0 \\ 1 \end{bmatrix}$$

Since A has three linearly independent eigenvectors, it is diagonalizable. Let

$$P = \begin{bmatrix} 1 & 1 & -4 \\ 0 & 1 & 0 \\ 0 & 0 & 1 \end{bmatrix}$$

Then

$$P^{-1}AP = D = \begin{bmatrix} 1 & 0 & 0 \\ 0 & -1 & 0 \\ 0 & 0 & -1 \end{bmatrix}$$

Thus, $A^k = PD^kP^{-1}$ for any positive integer k, or

$$A^k = \begin{cases} PIP^{-1} = PP^{-1} = I & \text{if } k \text{ is even} \\ PDP^{-1} = A & \text{if } k \text{ is odd} \end{cases}$$

Finally, since D is its own inverse, we have

$$A^{-1} = (P D P^{-1})^{-1} = P D^{-1} P^{-1} = A$$

so that

$$A^{-k} = (A^{-1})^k = A^k$$

Of course, all of the above could have been greatly simplified if we had been clever enough to notice at the beginning that $A^2 = I$.

18. The characteristic equation of A is $(\lambda - 1)(\lambda - 3)(\lambda - 4) = 0$ so that the eigenvalues are $\lambda = 1$, 3, and 4. Corresponding eigenvectors are $[1 \;\; 2 \;\; 1]^t$, $[1 \;\; 0 \; -1]^t$, and $[1 \; -1 \;\; 1]^t$, respectively, so we let

$$P = \begin{bmatrix} 1 & 1 & 1 \\ 2 & 0 & -1 \\ 1 & -1 & 1 \end{bmatrix}$$

Hence

$$P^{-1} = \begin{bmatrix} 1/6 & 1/3 & 1/6 \\ 1/2 & 0 & -1/2 \\ 1/3 & -1/3 & 1/3 \end{bmatrix}$$

and therefore

$$A^n = \begin{bmatrix} 1 & 1 & 1 \\ 2 & 0 & -1 \\ 1 & -1 & 1 \end{bmatrix} \begin{bmatrix} 1^n & 0 & 0 \\ 0 & 3^n & 0 \\ 0 & 0 & 4^n \end{bmatrix} \begin{bmatrix} 1/6 & 1/3 & 1/6 \\ 1/2 & 0 & -1/2 \\ 1/3 & -1/3 & 1/3 \end{bmatrix}$$

19. **(a)** By Theorem 6.2.3, we can diagonalize A if it has two distinct eigenvalues. By Exercise 18 of Section 6.1, this will occur if and only if $(a - d)^2 + 4bc > 0$.

(b) See Part (a). If the discriminant, $(a - d)^2 + 4bc$, of the characteristic equation is negative, then there are no real roots and hence no real eigenvalues or eigenvectors. Thus, by Theorem 6.2.1, A is not diagonalizable.

21. Since A is diagonalizable, we have $A = PDP^{-1}$ where D is a diagonal matrix with the eigenvalues of A as its diagonal elements. Thus the rank of D is the number of nonzero eigenvalues of A. Since P and P^{-1} are invertible, they are products of elementary matrices. But multiplication on the left or on the right by an elementary matrix does not change the rank of the product. Therefore rank (A) = rank (D), and we are done.

EXERCISE SET 6.3

1. **(a)** The characteristic equation is $\lambda(\lambda - 5) = 0$. Thus each eigenvalue is repeated once and hence each eigenspace is 1-dimensional.

   (c) The characteristic equation is $\lambda^2(\lambda - 3) = 0$. Thus the eigenspace corresponding to $\lambda = 0$ is 2-dimensional and that corresponding to $\lambda = 3$ is 1-dimensional.

   (e) The characteristic equation is $\lambda^3(\lambda - 8) = 0$. Thus the eigenspace corresponding to $\lambda = 0$ is 3-dimensional and that corresponding to $\lambda = 8$ is 1-dimensional.

2. The eigenvalues of A are $\lambda = 4$ and $\lambda = 2$. The eigenspaces are spanned by the orthogonal vectors

$$\begin{bmatrix} 1 \\ 1 \end{bmatrix} \quad \text{and} \quad \begin{bmatrix} -1 \\ 1 \end{bmatrix}$$

respectively, If we normalize these vectors, we obtain

$$\begin{bmatrix} \frac{1}{\sqrt{2}} \\ \frac{1}{\sqrt{2}} \end{bmatrix} \quad \text{and} \quad \begin{bmatrix} \frac{-1}{\sqrt{2}} \\ \frac{1}{\sqrt{2}} \end{bmatrix}$$

Thus

$$P = \begin{bmatrix} \frac{1}{\sqrt{2}} & -\frac{1}{\sqrt{2}} \\ \frac{1}{\sqrt{2}} & \frac{1}{\sqrt{2}} \end{bmatrix}$$

and

$$P^{-1}AP = \begin{bmatrix} 4 & 0 \\ 0 & 2 \end{bmatrix}$$

If we interchange the order of the vectors which span the eigenspaces, we obtain

$$P = \begin{bmatrix} -\frac{1}{\sqrt{2}} & \frac{1}{\sqrt{2}} \\ \frac{1}{\sqrt{2}} & \frac{1}{\sqrt{2}} \end{bmatrix}$$

and

$$P^{-1}AP = \begin{bmatrix} 2 & 0 \\ 0 & 4 \end{bmatrix}$$

6. The eigenvalues of A are $\lambda = 2$ and $\lambda = 0$. The eigenspace associated with $\lambda = 2$ is spanned by $\begin{bmatrix} 1 \\ 1 \\ 0 \end{bmatrix}$; the eigenspace associated with $\lambda = 0$ is spanned by $\begin{bmatrix} 1 \\ -1 \\ 0 \end{bmatrix}$ and $\begin{bmatrix} 0 \\ 0 \\ 1 \end{bmatrix}$. If we normalize these three orthogonal vectors, we obtain

$$\begin{bmatrix} \frac{1}{\sqrt{2}} \\ \frac{1}{\sqrt{2}} \\ 0 \end{bmatrix} \quad \begin{bmatrix} \frac{1}{\sqrt{2}} \\ -\frac{1}{\sqrt{2}} \\ 0 \end{bmatrix} \quad \begin{bmatrix} 0 \\ 0 \\ 1 \end{bmatrix}$$

Thus

$$P = \begin{bmatrix} \frac{1}{\sqrt{2}} & \frac{1}{\sqrt{2}} & 0 \\ \frac{1}{\sqrt{2}} & -\frac{1}{\sqrt{2}} & 0 \\ 0 & 0 & 1 \end{bmatrix}$$

and

$$P^{-1}AP = \begin{bmatrix} 2 & 0 & 0 \\ 0 & 0 & 0 \\ 0 & 0 & 0 \end{bmatrix}$$

Another possibility (which results from changing the order of the spanning vectors) is

$$P = \begin{bmatrix} \frac{1}{\sqrt{2}} & 0 & \frac{1}{\sqrt{2}} \\ -\frac{1}{\sqrt{2}} & 0 & \frac{1}{\sqrt{2}} \\ 0 & 1 & 0 \end{bmatrix}$$

so that

$$P^{-1}AP = \begin{bmatrix} 0 & 0 & 0 \\ 0 & 0 & 0 \\ 0 & 0 & 2 \end{bmatrix}$$

8. The eigenvalues of A are $\lambda = 0$, $\lambda = 4$, and $\lambda = 2$. The eigenspace associated with $\lambda = 0$ is spanned by $\begin{bmatrix} 0 \\ 0 \\ 1 \\ 0 \end{bmatrix}$ and $\begin{bmatrix} 0 \\ 0 \\ 0 \\ 1 \end{bmatrix}$; the eigenspaces associated with $\lambda = 4$ and $\lambda = 2$ are spanned by $\begin{bmatrix} 1 \\ 1 \\ 0 \\ 0 \end{bmatrix}$ and $\begin{bmatrix} 1 \\ -1 \\ 0 \\ 0 \end{bmatrix}$ respectively. Hence, the eigenspaces of A are spanned by the orthogonal vectors

$$\begin{bmatrix} 0 \\ 0 \\ 1 \\ 0 \end{bmatrix} \quad \begin{bmatrix} 0 \\ 0 \\ 0 \\ 1 \end{bmatrix} \quad \begin{bmatrix} 1 \\ 1 \\ 0 \\ 0 \end{bmatrix} \quad \begin{bmatrix} 1 \\ -1 \\ 0 \\ 0 \end{bmatrix}$$

Normalizing each of these vectors yields

$$\begin{bmatrix} 0 \\ 0 \\ 1 \\ 0 \end{bmatrix} \quad \begin{bmatrix} 0 \\ 0 \\ 0 \\ 1 \end{bmatrix} \quad \begin{bmatrix} 1/\sqrt{2} \\ 1/\sqrt{2} \\ 0 \\ 0 \end{bmatrix} \quad \begin{bmatrix} 1/\sqrt{2} \\ -1/\sqrt{2} \\ 0 \\ 0 \end{bmatrix}$$

Thus

$$P = \begin{bmatrix} 0 & 0 & 1/\sqrt{2} & 1/\sqrt{2} \\ 0 & 0 & 1/\sqrt{2} & -1/\sqrt{2} \\ 1 & 0 & 0 & 0 \\ 0 & 1 & 0 & 0 \end{bmatrix}$$

$$P^{-1} = \begin{bmatrix} 0 & 0 & 1 & 0 \\ 0 & 0 & 0 & 1 \\ 1/\sqrt{2} & 1/\sqrt{2} & 0 & 0 \\ 1/\sqrt{2} & -1/\sqrt{2} & 0 & 0 \end{bmatrix}$$

and

$$P^{-1}AP = \begin{bmatrix} 0 & 0 & 0 & 0 \\ 0 & 0 & 0 & 0 \\ 0 & 0 & 4 & 0 \\ 0 & 0 & 0 & 2 \end{bmatrix}$$

Changing the order of the spanning vectors may change the position of the eigenvalues on the diagonal of $P^{-1}AP$.

10. The characteristic equation is $\lambda^2 - 2a\lambda + a^2 - b^2 = 0$. By the quadratic formula, $\lambda = a \pm |b|$, or $\lambda = a \pm b$. Since $b \neq 0$, there are 2 distinct eigenvalues with eigenspaces spanned by the orthogonal vectors

$$\begin{bmatrix} 1 \\ 1 \end{bmatrix} \quad \text{and} \quad \begin{bmatrix} -1 \\ 1 \end{bmatrix}$$

Thus

$$\begin{bmatrix} \frac{1}{\sqrt{2}} & -\frac{1}{\sqrt{2}} \\ \frac{1}{\sqrt{2}} & \frac{1}{\sqrt{2}} \end{bmatrix}$$

will orthogonally diagonalize the given matrix.

12. **(a)** By Theorem 6.3.1, it will suffice to show that $I - \mathbf{v}\mathbf{v}^t$ is symmetric. This follows from properties of the transpose.

12. **(b)** In this case,

$$I - \mathbf{v}\mathbf{v}^t = \begin{bmatrix} 1 & 0 & 0 \\ 0 & 1 & 0 \\ 0 & 0 & 1 \end{bmatrix} - \begin{bmatrix} 1 \\ 0 \\ 1 \end{bmatrix} [1 \quad 0 \quad 1]$$

$$= \begin{bmatrix} 0 & 0 & -1 \\ 0 & 1 & 0 \\ -1 & 0 & 0 \end{bmatrix}$$

The characteristic equation is $(\lambda - 1)(\lambda - 1)(\lambda + 1) = 0$, so the eigenvalues are $\lambda = \pm 1$. The eigenvectors corresponding to $\lambda = 1$ are $\begin{bmatrix} 1 \\ 0 \\ -1 \end{bmatrix}$ and $\begin{bmatrix} 0 \\ 1 \\ 0 \end{bmatrix}$. The eigenvector corresponding to $\lambda = -1$ is $\begin{bmatrix} 1 \\ 0 \\ 1 \end{bmatrix}$. Thus the matrix

$$P = \begin{bmatrix} 1/\sqrt{2} & 0 & 1/\sqrt{2} \\ 0 & 1 & 0 \\ -1/\sqrt{2} & 0 & 1/\sqrt{2} \end{bmatrix}$$

will orthogonally diagonalize $I - \mathbf{v}\mathbf{v}^t$.

13. Let A be a symmetric matrix and let B be orthogonally similar to A. Then $B = P^tAP$ and

$$\begin{aligned} B^t &= (P^tAP)^t \\ &= P^tA^t(P^t)^t \\ &= P^tAP \qquad (A \text{ is symmetric}) \\ &= B \end{aligned}$$

Hence B is also symmetric.

14. If A is a 2×2 symmetric matrix, then A has the form of the matrix in Exercise 10. In case $b \neq 0$, then, as in Exercise 10, there are two eigenspaces, each spanned by a single vector, and these vectors are orthogonal. Thus eigenvalues from different eigenspaces are orthogonal. In case $b = 0$, there is only one eigenvalue, $\lambda = a$, and thus only one eigenspace.

SUPPLEMENTARY EXERCISES 6

1. **(a)** The characteristic equation of A is $\lambda^2 - 2\cos\theta + 1 = 0$. The discriminant of this equation is $4(\cos^2\theta - 1)$, which is negative unless $\cos^2\theta = 1$. Thus A can have no real eigenvalues or eigenvectors in case $0 < \theta < \pi$.

2. **(a)** If

$$D = \begin{bmatrix} a_1 & 0 & \cdots & 0 \\ 0 & a_2 & \cdots & 0 \\ \vdots & \vdots & & \vdots \\ 0 & 0 & \cdots & a_n \end{bmatrix}$$

then $D = S^2$, where

$$S = \begin{bmatrix} \sqrt{a_1} & 0 & \cdots & 0 \\ 0 & \sqrt{a_2} & \cdots & 0 \\ \vdots & \vdots & & \vdots \\ 0 & 0 & \cdots & \sqrt{a_n} \end{bmatrix}$$

Of course, this makes sense only if $a_1 \geq 0, \ldots, a_n \geq 0$.

(b) If A is diagonalizable, then there are matrices P and D such that D is diagonal and $D = P^{-1}AP$. Moreover, if A has nonnegative eigenvalues, then the entries of D are nonnegative since they are all eigenvalues. Thus there is a matrix T, by virtue of Part (a), such that $D = T^2$. Therefore,

$$\begin{aligned} A &= PDP^{-1} \\ &= PT^2P^{-1} \\ &= PTP^{-1}PTP^{-1} \\ &= (PTP^{-1})^2 \end{aligned}$$

That is, if we let $S = PTP^{-1}$, then $A = S^2$.

3. **(c)** The eigenvalues of A are $\lambda = 9$, $\lambda = 1$, and $\lambda = 4$. The eigenspaces are spanned by the vectors

$$\begin{bmatrix} 1 \\ 2 \\ 2 \end{bmatrix} \quad \begin{bmatrix} 1 \\ 0 \\ 0 \end{bmatrix} \quad \begin{bmatrix} 1 \\ 1 \\ 0 \end{bmatrix}$$

Thus, we have

$$P = \begin{bmatrix} 1 & 1 & 1 \\ 2 & 0 & 1 \\ 2 & 0 & 0 \end{bmatrix} \quad \text{and} \quad P^{-1} = \begin{bmatrix} 0 & 0 & 1/2 \\ 1 & -1 & 1/2 \\ 0 & 1 & -1 \end{bmatrix}$$

while

$$D = \begin{bmatrix} 9 & 0 & 0 \\ 0 & 1 & 0 \\ 0 & 0 & 4 \end{bmatrix} \quad \text{and} \quad T = \begin{bmatrix} 3 & 0 & 0 \\ 0 & 1 & 0 \\ 0 & 0 & 2 \end{bmatrix}$$

Therefore

$$S = PTP^{-1} = \begin{bmatrix} 1 & 1 & 0 \\ 0 & 2 & 1 \\ 0 & 0 & 3 \end{bmatrix}$$

4. **(a)** Since $(\lambda I - A)^t = \lambda I - A^t$, we have

$$\begin{aligned} \det(\lambda I - A) &= \det((\lambda I - A)^t) \\ &= \det(\lambda I - A^t) \end{aligned}$$

Thus A and A^t have the same characteristic polynomials, and hence the same eigenvalues.

(b) By the result of Exercise 19, Section 6.1, if A has distinct eigenvalues λ_1 and λ_2, and if $bc \neq 0$, then the eigenvectors of A are

$$\begin{bmatrix} -b \\ a - \lambda_1 \end{bmatrix} \quad \text{and} \quad \begin{bmatrix} -b \\ a - \lambda_2 \end{bmatrix}$$

Corresponding eigenvectors of A^t are

$$\begin{bmatrix} -c \\ a - \lambda_1 \end{bmatrix} \quad \text{and} \quad \begin{bmatrix} -c \\ a - \lambda_2 \end{bmatrix}$$

Thus we need only look for a matrix with two distinct eigenvalues, with $bc \neq 0$, and with $b \neq c$. For instance, if we let

$$A = \begin{bmatrix} 1 & 1 \\ 3 & -1 \end{bmatrix}$$

then the resulting pairs of eigenvectors will be different.

5. Since $\det(\lambda I - A)$ is a sum of signed elementary products, we ask which terms involve λ^{n-1}. Obviously the signed elementary product

$$\begin{aligned} q &= (\lambda - a_{11})(\lambda - a_{22})\cdots(\lambda - a_{nn}) \\ &= \lambda^n - (a_{11} + a_{22} + \cdots + a_{nn})\lambda^{n-1} \\ &\qquad + \text{terms involving } \lambda^r \text{ where } r < n - 1 \end{aligned}$$

has a term $-$ (trace of A)λ^{n-1}. Any elementary product containing fewer than n-1 factors of the form $\lambda - a_{ii}$ cannot have a term which contains λ^{n-1}. But there is no elementary product which contains exactly n-1 of these factors (Why?). Thus the coefficient of λ^{n-1} is minus the trace of A.

6. The only eigenvalue is $\lambda = a$ and if $b \neq 0$, then the eigenspace is spanned by the vector $\begin{bmatrix} 1 \\ 0 \end{bmatrix}$. Thus, by Theorem 6.2.1, A is not diagonalizable.

7. **(b)** The characteristic equation is

$$p(\lambda) = -1 + 3\lambda - 3\lambda^2 + \lambda^3$$

Moreover,

$$A^2 = \begin{bmatrix} 0 & 0 & 1 \\ 1 & -3 & 3 \\ 3 & -8 & 6 \end{bmatrix}$$

and

$$A^3 = \begin{bmatrix} 1 & -3 & 3 \\ 3 & -8 & 6 \\ 6 & -15 & 10 \end{bmatrix}$$

It then follows that

$$p(A) = -I + 3A - 3A^2 + A^3 = 0$$

9. Since $c_0 = 0$ and $c_1 = -5$, we have $A^2 = 5A$, and, in general, $A^n = 5^{n-1}A$.

11. Call the matrix A and show that $A^2 = (c_1 + c_2 + \cdots + c_n)A = [\text{tr}(A)]A$. Thus $A^k = [\text{tr}(A)]^{k-1}A$ for $k = 2,3,\ldots$ Now if λ is an eigenvalue of A, then λ^k is an eigenvalue of A^k, so that in case $\text{tr}(A) \neq 0$, we have that $\lambda^k/[\text{tr}(A)]^{k-1} = [\lambda/\text{tr}(A)]^{k-1}\lambda$ is an eigenvalue of A for k = 2,3,... Why? We know that A has at most n eigenvalues, so that this expression can take on only finitely many values. This means that either $\lambda = 0$ or $\lambda = \text{tr}(A)$. Why? In case $\text{tr}(A) = 0$, then all of the eigenvalues of A are 0. Why? Thus the only possible eigenvalues of A are zero and $\text{tr}(A)$. It is easy to check that each of these is, in fact, an eigenvalue of A.

Alternatively, we could try to evaluate $\det(I\lambda - A)$ by brute force. If we add Column 1 to Column 2, the new Column 2 to Column 3, the new Column 3 to Column 4, and so on, we obtain the determinant

$$\det(I\lambda - A) = \det\begin{bmatrix} \lambda - c_1 & \lambda - c_1 - c_2 & \lambda - c_1 - c_2 - c_3 & \cdots & \lambda - c_1 - c_2 - \cdots - c_n \\ -c_1 & \lambda - c_1 - c_2 & \lambda - c_1 - c_2 - c_3 & \cdots & \lambda - c_1 - c_2 - \cdots - c_n \\ -c_1 & -c_1 - c_2 & \lambda - c_1 - c_2 - c_3 & \cdots & \lambda - c_1 - c_2 - \cdots - c_n \\ \vdots & \vdots & \vdots & & \vdots \\ -c_1 & -c_1 - c_2 & -c_1 - c_2 - c_3 & \cdots & \lambda - c_1 - c_2 - \cdots - c_n \end{bmatrix}$$

If we subtract Row 1 from each of the other rows and then expand by cofactors along the nth column, we have

$$\det(I\lambda - A) = \det\begin{bmatrix} \lambda - c_1 & \lambda - c_1 - c_2 & \lambda - c_1 - c_2 - c_3 & \cdots & \lambda - \operatorname{tr}(A) \\ -\lambda & 0 & 0 & \cdots & 0 \\ -\lambda & -\lambda & 0 & \cdots & 0 \\ \vdots & \vdots & \vdots & & \vdots \\ -\lambda & -\lambda & -\lambda & \cdots & 0 \end{bmatrix}$$

$$= (-1)^{n+1}(\lambda - \operatorname{tr}(A))\det\begin{bmatrix} -\lambda & 0 & 0 & \cdots & 0 \\ -\lambda & -\lambda & 0 & \cdots & 0 \\ -\lambda & -\lambda & -\lambda & \cdots & 0 \\ \vdots & \vdots & \vdots & & \vdots \\ -\lambda & -\lambda & -\lambda & \cdots & -\lambda \end{bmatrix}$$

$$= (-1)^{n+1}(\lambda - \operatorname{tr}(A))(-\lambda)^{n-1}$$ because the above matrix is triangular

$$= (-1)^{2n}(\lambda - \operatorname{tr}(A))\lambda^{n-1}$$

$$= \lambda^{n-1}(\lambda - \operatorname{tr}(A))$$

Thus $\lambda = \operatorname{tr}(A)$ and $\lambda = 0$ are the eigenvalues, with $\lambda = 0$ repeated $n - 1$ times.

12. **(a)** The characteristic polynomial of the given matrix is

$$p(\lambda) = \det\begin{bmatrix} \lambda & 0 & 0 & \cdots & 0 & c_1 \\ -1 & \lambda & 0 & \cdots & 0 & c_2 \\ 0 & -1 & \lambda & \cdots & 0 & c_3 \\ \vdots & \vdots & \vdots & & \vdots & \vdots \\ 0 & 0 & 0 & \cdots & -1 & \lambda + c_{n-1} \end{bmatrix}$$

If we add λ times the second row to the first row, this becomes

$$p(\lambda) = \det\begin{bmatrix} 0 & \lambda^2 & 0 & \cdots & 0 & c_1 + c_2\lambda \\ -1 & \lambda & 0 & \cdots & 0 & c_2 \\ 0 & -1 & \lambda & \cdots & 0 & c_3 \\ \vdots & \vdots & \vdots & & \vdots & \vdots \\ 0 & 0 & 0 & \cdots & -1 & \lambda + c_{n-1} \end{bmatrix}$$

Expanding by cofactors along the first column gives

$$p(\lambda) = \det\begin{bmatrix} \lambda^2 & 0 & 0 & \cdots & 0 & c_1 + c_2\lambda \\ -1 & \lambda & 0 & \cdots & 0 & c_3 \\ 0 & -1 & \lambda & \cdots & 0 & c_4 \\ \vdots & \vdots & \vdots & & \vdots & \vdots \\ 0 & 0 & 0 & \cdots & -1 & \lambda + c_{n-1} \end{bmatrix}$$

Adding λ^2 times the second row to the first row yields

$$p(\lambda) = \det\begin{bmatrix} 0 & \lambda^3 & 0 & \cdots & 0 & c_1 + c_2\lambda + c_3\lambda^2 \\ -1 & \lambda & 0 & \cdots & 0 & c_3 \\ 0 & -1 & \lambda & \cdots & 0 & c_4 \\ \vdots & \vdots & \vdots & & \vdots & \vdots \\ 0 & 0 & 0 & \cdots & -1 & \lambda + c_{n-1} \end{bmatrix}$$

Expanding by cofactors along the first column gives

$$p(\lambda) = \det\begin{bmatrix} \lambda^3 & 0 & 0 & \cdots & 0 & c_1 + c_2\lambda + c_3\lambda^2 \\ -1 & \lambda & 0 & \cdots & 0 & c_4 \\ 0 & -1 & \lambda & \cdots & 0 & c_5 \\ \vdots & \vdots & \vdots & & \vdots & \vdots \\ 0 & 0 & 0 & \cdots & -1 & \lambda + c_{n-1} \end{bmatrix}$$

Eventually this procedure will yield

$$p(\lambda) = \det \begin{bmatrix} \lambda^{n-2} & c_1 + c_2\lambda + \cdots + c_{n-2}\lambda^{n-3} \\ -1 & \lambda + c_{n-1} \end{bmatrix}$$

Adding λ^{n-2} times the second row to the first row gives

$$p(\lambda) = \det \begin{bmatrix} 0 & c_1 + c_2\lambda + \cdots + c_{n-1}\lambda^{n-2} + \lambda^{n-1} \\ -1 & \lambda + c_{n-1} \end{bmatrix}$$

or

$$p(\lambda) = c_1 + c_2\lambda + \cdots + c_{n-1}\lambda^{n-2} + \lambda^{n-1}$$

which is the desired result.

14. The characteristic equation of an $n \times n$ matrix has degree n and therefore n roots. Since the complex roots of a polynomial equation with real coefficients come in conjugate pairs, any such equation of odd degree must have at least one real root. Hence A has at least one real eigenvalue.

15. Suppose that $A\mathbf{x} = \lambda\mathbf{x}$ where A is invertible. Then

$$\mathbf{x} = A^{-1}A\mathbf{x} = A^{-1}\lambda\mathbf{x} = \lambda A^{-1}\mathbf{x}$$

or

$$A^{-1}\mathbf{x} = \frac{1}{\lambda}\,\mathbf{x}$$

That is, $1/\lambda$ is an eigenvalue of A^{-1} and $\mathbf{x}$ is a corresponding eigenvector.

16. Suppose that $A\mathbf{x} = \lambda\mathbf{x}$. Then

$$(A - sI)\mathbf{x} = A\mathbf{x} - sI\mathbf{x} = \lambda\mathbf{x} - s\mathbf{x} = (\lambda - s)\mathbf{x}$$

That is, $\lambda - s$ is an eigenvalue of $A - sI$ and $\mathbf{x}$ is a corresponding eigenvector.

17. The characteristic equation of A is $\lambda^3 - 6\lambda^2 + 11\lambda - 6 = 0$. The roots of this equation are $\lambda = 1, 2, 3$, so that these are the eigenvalues of A. If $\lambda = 1$, then $\lambda I - A$ reduces to

$$\begin{bmatrix} 1 & 0 & -1 \\ 0 & 1 & 0 \\ 0 & 0 & 0 \end{bmatrix}$$

Thus if $\mathbf{x} = [x_1 \quad x_2 \quad x_3]^t$ is an eigenvector of A corresponding to $\lambda = 1$, then $x_1 = x_3$ and $x_2 = 0$, so that

$$\mathbf{x} = \begin{bmatrix} r \\ 0 \\ r \end{bmatrix} = r\begin{bmatrix} 1 \\ 0 \\ 1 \end{bmatrix}$$

If $\lambda = 2$, then $\lambda I - A$ reduces to

$$\begin{bmatrix} 1 & -1/2 & 0 \\ 0 & 0 & 1 \\ 0 & 0 & 0 \end{bmatrix}$$

Thus any eigenvector $\mathbf{x}$ corresponding to $\lambda = 2$ is of the form

$$\mathbf{x} = \begin{bmatrix} s \\ s/2 \\ 0 \end{bmatrix} = s\begin{bmatrix} 1 \\ 1/2 \\ 0 \end{bmatrix}$$

If $\lambda = 3$, then $\lambda I - A$ reduces to

$$\begin{bmatrix} 1 & 0 & -1 \\ 0 & 1 & -1 \\ 0 & 0 & 0 \end{bmatrix}$$

Thus any eigenvector **x** corresponding to $\lambda = 2$ is of the form

$$\mathbf{x} = \begin{bmatrix} t \\ t \\ t \end{bmatrix} = t\begin{bmatrix} 1 \\ 1 \\ 1 \end{bmatrix}$$

17. **(a)** By Exercise 15, the eigenvalues of A^{-1} are 1, 1/2, and 1/3. The corresponding eigenspaces are spanned by the vectors.

$$\begin{bmatrix} 1 \\ 0 \\ 1 \end{bmatrix} \quad \begin{bmatrix} 1 \\ 1/2 \\ 0 \end{bmatrix} \quad \begin{bmatrix} 1 \\ 1 \\ 1 \end{bmatrix}$$

respectively.

18. Since every odd power of A is again A, we have that every odd power of an eigenvalue of A is again an eigenvalue of A. Thus the only possible eigenvalues of A are $\lambda = 0, \pm 1$.

EXERCISE SET 7.1

1. **(a)** Since

$$\begin{aligned} F((x_1,y_1) + (x_2,y_2)) &= (2(x_1 + x_2),\ y_1 + y_2) \\ &= (2x_1,y_1) + (2x_2,y_2) \\ &= F(x_1,y_1) + F(x_2,y_2) \end{aligned}$$

and

$$\begin{aligned} F(k(x,y)) &= F(kx,ky) = (2kx,ky) \\ &= k(2x,y) = kF(x,y) \end{aligned}$$

F is linear.

(b) Since

$$\begin{aligned} F((x_1,y_1) + (x_2,y_2)) &= \left(\left[x_1 + x_2\right]^2,\ y_1 + y_2\right) \\ &= (x_1^2,y_1) + (2x_1x_2,0) + (x_2^2,y_2) \\ &= F(x_1,y_1) + F(x_2,y_2) + (2x_1x_2,0) \\ &\neq F(x_1,y_1) + F(x_2,y_2) \text{ if } x_1x_2 \neq 0 \end{aligned}$$

and

$$\begin{aligned} F(k(x,y)) &= (k^2x^2,ky) = k(kx^2,y) \\ &\neq kF(x,y) \text{ if } k \neq 1 \text{ or } x \neq 0 \end{aligned}$$

F is nonlinear.

2. **(a)** Since

$$\begin{aligned} F((x_1,y_1) + (x_2,y_2)) &= (2(x_1 + x_2) + (y_1 + y_2),\ (x_1 + x_2) - (y_1 + y_2)) \\ &= (2x_1 + y_1,\ x_1 - y_1) + (2x_2 + y_2,\ x_2 - y_2) \\ &= F(x_1,y_1) + F(x_2,y_2) \end{aligned}$$

and

$$\begin{aligned} F(k(x,y)) &= (2kx + ky,\ kx - ky) \\ &= k(2x + y,\ x - y) = kF(x,y) \end{aligned}$$

F is linear.

(b) Since

$$\begin{aligned} F((x_1,y_1) + (x_2,y_2)) &= (x_1 + x_2 + 1,\ y_1 + y_2) \\ &= (\ x_1 + 1,\ y_1) + (x_2,y_2) \\ &\neq F(x_1,y_1) + F(x_2,y_2) \end{aligned}$$

and $F(k(x,y)) = (kx + 1,\ ky) \neq kF(x,y)$ unless $k = 1$, F is nonlinear.

3. **(a)** Since

$$\begin{aligned} F((x_1,y_1,z_1) + (x_2,y_2,z_2)) &= (x_1 + x_2,\ x_1 + x_2 + y_1 + y_2 + z_1 + z_2) \\ &= (x_1,\ x_1 + y_1 + z_1) + (x_2,\ x_2 + y_2 + z_2) \\ &= F(x_1,y_1,z_1) + F(x_2,y_2,z_2) \end{aligned}$$

and

$$\begin{aligned} F(k(x,y,z)) &= (kx,\ kx + ky + kz) \\ &= kF(x,y,z) \end{aligned}$$

F is linear.

(b) Since

$$F(\mathbf{u} + \mathbf{v}) = (1,1) \neq F(\mathbf{u}) + F(\mathbf{v}) = (2,2)$$

and

$$F(k\mathbf{u}) = (1,1) \neq kF(\mathbf{u}) = (k,k) \text{ if } k \neq 1$$

F is nonlinear.

5. **(b)** Since both properties fail, F is nonlinear. For instance,

$$F\left(k\begin{bmatrix} a & b \\ c & d \end{bmatrix}\right) = \det\begin{bmatrix} ka & kb \\ kc & kd \end{bmatrix} = k^2(ad - bc)$$

$$\neq kF\left(\begin{bmatrix} a & b \\ c & d \end{bmatrix}\right) \text{ in general.}$$

6. **(b)** Since both properties fail, F is nonlinear. For instance,

$$F\left(k\begin{bmatrix} a & b \\ c & d \end{bmatrix}\right) = k^2a^2 + k^2b^2$$

$$\neq kF\left(\begin{bmatrix} a & b \\ c & d \end{bmatrix}\right) \text{ in general.}$$

7. **(b)** Since $F\left(\left[a_0 + a_1x + a_2x^2\right] + \left[b_0 + b_1x + b_2x^2\right]\right)$

$$= (a_0 + b_0) + (a_1 + b_1)(x + 1) + (a_2 + b_2)(x + 1)^2$$

$$= F\left(a_0 + a_1x + a_2x^2\right) + F\left(b_0 + b_1x + b_2x^2\right)$$

and

$$F\left(k\left[a_0 + a_1x + a_2x^2\right]\right) = ka_0 + ka_1(x + 1) + ka_2(x + 1)^2$$

$$= kF\left(a_0 + a_1x + a_2x^2\right)$$

F is linear.

10. We observe that

$$T(A_1 + A_2) = (A_1 + A_2)B = A_1B + A_2B = T(A_1) + T(A_2)$$

and

$$T(kA) = (kA)B = k(AB) = kT(A)$$

Hence, T is linear.

11. **(a)** We have

$$T\left(\begin{bmatrix} 1 \\ 3 \\ 8 \end{bmatrix}\right) = T\left(\begin{bmatrix} 1 \\ 0 \\ 0 \end{bmatrix}\right) + 3T\left(\begin{bmatrix} 0 \\ 1 \\ 0 \end{bmatrix}\right) + 8T\left(\begin{bmatrix} 0 \\ 0 \\ 1 \end{bmatrix}\right)$$

$$= \begin{bmatrix} 1 \\ 1 \end{bmatrix} + 3\begin{bmatrix} 3 \\ 0 \end{bmatrix} + 8\begin{bmatrix} 4 \\ -7 \end{bmatrix}$$

$$= \begin{bmatrix} 42 \\ -55 \end{bmatrix}$$

(b) Similarly,

$$T\left(\begin{bmatrix} x \\ y \\ z \end{bmatrix}\right) = x\begin{bmatrix} 1 \\ 1 \end{bmatrix} + y\begin{bmatrix} 3 \\ 0 \end{bmatrix} + z\begin{bmatrix} 4 \\ -7 \end{bmatrix}$$

$$= \begin{bmatrix} x + 3y + 4z \\ x + 0y - 7z \end{bmatrix}$$

(c) Therefore

$$A = \begin{bmatrix} 1 & 3 & 4 \\ 1 & 0 & -7 \end{bmatrix}$$

12. (a) This is similar to Example 8. The vectors $\mathbf{w}_1 = (1,0,0)$ and $\mathbf{w}_2 = (0,0,1)$ form an orthonormal basis for the xz-plane. If we let $\mathbf{v} = (x,y,z)$, then

$$T(x,y,z) = (\mathbf{v} \cdot \mathbf{w}_1)\mathbf{w}_1 + (\mathbf{v} \cdot \mathbf{w}_2)\mathbf{w}_2 = (x,0,z)$$

13. (a) The vectors $(1,-1,0)$ and $(0,1,-1)$ form a basis for W since they both lie in the plane and are linearly independent. If we apply the Gram-Schmidt process, we obtain the orthonormal basis $\{\mathbf{w}_1, \mathbf{w}_2\} = \{(1/\sqrt{2}, -1/\sqrt{2}, 0), (1/\sqrt{6}, 1/\sqrt{6}, -2/\sqrt{6})\}$. Then, as in Example 7, if $\mathbf{v} = (x,y,z)$, we have

$$T(\mathbf{v}) = (\mathbf{v} \cdot \mathbf{w}_1)\mathbf{w}_1 + (\mathbf{v} \cdot \mathbf{w}_2)\mathbf{w}_2$$

or

$$\begin{aligned} T(x,y,z) &= \frac{x - y}{2}(1,-1,0) + \frac{x + y - 2z}{6}(1,1,-2) \\ &= \left(\frac{2x - y - z}{3}, \frac{-x + 2y - z}{3}, \frac{-x - y + 2z}{3}\right) \end{aligned}$$

14. (a) Let $\theta = \frac{\pi}{4}$ in the matrix of Example 2. Then

$$A = \begin{bmatrix} \cos\frac{\pi}{4} & -\sin\frac{\pi}{4} \\ \sin\frac{\pi}{4} & \cos\frac{\pi}{4} \end{bmatrix} = \begin{bmatrix} 1/\sqrt{2} & -1/\sqrt{2} \\ 1/\sqrt{2} & 1/\sqrt{2} \end{bmatrix}$$

Hence

$$T(x,y) = A\begin{bmatrix} x \\ y \end{bmatrix} = \begin{bmatrix} (1/\sqrt{2})x - (1/\sqrt{2})y \\ (1/\sqrt{2})x + (1/\sqrt{2})y \end{bmatrix}$$

and so

$$T(-1,2) = (-3/\sqrt{2},\ 1/\sqrt{2})$$

15. **(c)** Since the vectors $\mathbf{v}_1$ and $\mathbf{v}_2$ are orthonormal, we have

$$T(x,y,z) = \left[\frac{1}{3}x + \frac{2}{3}y + \frac{2}{3}z\right]\mathbf{v}_1 + \left[-\frac{1}{\sqrt{2}}y + \frac{1}{\sqrt{2}}z\right]\mathbf{v}_2$$

$$= \left[\frac{1}{9}x + \frac{2}{9}y + \frac{2}{9}z,\ \frac{2}{9}x + \frac{17}{18}y - \frac{1}{18}z,\ \frac{2}{9}x - \frac{1}{18}y + \frac{17}{18}z\right]$$

17. **(a)** Since T_1 is defined on all of R^2, the domain of $T_2 \circ T_1$ is R^2. We have $T_2 \circ T_1(x,y) = T_2(T_1(x,y)) = T_2(2x,3y) = (2x - 3y,\ 2x + 3y)$. Since the system of equations

$$2x - 3y = a$$

$$2x + 3y = b$$

can be solved for all values of a and b, the image space is also all of R^2.

18. **(b)** Since T_1 is defined on all of R^2, the domain of $T_2 \circ T_1$ is R^2. We have

$$T_2\ (T_1(x,y)) = T_2(x - y,\ y + z,\ x - z) = (0,2x)$$

Thus the image space of $T_2 \circ T_1$ is the y-axis.

19. **(a)** Since T_1 is defined on all of R^2, the domain of $T_3 \circ T_2 \circ T_1$ is R^2. We have

$$\begin{aligned} T_3 \circ T_2 \circ T_1(x,y) &= T_3 \circ T_2(-2y,\ 3x,\ x-2y) \\ &= T_3(3x,\ x-2y,\ -2y) \\ &= (3x-2y,\ x) \end{aligned}$$

This vector can assume all possible values in R^2, so the image space is R^2.

21. We have

$$\begin{aligned} T_2 \circ T_1\left[a_0 + a_1x + a_2x^2\right] &= T_2\left[a_0 + a_1(x+1) + a_2(x+1)^2\right] \\ &= T_2\left[(a_0 + a_1 + a_2) + (a_1 + 2a_2)x + a_2x^2\right] \\ &= (a_0 + a_1 + a_2)x + (a_1 + 2a_2)x^2 + a_2x^3 \end{aligned}$$

23. We have

$$\begin{aligned} T(2\mathbf{v}_1 - 3\mathbf{v}_2 + 4\mathbf{v}_3) &= 2T(\mathbf{v}_1) - 3T(\mathbf{v}_2) + 4T(\mathbf{v}_3) \\ &= (2,-2,4) + (0,-9,-6) + (-12,4,8) \\ &= (-10,-7,6) \end{aligned}$$

25. **(b)** We have

$$\begin{aligned} (T_3 \circ T_2) \circ T_1(\mathbf{v}) &= (T_3 \circ T_2)(T_1(\mathbf{v})) \\ &= T_3(T_2(T_1(\mathbf{v}))) \\ &= T_3 \circ T_2 \circ T_1(\mathbf{v}) \end{aligned}$$

26. Since (1,0,0) and (0,1,0) form an orthonormal basis for the xy-plane, we have $T(x,y,z) = (x,0,0) + (0,y,0) = (x,y,0)$, which can also be arrived at by inspection. Then $T(T(x,y,z)) = T(x,y,0) = (x,y,0) = T(x,y,z)$. This says that T leaves every point in the x-y plane unchanged.

27. **(a)** From the linearity of T and the definition of kT, we have

$$(kT)(\mathbf{u} + \mathbf{v}) = k(T(\mathbf{u} + \mathbf{v})) = k(T(\mathbf{u}) + T(\mathbf{v}))$$

$$= kT(\mathbf{u}) + kT(\mathbf{v}) = (kT)(\mathbf{u}) + (kT)(\mathbf{v})$$

and

$$(kT)(k'\mathbf{u}) = k(T(k'\mathbf{u})) = kk'T(\mathbf{u}) = k'(kT)(\mathbf{u})$$

Therefore, kT is linear.

29. Following the hint, we have $T(\mathbf{u}) = \mathbf{v}_0 \times \mathbf{u}$. By properties of the cross product,

$$T(\mathbf{u} + \mathbf{v}) = \mathbf{v}_0 \times (\mathbf{u} + \mathbf{v}) = (\mathbf{v}_0 \times \mathbf{u}) + (\mathbf{v}_0 \times \mathbf{v}) = T(\mathbf{u}) + T(\mathbf{v})$$

and

$$T(k\mathbf{u}) = \mathbf{v}_0 \times (k\mathbf{u}) = k(\mathbf{v}_0 \times \mathbf{u}) = kT(\mathbf{u})$$

Thus T is linear.

31. **(b)** We have

$$(J \circ D)(\sin x) = \int_0^1 (\sin x)'\,dx = \sin(1) - \sin(0) = \sin(1)$$

32. Write $\mathbf{u} - \mathbf{v}$ as $\mathbf{u} + (-1)\mathbf{v}$. Since T is linear,

$$T(\mathbf{u} - \mathbf{v}) = T(\mathbf{u}) + T((-1)\mathbf{v}) = T(\mathbf{u}) - T(\mathbf{v})$$

34. Let $\mathbf{v} = a_1\mathbf{v}_1 + \cdots + a_n\mathbf{v}_n$ be any vector in V. Then

$$\begin{aligned} T(\mathbf{v}) &= a_1 T(\mathbf{v}_1) + \cdots + a_n T(\mathbf{v}_n) \\ &= a_1\mathbf{v}_1 + \cdots + a_n\mathbf{v}_n \\ &= \mathbf{v} \end{aligned}$$

Hence T is the identity transformation on V.

EXERCISE SET 7.2

1. **(a)** If (1,–4) is in $R(T)$, then there must be a vector (x,y) such that $T(x,y) = (2x - y, -8x + 4y) = (1,-4)$. If we equate components, we find that $2x - y = 1$ or $y = t$ and $x = (1 + t)/2$. Thus T maps infinitely many vectors into (1,–4).

   (b) Proceeding as above, we obtain the system of equations

$$\begin{aligned} 2x - y &= 5 \\ -8x + 4y &= 0 \end{aligned}$$

   Since $2x - y = 5$ implies that $-8x + 4y = -20$, this system has no solution. Hence (5,0) is not in $R(T)$.

2. **(a)** The vector (5,10) is in $\ker(T)$ since

$$T(5,10) = (2(5) - 10, -8(5) + 4(10)) = (0,0)$$

   (b) The vector (3,2) is not in $\ker(T)$ since $T(3,2) = (4,-16) \neq (0,0)$.

3. **(b)** The vector (1,3,0) is in $R(T)$ if and only if the following system of equations has a solution:

$$\begin{aligned} 4x + y - 2z - 3w &= 1 \\ 2x + y + z - 4w &= 3 \\ 6x - 9z + 9w &= 0 \end{aligned}$$

   This system has infinitely many solutions $x = (3/2)(t - 1)$, $y = 10 - 4t$, $z = t$, $w = 1$ where t is arbitrary. Thus (1,3,0) is in $R(T)$.

4. **(b)** Since $T(0,0,0,1) = (-3,-4,9) \neq (0,0,0)$, the vector $(0,0,0,1)$ is not in $\ker(T)$.

5. **(a)** Since $T(x^2) = x^3 \neq 0$, the polynomial x^2 is not in $\ker(T)$.

6. **(a)** Since $T(1 + x) = x + x^2$, the polynomial $x + x^2$ is in $R(T)$.

   (c) Since there doesn't exist a polynomial $p(x)$ such that $xp(x) = 3 - x^2$, then $3 - x^2$ is not in $R(T)$.

7. **(a)** We look for conditions on x and y such that $2x - y = 0$ and $-8x + 4y = 0$. Since these equations are satisfied if and only if $y = 2x$, the kernel will be spanned by the vector $(1,2)$, which is then a basis.

   (c) Since the only vector which is mapped to zero is the zero vector, the kernel is $\{\mathbf{0}\}$ and has dimension zero and hence no basis.

8. **(a)** Since $-8x + 4y = -4(2x - y)$ and $2x - y$ can assume any real value, the range of T is the set of vectors $(x,-4x)$ or the line $y = -4x$. The vector $(1,-4)$ is a basis for this space.

   (c) The range of T is just the set of all polynomials in P_3 with constant term zero. The set $\{x,x^2,x^3\}$ is a basis for this space.

9. **(a)** Here $n = \dim(R^2) = 2$, $\text{rank}(T) = 1$ by Exercise 8(a), and $\text{nullity}(T) = 1$ by Exercise 7(a). Recall that $1 + 1 = 2$.

   (c) Here $n = \dim(P_2) = 3$, $\text{rank}(T) = 3$ by Exercise 8(c), and $\text{nullity}(T) = 0$ by Exercise 7(c). Thus we have $3 = 3 + 0$.

10. **(a)** We know from Theorem 7.2.2 that the range of T is the column space of A. Using elementary column operations, we reduce A to the matrix

$$\begin{bmatrix} 1 & 0 & 0 \\ 5 & 1 & 0 \\ 7 & 1 & 0 \end{bmatrix} \quad \text{or} \quad \begin{bmatrix} 1 & 0 & 0 \\ 0 & 1 & 0 \\ 2 & 1 & 0 \end{bmatrix}$$

Thus $\begin{bmatrix} 1 \\ 5 \\ 7 \end{bmatrix}$ and $\begin{bmatrix} 0 \\ 1 \\ 1 \end{bmatrix}$ as well as $\begin{bmatrix} 1 \\ 0 \\ 2 \end{bmatrix}$ and $\begin{bmatrix} 0 \\ 1 \\ 1 \end{bmatrix}$ form a basis for the column space and therefore also for the range of T.

(b) To investigate the solution space of A, we look for conditions on x, y, and z such that

$$\begin{bmatrix} 1 & -1 & 3 \\ 5 & 6 & -4 \\ 7 & 4 & 2 \end{bmatrix} \begin{bmatrix} x \\ y \\ z \end{bmatrix} = \begin{bmatrix} 0 \\ 0 \\ 0 \end{bmatrix}$$

The solution of the resulting system of equations is $x = -14t$, $y = 19t$, and $z = 11t$ where t is arbitrary. Thus one basis for $\ker(T)$ is the vector $\begin{bmatrix} -14 \\ 19 \\ 11 \end{bmatrix}$.

(c) By parts (a) and (b), we have $\text{rank}(T) = 2$ and $\text{nullity}(T) = 1$.

(d) Since $\text{rank}(A) = 2$ and A has 3 columns, $\text{nullity}(A) = 3 - 2 = 1$.

12. **(a)** Since the range of T is the column space of A, we reduce A using column operations. This yields

$$\begin{bmatrix} 1 & 0 & 0 & 0 \\ 0 & 1 & 0 & 0 \end{bmatrix}$$

Thus $\begin{bmatrix} 1 \\ 0 \end{bmatrix}$ and $\begin{bmatrix} 0 \\ 1 \end{bmatrix}$ form a basis for the range of T.

(b) If

$$\begin{bmatrix} 4 & 1 & 5 & 2 \\ 1 & 2 & 3 & 0 \end{bmatrix}\begin{bmatrix} x \\ y \\ z \\ w \end{bmatrix} = \begin{bmatrix} 0 \\ 0 \end{bmatrix}$$

then $x = -s - 4r$, $y = -s + 2r$, $z = s$, and $w = 7r$ where s and r are arbitrary. That is,

$$\begin{bmatrix} x \\ y \\ z \\ w \end{bmatrix} = \begin{bmatrix} -1 \\ -1 \\ 1 \\ 0 \end{bmatrix} s + \begin{bmatrix} -4 \\ 2 \\ 0 \\ 7 \end{bmatrix} r$$

Thus, the set $\{(-1,-1,1,0)^t, (-4,2,0,7)^t\}$ is one basis for $\ker(T)$.

(c) By Parts (a) and (b), $\text{rank}(T) = \text{nullity}(T) = 2$.

(d) Since A has 4 columns and $\text{rank}(A) = 2$, we have $\text{nullity}(A) = 4 - 2 = 2$.

14. **(a)** The orthogonal projection on the xz-plane maps R^3 to the entire xz-plane. Hence its range is the xz-plane. It maps only the y-axis to $(0,0,0)$, so that its nullspace is the y-axis.

16. **(a)** The nullity of T is $5 - 3 = 2$.

(c) The nullity of T is $6 - 3 = 3$.

18. **(a)** The dimension of the solution space of $A\mathbf{x} = \mathbf{0}$ is, by the Dimension Theorem, $7 - 4 = 3$.

(b) Since the range space of A has dimension 4, the range cannot be R^5, which has dimension 5.

19. By Theorem 7.2.1, the kernel of T is a subspace of R^3. Now refer to Exercise 24 of Section 4.5.

21. **(a)** If

$$\begin{bmatrix} 1 & 3 & 4 \\ 3 & 4 & 7 \\ -2 & 2 & 0 \end{bmatrix} \begin{bmatrix} x \\ y \\ z \end{bmatrix} = \begin{bmatrix} 0 \\ 0 \\ 0 \end{bmatrix}$$

then $x = -t$, $y = -t$, $z = t$. These are parametric equations for a line through the origin.

(b) Using elementary column operations, we reduce the given matrix to

$$\begin{bmatrix} 1 & 0 & 0 \\ 3 & -5 & 0 \\ -2 & 8 & 0 \end{bmatrix}$$

Thus, $(1,3,-2)^t$ and $(0,-5,8)^t$ form a basis for the range space. That range space, which we can interpret as a subspace of R^3, is a plane through the origin. To find a normal to that plane, we compute

$$(1,3,-2) \times (0,-5,8) = (14,-8,-5)$$

Therefore, an equation for the plane is

$$14x - 8y - 5z = 0$$

Alternatively, but more painfully, we can use elementary row operations to reduce the matrix

$$\begin{bmatrix} 1 & 3 & 4 & x \\ 3 & 4 & 7 & y \\ -2 & 2 & 0 & z \end{bmatrix}$$

to the matrix

$$\begin{bmatrix} 1 & 0 & 1 & (-4x+3y)/5 \\ 0 & 1 & 1 & (3x-y)/5 \\ 0 & 0 & 0 & 14x-8y-5z \end{bmatrix}$$

Thus the vector $(x,y,z)^t$ is in the range of T if and only if $14x - 8y - 5z = 0$.

22. Suppose that $\mathbf{v}$ is any vector in V and write

(*) $$\mathbf{v} = a_1\mathbf{v}_1 + a_2\mathbf{v}_2 + \cdots + a_n\mathbf{v}_n$$

Now define a transformation $T: V \longrightarrow W$ by

$$T(\mathbf{v}) = a_1\mathbf{w}_1 + a_2\mathbf{w}_2 + \cdots + a_n\mathbf{w}_n$$

Note that T is well-defined because, by Theorem 4.5.1, the constants $a_1, a_2, \ldots, a_n$ in (*) are unique.

In order to complete the problem, you must show that (i) T is linear and (ii) $T(\mathbf{v}_i) = \mathbf{w}_i$ for $i = 1,2,\ldots,n$.

23. **(a)** Suppose that $\{\mathbf{v}_1, \mathbf{v}_2, \ldots, \mathbf{v}_n\}$ is a basis for V and let

(*) $$\mathbf{v} = a_1\mathbf{v}_1 + a_2\mathbf{v}_2 + \cdots + a_n\mathbf{v}_n$$

be an arbitrary vector in V. Since T is linear, we have

(**) $$T(\mathbf{v}) = a_1T(\mathbf{v}_1) + \cdots + a_nT(\mathbf{v}_n)$$

The hypothesis that $\dim(\ker(T)) = 0$ implies that $T(\mathbf{v}) = \mathbf{0}$ if and only if $\mathbf{v} = \mathbf{0}$. Since $\{\mathbf{v}_1, \ldots, \mathbf{v}_n\}$ is a linearly independent set, by (*), $\mathbf{v} = \mathbf{0}$ if and only if

$$a_1 = a_2 = \cdots = a_n = 0$$

Thus, by (**), $T(\mathbf{v}) = \mathbf{0}$ if and only if

$$a_1 = a_2 = \cdots = a_n = 0$$

That is, the vectors $T(\mathbf{v}_1)$, $T(\mathbf{v}_2), \ldots,$ $T(\mathbf{v}_n)$ form a linearly independent set in $R(T)$. Since they also span $R(T)$, then $\dim(R(T)) = n$.

24. If $R(T) = V$, then, by the dimension theorem, $\text{nullity}(T) = 0$. Thus, $\ker(T) = \{0\}$. Conversely, if $\ker(T) = \{0\}$, then $\text{nullity}(T) = 0$, so that $\text{rank}(T) = \dim(V)$. But the only subspace of V which has the same dimension as V is V itself. (Why?) Thus $R(T) = V$.

EXERCISE SET 7.3

1. **(a)** Clearly $\ker(T) = \{(0,0)\}$, so T is one-to-one.

   (c) Since $T(x,y) = (0,0)$ if and only if $x = y$ and $x = -y$, the kernel is $(0,0)$ and T is one-to-one.

   (e) Here $T(x,y) = (0,0,0)$ if and only if x and y satisfy the equations $x - y = 0$, $-x + y = 0$, and $2x - 2y = 0$. That is, (x,y) is in $\ker(T)$ if and only if $x = y$, so the kernel of T is this line and T is not one-to-one.

2. **(a)** Since $\det(A) = 1 \neq 0$, then T has an inverse. By direct calculation

$$A^{-1} = \begin{bmatrix} 1 & -2 \\ -2 & 5 \end{bmatrix}$$

   and so

$$T^{-1}\begin{bmatrix} x_1 \\ x_2 \end{bmatrix} = A^{-1}\begin{bmatrix} x_1 \\ x_2 \end{bmatrix} = \begin{bmatrix} x_1 - 2x_2 \\ -2x_1 + 5x_2 \end{bmatrix}$$

3. **(a)** Since $\det(A) = 0$, or equivalently, $\operatorname{rank}(A) < 3$, T has no inverse.

3. **(c)** Since A is invertible, we have

$$T^{-1}\begin{bmatrix} x_1 \\ x_2 \\ x_3 \end{bmatrix} = A^{-1}\begin{bmatrix} x_1 \\ x_2 \\ x_3 \end{bmatrix} = \begin{bmatrix} \frac{1}{2} & -\frac{1}{2} & \frac{1}{2} \\ -\frac{1}{2} & \frac{1}{2} & \frac{1}{2} \\ \frac{1}{2} & \frac{1}{2} & -\frac{1}{2} \end{bmatrix}\begin{bmatrix} x_1 \\ x_2 \\ x_3 \end{bmatrix}$$

$$= \begin{bmatrix} \frac{1}{2}\,(\ x_1 - x_2 + x_3) \\ \frac{1}{2}\,(-x_1 + x_2 + x_3) \\ \frac{1}{2}\,(\ x_1 + x_2 - x_3) \end{bmatrix}$$

4. **(a)** Since the kernel of this transformation is the line $x = 2y$, the transformation is not one-to-one.

   (c) Since the kernel of the transformation is $(0,0)$, the transformation is one-to-one.

5. **(a)** The kernel of T is the line $y = -x$ since all points on this line (and only those points) map to the origin.

   (b) Since the kernel is not $(0,0)$, the transformation is not one-to-one.

7. **(b)** Since $\text{nullity}(T) = n - \text{rank}(T) = 1$, T is not one-to-one.

   (c) Here T cannot be one-to-one since $\text{rank}(T) \leq n < m$, so $\text{nullity}(T) \geq 1$.

8. **(b)** If $p(x) = a_0 + a_1x + a_2x^2$, then

$$\begin{aligned} T(p(x)) &= a_0 + a_1(x + 1) + a_2(x + 1)^2 \\ &= (a_0 + a_1 + a_2) + (a_1 + 2a_2)x + a_2x^2 \end{aligned}$$

Thus $T(p(x)) = 0$ if and only if $a_2 = a_1 + 2a_2 = a_0 + a_1 + a_2 = 0$, or $a_2 = a_1 = a_0 = 0$. That is, the nullity of T is zero so T is one-to-one.

10. **(a)** Since infinitely many vectors $(0,0,\ldots,0,x_n)$ map to the zero vector, T is not one-to-one.

(c) Here T is one-to-one and $T^{-1}(x_1,x_2,\ldots,x_n) = (x_n,x_1,\ldots,x_{n-1})$.

11. **(a)** We know that T will have an inverse if and only if its kernel is the zero vector, which means if and only if none of the numbers $a_i = 0$.

12. **(a)** Since T_1 and T_2 are one-to-one, their kernels are the zero vector in V. Therefore, $T_2 \circ T_1$ also must have nullity zero, and so be one-to-one.

(b) We have

$$(T_2 \circ T_1) \circ \left[T_1^{-1} \circ T_2^{-1}\right](\mathbf{v}) = T_2\left[T_1\left[T_1^{-1}\left[T_2^{-1}(\mathbf{v})\right]\right]\right] = T_2\left[T_2^{-1}(\mathbf{v})\right] = \mathbf{v}$$

15. If $T(x,y) = (x + ky, -y) = (0,0)$, then $x = y = 0$. Hence $\ker(T) = \{(0,0)\}$ and therefore T is one-to-one. Since $T(T(x,y)) = T(x + ky, -y) = (x + ky + k(-y), -(-y)) = (x,y)$, we have $T^{-1} = T$.

16. Suppose that $\mathbf{w}_1$ and $\mathbf{w}_2$ are in $R(T)$. We must show that

$$T^{-1}(\mathbf{w}_1 + \mathbf{w}_2) = T^{-1}(\mathbf{w}_1) + T^{-1}(\mathbf{w}_2)$$

and

$$T^{-1}(k\mathbf{w}_1) = kT^{-1}(\mathbf{w}_1)$$

Because T is one-to-one, the above equalities will hold if and only if the results of applying T to both sides are indeed valid equalities. This follows immediately from the linearity of T.

17. Since the derivative of any constant function is zero, the kernel of D is not the zero polynomial. Thus, D is not one-to-one.

EXERCISE SET 7.4

2. Suppose that T is multiplication by the matrix

$$A = \begin{bmatrix} a & b & c \\ d & e & f \end{bmatrix}$$

Then

$$A\begin{bmatrix} 1 \\ 2 \\ 3 \end{bmatrix} = \begin{bmatrix} 1 \\ 0 \end{bmatrix} \qquad A\begin{bmatrix} 2 \\ 5 \\ 3 \end{bmatrix} = \begin{bmatrix} 1 \\ 0 \end{bmatrix} \qquad A\begin{bmatrix} 1 \\ 0 \\ 10 \end{bmatrix} = \begin{bmatrix} 0 \\ 1 \end{bmatrix}$$

Thus we have the system of equations

$$\begin{aligned} a + 2b + 3c \qquad\qquad\qquad\qquad & = 1 \\ d + 2e + 3f & = 0 \\ 2a + 5b + 3c \qquad\qquad\qquad\qquad & = 1 \\ 2d + 5e + 3f & = 0 \\ a \qquad + 10c \qquad\qquad\qquad\qquad & = 0 \\ d \qquad + 10f & = 1 \end{aligned}$$

The solution to this system is $a = 30$, $b = -10$, $c = -3$, $d = -9$, $e = 3$, and $f = 1$. Thus

$$A = \begin{bmatrix} 30 & -10 & -3 \\ -9 & 3 & 1 \end{bmatrix}$$

and hence

$$T(x_1,x_2,x_3) = (30x_1 - 10x_2 - 3x_3,\ -9x_1 + 3x_2 + x_3)$$

From the above equation, we see that

$$T(1,1,1) = (17,-5)$$

<u>Alternate Solution</u>: We ask how to write an arbitrary vector (x_1,x_2,x_3) as a linear combination of the basis vectors $\mathbf{v}_1$, $\mathbf{v}_2$, and $\mathbf{v}_3$. If $(x_1,x_2,x_3) = a\mathbf{v}_1 + b\mathbf{v}_2 + c\mathbf{v}_3$, then

$$\begin{aligned} a + 2b + c &= x_1 \\ 2a + 5b &= x_2 \\ 3a + 3b + 10c &= x_3 \end{aligned}$$

The solution to this system is

$$\begin{aligned} a &= 50x_1 - 17x_2 - 5x_3 \\ b &= -20x_1 + 7x_2 + 2x_3 \\ c &= -9x_1 + 3x_2 + x_3 \end{aligned}$$

Thus

$$\begin{aligned} T(x_1,x_2,x_3) &= aT(\mathbf{v}_1) + bT(\mathbf{v}_2) + cT(\mathbf{v}_3) \\ &= a(1,0) + b(1,0) + c(0,1) \\ &= (30x_1 - 10x_2 - 3x_3,\ -9x_1 + 3x_2 + x_3) \end{aligned}$$

3. Since

$$\begin{aligned} T(a_0 + a_1x + a_2x^2) &= a_0T(1) + a_1T(x) + a_2T(x^2) \\ &= a_0(1 + x) + a_1(3 - x^2) + a_2(4 + 2x - 3x^2) \\ &= (a_0 + 3a_1 + 4a_2) + (a_0 + 2a_2)x + (-a_1 - 3a_2)x^2 \end{aligned}$$

we have

$$T(2 - 2x + 3x^2) = 8 + 8x - 7x^2$$

Alternate Solution: Since $T: P_2 \rightarrow P_2$, we look for the matrix (necessarily 3 × 3) such that $A\mathbf{x} = T(\mathbf{x})$. The equations

$$A\begin{bmatrix}1\\0\\0\end{bmatrix} = \begin{bmatrix}1\\1\\0\end{bmatrix}, \quad A\begin{bmatrix}0\\1\\0\end{bmatrix} = \begin{bmatrix}3\\0\\-1\end{bmatrix}, \quad \text{and} \quad A\begin{bmatrix}0\\0\\1\end{bmatrix} = \begin{bmatrix}4\\2\\-3\end{bmatrix}$$

yield

$$A = \begin{bmatrix}1 & 3 & 4\\1 & 0 & 2\\0 & -1 & -3\end{bmatrix}$$

which is equivalent to the previous answer.

5. **(a)** Since $T(1,0) = (2,1)$ and $T(0,1) = (-1,1)$, the standard matrix is

$$\begin{bmatrix}2 & -1\\1 & 1\end{bmatrix}$$

6. **(a)** Since $T(1,0) = (0,-1,1,1)$ and $T(0,1) = (1,0,3,-1)$, the standard matrix is

$$\begin{bmatrix}0 & 1\\-1 & 0\\1 & 3\\1 & -1\end{bmatrix}$$

9. **(a)** Since the determinant of the matrix

$$[T] = \begin{bmatrix}1 & 2\\-1 & 1\end{bmatrix}$$

is 3, $[T]$ is invertible with

$$[T]^{-1} = \begin{bmatrix} \frac{1}{3} & -\frac{2}{3} \\ \frac{1}{3} & \frac{1}{3} \end{bmatrix}$$

Thus $T^{-1}(x_1,x_2) = \left(\frac{1}{3}x_1 - \frac{2}{3}x_2, \frac{1}{3}x_1 + \frac{1}{3}x_2\right)$.

9. **(b)** Since the determinant of the matrix

$$[T] = \begin{bmatrix} 4 & -6 \\ -2 & 3 \end{bmatrix}$$

is zero, T is not one-to-one.

10. **(a)** Since the matrix

$$[T] = \begin{bmatrix} 1 & -2 & 2 \\ 2 & 1 & 1 \\ 1 & 0 & 1 \end{bmatrix}$$

is invertible, T is one-to-one. Since

$$[T]^{-1} = \begin{bmatrix} 1 & 2 & -4 \\ -1 & -1 & 3 \\ -1 & -2 & 5 \end{bmatrix}$$

$$T^{-1}(x_1,x_2,x_3) = (x_1 + 2x_2 - 4x_3, -x_1 - x_2 + 3x_3, -x_1 - 2x_2 + 5x_3).$$

(b) Since the matrix

$$[T] = \begin{bmatrix} 1 & -3 & 4 \\ -1 & 1 & 1 \\ 0 & -2 & 5 \end{bmatrix}$$

is not invertible, T is not one-to-one.

11. **(a)** The standard matrix for T is

$$\begin{bmatrix} 1 & -2 & 1 & 3 \\ 2 & 1 & 0 & 4 \\ -1 & -3 & 1 & -1 \\ 3 & 3 & 1 & 7 \end{bmatrix}$$

The reduced row-echelon form of this matrix is

$$\begin{bmatrix} 1 & 0 & 0 & 2 \\ 0 & 1 & 0 & 0 \\ 0 & 0 & 1 & 1 \\ 0 & 0 & 0 & 0 \end{bmatrix}$$

Since the rank of this 4×4 matrix is 3, we have $\text{rank}(T) = 3$ and $\text{nullity}(T) = 4 - 3 = 1$.

12. **(a)** Since $T_2(T_1(x_1,x_2)) = (3(x_1 + x_2), 2(x_1 + x_2) + 4(x_1 - x_2)) = (3x_1 + 3x_2, 6x_1 - 2x_2)$, we have

$$[T_2 \circ T_1] = \begin{bmatrix} 3 & 3 \\ 6 & -2 \end{bmatrix}$$

We also have

$$[T_2]\ [T_1] = \begin{bmatrix} 3 & 0 \\ 2 & 4 \end{bmatrix} \begin{bmatrix} 1 & 1 \\ 1 & -1 \end{bmatrix} = \begin{bmatrix} 3 & 3 \\ 6 & -2 \end{bmatrix}$$

13. Call the basis $\mathbf{v}_1,\ldots,\mathbf{v}_n$, so that $A\mathbf{v}_i = B\mathbf{v}_i$ for $i = 1,\ldots,n$. If $\mathbf{v}$ is any vector in R^n, we can write $\mathbf{v}$ as a linear combination of basis vectors — say $\mathbf{v} = a_1\mathbf{v}_1 + \cdots + a_n\mathbf{v}_n$. Then, since A and B represent linear transformations, we have

$$\begin{aligned} A\mathbf{v} &= a_1A\mathbf{v}_1 + \cdots + a_nA\mathbf{v}_n \\ &= a_1B\mathbf{v}_1 + \cdots + a_nB\mathbf{v}_n \\ &= B\mathbf{v} \end{aligned}$$

Therefore, by Theorem 7.4.2, $A = B$.

15. **(a)** If T is one-to-one, then nullity(T) = 0, so that rank(T) = n. Consequently, rank([T]) = n, so that [T] is invertible.

(b) If [T] is invertible, then rank([T]) = n. Consequently, nullity([T]) = nullity(T) = 0, so that T is one-to-one.

17. **(a)** Note that

$$T(x,y,z) = \mathbf{i}\begin{vmatrix} b_0 & c_0 \\ y & z \end{vmatrix} - \mathbf{j}\begin{vmatrix} a_0 & c_0 \\ x & z \end{vmatrix} + \mathbf{k}\begin{vmatrix} a_0 & b_0 \\ x & y \end{vmatrix}$$

$$= \left[\begin{vmatrix} b_0 & c_0 \\ y & z \end{vmatrix}, -\begin{vmatrix} a_0 & c_0 \\ x & z \end{vmatrix}, \begin{vmatrix} a_0 & b_0 \\ x & y \end{vmatrix}\right]$$

Thus we have

$$T(1,0,0) = (\ 0\ ,\ c_0,\ -b_0)$$
$$T(0,1,0) = (-c_0,\ 0\ ,\ a_0)$$
$$T(0,0,1) = (\ b_0,\ -a_0,\ 0\ )$$

so that

$$[T] = \begin{bmatrix} 0 & -c_0 & b_0 \\ c_0 & 0 & -a_0 \\ -b_0 & a_0 & 0 \end{bmatrix}$$

(b) Thus $T(x,y,z) = (-c_0y + b_0z,\ c_0x - a_0z,\ -b_0x + a_0y)$. This, fortunately, is the same result as that obtained by expanding the three 2×2 determinants in Part (a).

EXERCISE SET 7.5

2. (a) Let $B = \{1, x, x^2\} = \{\mathbf{u}_1, \mathbf{u}_2, \mathbf{u}_3\}$ and $B' = \{1, x\} = \{\mathbf{v}_1, \mathbf{v}_2\}$. Observe that

$$\begin{aligned} T(\mathbf{u}_1) &= T(1) &&= 1 &&= 1\mathbf{v}_1 \\ T(\mathbf{u}_2) &= T(x) &&= 1 - 2x &&= 1\mathbf{v}_1 - 2\mathbf{v}_2 \\ T(\mathbf{u}_3) &= T(x^2) &&= -3x &&= -3\mathbf{v}_2 \end{aligned}$$

Hence the matrix of T with respect to B and B' is

$$\begin{bmatrix} 1 & 1 & 0 \\ 0 & -2 & -3 \end{bmatrix}$$

(b) We have

$$[T]_{B',B}[\mathbf{x}]_B = \begin{bmatrix} 1 & 1 & 0 \\ 0 & -2 & -3 \end{bmatrix} \begin{bmatrix} c_0 \\ c_1 \\ c_2 \end{bmatrix} = \begin{bmatrix} c_0 + c_1 \\ -2c_1 - 3c_2 \end{bmatrix}$$

and

$$\left[T\left[c_0 + c_1 x + c_2 x^2\right]\right]_{B'} = [(c_0 + c_1) - (2c_1 + 3c_2)x]_{B'}$$

$$= \begin{bmatrix} c_0 + c_1 \\ -2c_1 - 3c_2 \end{bmatrix}$$

4. **(a)** Since

$$T(\mathbf{u}_1) = \begin{bmatrix} 0 \\ 2 \end{bmatrix} = 2\mathbf{u}_1 + 2\mathbf{u}_2$$

and

$$T(\mathbf{u}_2) = \begin{bmatrix} -1 \\ -1 \end{bmatrix} = -\mathbf{u}_1$$

we have

$$[T]_B = \begin{bmatrix} 2 & -1 \\ 2 & 0 \end{bmatrix}$$

(b) Since

$$\begin{aligned} [T]_B[x]_B &= \begin{bmatrix} 2 & -1 \\ 2 & 0 \end{bmatrix} \begin{bmatrix} x_1 \\ x_2 \end{bmatrix}_B \\ &= \begin{bmatrix} 2 & -1 \\ 2 & 0 \end{bmatrix} [x_1(-\mathbf{u}_2) + x_2(\mathbf{u}_1 + \mathbf{u}_2)]_B \\ &= \begin{bmatrix} 2 & -1 \\ 2 & 0 \end{bmatrix} \begin{bmatrix} x_2 \\ x_2 - x_1 \end{bmatrix} = \begin{bmatrix} x_1 + x_2 \\ 2x_2 \end{bmatrix} \end{aligned}$$

and

$$[T(\mathbf{x})]_B = \begin{bmatrix} x_1 - x_2 \\ x_1 + x_2 \end{bmatrix}_B = (x_1 + x_2)\mathbf{u}_1 + 2x_2\mathbf{u}_2$$

Formula (5a) does, indeed, hold.

6. **(a)** Observe that

$$T(\mathbf{v}_1) = (1,-1,0) = \mathbf{v}_1 - \mathbf{v}_2$$

$$T(\mathbf{v}_2) = (-1,1,-1) = -\frac{3}{2}\mathbf{v}_1 + \frac{1}{2}\mathbf{v}_2 + \frac{1}{2}\mathbf{v}_3$$

$$T(\mathbf{v}_3) = (0,0,1) = \frac{1}{2}\mathbf{v}_1 + \frac{1}{2}\mathbf{v}_2 - \frac{1}{2}\mathbf{v}_3$$

Thus the matrix of T with respect to B is

$$\begin{bmatrix} 1 & -\frac{3}{2} & \frac{1}{2} \\ -1 & \frac{1}{2} & \frac{1}{2} \\ 0 & \frac{1}{2} & -\frac{1}{2} \end{bmatrix}$$

(b) If we solve for $(x_1,x_2,x_3)_B = a_1\mathbf{v}_1 + a_2\mathbf{v}_2 + a_3\mathbf{v}_3$, we find that

$$[\mathbf{x}]_B = \left[\frac{x_1 - x_2 + x_3}{2} \quad \frac{-x_1 + x_2 + x_3}{2} \quad \frac{x_1 + x_2 - x_3}{2}\right]^t$$

Thus we have

$$[T]_B[\mathbf{x}]_B = \begin{bmatrix} 1 & -\frac{3}{2} & \frac{1}{2} \\ -1 & \frac{1}{2} & \frac{1}{2} \\ 0 & \frac{1}{2} & -\frac{1}{2} \end{bmatrix} \begin{bmatrix} \frac{x_1 - x_2 + x_3}{2} \\ \frac{-x_1 + x_2 + x_3}{2} \\ \frac{x_1 + x_2 - x_3}{2} \end{bmatrix}$$

$$= \begin{bmatrix} \frac{3x_1 - 2x_2 - x_3}{2} \\ \frac{-x_1 + 2x_2 - x_3}{2} \\ \frac{-x_1 + x_3}{2} \end{bmatrix}$$

and

$$[T(\mathbf{x})]_B = [(x_1 - x_2,\ x_2 - x_1,\ x_1 - x_3)]_B$$

$$= \begin{bmatrix} \dfrac{(x_1 - x_2) - (x_2 - x_1) + (x_1 - x_3)}{2} \\ \dfrac{-(x_1 - x_2) + (x_2 - x_1) + (x_1 - x_3)}{2} \\ \dfrac{(x_1 - x_2) + (x_2 - x_1) - (x_1 - x_3)}{2} \end{bmatrix}$$

$$= \begin{bmatrix} \dfrac{3x_1 - 2x_2 - x_3}{2} \\ \dfrac{-x_1 + 2x_2 - x_3}{2} \\ \dfrac{-x_1 + x_3}{2} \end{bmatrix}$$

8. **(a)** Since

$$T(1) = x$$

$$T(x) = x(x - 3) = -3x + x^2$$

$$T(x^2) = x(x - 3)^2 = 9x - 6x^2 + x^3$$

we have

$$[T]_{B',B} = \begin{bmatrix} 0 & 0 & 0 \\ 1 & -3 & 9 \\ 0 & 1 & -6 \\ 0 & 0 & 1 \end{bmatrix}$$

(b) Since we are working with standard bases, we have

$$[T]_{B',B}[x]_B = \begin{bmatrix} 0 & 0 & 0 \\ 1 & -3 & 9 \\ 0 & 1 & -6 \\ 0 & 0 & 1 \end{bmatrix} \begin{bmatrix} 1 \\ 1 \\ -1 \end{bmatrix} = \begin{bmatrix} 0 \\ -11 \\ 7 \\ -1 \end{bmatrix}$$

so that $T(1 + x - x^2) = -11x + 7x^2 - x^3$.

(c) We have $T(1 + x - x^2) = x(1 + (x - 3) - (x - 3)^2) = -11x + 7x^2 - x^3$.

9. **(a)** Since A is the matrix of T with respect to B, then we know that the first and second columns of A must be $[T(\mathbf{v}_1)]_B$ and $[T(\mathbf{v}_2)]_B$, respectively. That is

$$[T(\mathbf{v}_1)]_B = \begin{bmatrix} 1 \\ -2 \end{bmatrix}$$

$$[T(\mathbf{v}_2)]_B = \begin{bmatrix} 3 \\ 5 \end{bmatrix}$$

Alternatively, since $\mathbf{v}_1 = 1\mathbf{v}_1 + 0\mathbf{v}_2$ and $\mathbf{v}_2 = 0\mathbf{v}_1 + 1\mathbf{v}_2$, we have

$$[T(\mathbf{v}_1)]_B = A\begin{bmatrix} 1 \\ 0 \end{bmatrix} = \begin{bmatrix} 1 \\ -2 \end{bmatrix}$$

and

$$[T(\mathbf{v}_2)]_B = A\begin{bmatrix} 0 \\ 1 \end{bmatrix} = \begin{bmatrix} 3 \\ 5 \end{bmatrix}$$

(b) From Part (a),

$$T(\mathbf{v}_1) = \mathbf{v}_1 - 2\mathbf{v}_2 = \begin{bmatrix} 3 \\ -5 \end{bmatrix}$$

and

$$T(\mathbf{v}_2) = 3\mathbf{v}_1 + 5\mathbf{v}_2 = \begin{bmatrix} -2 \\ 29 \end{bmatrix}$$

9. **(c)** Since we already know $T(\mathbf{v}_1)$ and $T(\mathbf{v}_2)$, all we have to do is express $[x_1 \quad x_2]^t$ in terms of $\mathbf{v}_1$ and $\mathbf{v}_2$. If

$$\begin{bmatrix} x_1 \\ x_2 \end{bmatrix} = a\mathbf{v}_1 + b\mathbf{v}_2 = a\begin{bmatrix} 1 \\ 3 \end{bmatrix} + b\begin{bmatrix} -1 \\ 4 \end{bmatrix}$$

then

$$x_1 = a - b$$
$$x_2 = 3a + 4b$$

or

$$a = (4x_1 + x_2)/7$$
$$b = (-3x_1 + x_2)/7$$

Thus

$$T\left(\begin{bmatrix} x_1 \\ x_2 \end{bmatrix}\right) = \frac{4x_1 + x_2}{7}\begin{bmatrix} 3 \\ -5 \end{bmatrix} + \frac{-3x_1 + x_2}{7}\begin{bmatrix} -2 \\ 29 \end{bmatrix}$$

$$= \begin{bmatrix} \dfrac{18x_1 + x_2}{7} \\ \dfrac{-107x_1 + 24x_2}{7} \end{bmatrix}$$

(d) By the above formula,

$$T\left(\begin{bmatrix} 1 \\ 1 \end{bmatrix}\right) = \begin{bmatrix} 19/7 \\ -83/7 \end{bmatrix}$$

11. **(a)** The columns of A, by definition, are $[T(\mathbf{v}_1)]_B$, $[T(\mathbf{v}_2)]_B$, and $[T(\mathbf{v}_3)]_B$, respectively.

(b) From Part (a),

$$\begin{aligned} T(\mathbf{v}_1) &= \mathbf{v}_1 + 2\mathbf{v}_2 + 6\mathbf{v}_3 = 16 + 51x + 19x^2 \\ T(\mathbf{v}_2) &= 3\mathbf{v}_1 - 2\mathbf{v}_3 = -6 - 5x + 5x^2 \\ T(\mathbf{v}_3) &= -\mathbf{v}_1 + 5\mathbf{v}_2 + 4\mathbf{v}_3 = 7 + 40x + 15x^2 \end{aligned}$$

(c) Let $a_0 + a_1x + a_2x^2 = b_0\mathbf{v}_1 + b_1\mathbf{v}_2 + b_2\mathbf{v}_3$. Then

$$\begin{aligned} a_0 &= - b_1 + 3b_2 \\ a_1 &= 3b_0 + 3b_1 + 7b_2 \\ a_2 &= 3b_0 + 2b_1 + 2b_2 \end{aligned}$$

This system of equations has the solution

$$\begin{aligned} b_0 &= (a_0 - a_1 + 2a_2)/3 \\ b_1 &= (-5a_0 + 3a_1 - 3a_2)/8 \\ b_2 &= (a_0 + a_1 - a_2)/8 \end{aligned}$$

Thus

$$T(a_0 + a_1x + a_2x^2) = b_0T(\mathbf{v}_1) + b_1T(\mathbf{v}_2) + b_2T(\mathbf{v}_3)$$

$$= \frac{239a_0 - 161a_1 + 289a_2}{24} + \frac{201a_0 - 111a_1 + 247a_2}{8}x + \frac{61a_0 - 31a_1 + 107a_2}{12}x^2$$

11. **(d)** By the above formula,

$$T(1 + x^2) = 22 + 56x + 14x^2$$

13. **(a)** Since

$$T_1(1) = 2 \quad \text{and} \quad T_1(x) = -3x^2$$

$$T_2(1) = 3x \quad T_2(x) = 3x^2 \quad \text{and} \quad T_2(x^2) = 3x^3$$

$$T_2 \circ T_1(1) = 6x \quad \text{and} \quad T_2 \circ T_1(x) = -9x^3$$

we have

$$[T_1]_{B'',B} = \begin{bmatrix} 2 & 0 \\ 0 & 0 \\ 0 & -3 \end{bmatrix} \qquad [T_2]_{B',B''} = \begin{bmatrix} 0 & 0 & 0 \\ 3 & 0 & 0 \\ 0 & 3 & 0 \\ 0 & 0 & 3 \end{bmatrix}$$

and

$$[T_2 \circ T_1]_{B',B} = \begin{bmatrix} 0 & 0 \\ 6 & 0 \\ 0 & 0 \\ 0 & -9 \end{bmatrix}$$

(b) We observe that here

$$[T_2 \circ T_1]_{B',B} = [T_2]_{B',B''}\ [T_1]_{B'',B}.$$

15. If T is a contraction or a dilation of V, then T maps any basis $B = \{\mathbf{v}_1, \ldots, \mathbf{v}_n\}$ of V to $\{k\mathbf{v}_1, \ldots, k\mathbf{v}_n\}$ where k is a nonzero constant. Therefore the matrix of T with respect to B is

$$\begin{bmatrix} k & 0 & 0 & \cdots & 0 \\ 0 & k & 0 & \cdots & 0 \\ 0 & 0 & k & \cdots & 0 \\ \vdots & \vdots & \vdots & & \vdots \\ 0 & 0 & 0 & \cdots & k \end{bmatrix}$$

17. **(a)** Since $D(1) = 0$, $D(x) = 1$, and $D(x^2) = 2x$, then

$$\begin{bmatrix} 0 & 1 & 0 \\ 0 & 0 & 2 \\ 0 & 0 & 0 \end{bmatrix}$$

is the matrix of D with respect to B.

(b) Since $D(2) = 0$, $D(2 - 3x) = -3 = -\frac{3}{2}\,\mathbf{p}_1$ and $D(2 - 3x + 8x^2) = -3 + 16x = \frac{23}{6}\,\mathbf{p}_1 - \frac{16}{3}\,\mathbf{p}_2$, the matrix of D with respect to B is

$$\begin{bmatrix} 0 & -\frac{3}{2} & \frac{23}{6} \\ 0 & 0 & -\frac{16}{3} \\ 0 & 0 & 0 \end{bmatrix}$$

(c) Using the matrix of Part (a), we obtain

$$D(6 - 6x + 24x^2) = \begin{bmatrix} 0 & 1 & 0 \\ 0 & 0 & 2 \\ 0 & 0 & 0 \end{bmatrix}\begin{bmatrix} 6 \\ -6 \\ 24 \end{bmatrix} = \begin{bmatrix} -6 \\ 48 \\ 0 \end{bmatrix} = -6 + 48x$$

17. (d) Since $6 - 6x + 24x^2 = \mathbf{p}_1 - \mathbf{p}_2 + 3\mathbf{p}_3$, we have

$$[D(6 - 6x + 24x^2)]_B = \begin{bmatrix} 0 & -\frac{3}{2} & \frac{23}{6} \\ 0 & 0 & -\frac{16}{3} \\ 0 & 0 & 0 \end{bmatrix} \begin{bmatrix} 1 \\ -1 \\ 3 \end{bmatrix} = \begin{bmatrix} 13 \\ -16 \\ 0 \end{bmatrix}_B$$

or

$$D(6 - 6x + 24x^2) = 13(2) - 16(2 - 3x) = -6 + 48x$$

18. (c) Since $D(\mathbf{f}_1) = 2\mathbf{f}_1$, $D(\mathbf{f}_2) = \mathbf{f}_1 + 2\mathbf{f}_2$, and $D(\mathbf{f}_3) = 2\mathbf{f}_2 + 2\mathbf{f}_3$, we have the matrix

$$\begin{bmatrix} 2 & 1 & 0 \\ 0 & 2 & 2 \\ 0 & 0 & 2 \end{bmatrix}$$

19. The standard matrix for T is just the $m \times n$ matrix whose columns are the transforms of the standard basis vectors. But since B is indeed the standard basis for R^n, the matrices are the same. Moreover, since B' is the standard basis for R^m, the resulting transformation will yield vector components relative to the standard basis, rather than to some other basis.

EXERCISE SET 7.6

1. First, we find the matrix of T with respect to B. Since

$$T(\mathbf{u}_1) = \begin{bmatrix} 1 \\ 0 \end{bmatrix}$$

and

$$T(\mathbf{u}_2) = \begin{bmatrix} -2 \\ -1 \end{bmatrix}$$

then

$$A = [T]_B = \begin{bmatrix} 1 & -2 \\ 0 & -1 \end{bmatrix}$$

In order to find P, we note that $\mathbf{v}_1 = 2\mathbf{u}_1 + \mathbf{u}_2$ and $\mathbf{v}_2 = -3\mathbf{u}_1 + 4\mathbf{u}_2$. Hence the transition matrix from B' to B is

$$P = \begin{bmatrix} 2 & -3 \\ 1 & 4 \end{bmatrix}$$

Thus

$$P^{-1} = \begin{bmatrix} \frac{4}{11} & \frac{3}{11} \\ -\frac{1}{11} & \frac{2}{11} \end{bmatrix}$$

and therefore

$$A' = [T]_{B'} = P^{-1}[T]_B P = \frac{1}{11}\begin{bmatrix} 4 & 3 \\ -1 & 2 \end{bmatrix}\begin{bmatrix} 1 & -2 \\ 0 & -1 \end{bmatrix}\begin{bmatrix} 2 & -3 \\ 1 & 4 \end{bmatrix}$$

$$= \begin{bmatrix} -\frac{3}{11} & -\frac{56}{11} \\ -\frac{2}{11} & \frac{3}{11} \end{bmatrix}$$

2. In order to compute $A = [T]_B$, we note that

$$T(\mathbf{u}_1) = \begin{bmatrix} 16 \\ -2 \end{bmatrix} = (0.8)\mathbf{u}_1 + (3.6)\mathbf{u}_2$$

and

$$T(\mathbf{u}_2) = \begin{bmatrix} -3 \\ 16 \end{bmatrix} = (6.1)\mathbf{u}_1 + (-3.8)\mathbf{u}_2$$

Hence

$$A = [T]_B = \begin{bmatrix} 0.8 & 6.1 \\ 3.6 & -3.8 \end{bmatrix}$$

In order to find P, we note that

$$\mathbf{v}_1 = (1.3)\mathbf{u}_1 + (-0.4)\mathbf{u}_2$$
$$\mathbf{v}_2 = (-0.5)\mathbf{u}_1$$

Hence

$$P = \begin{bmatrix} 1.3 & -0.5 \\ -0.4 & 0 \end{bmatrix}$$

and

$$P^{-1} = \begin{bmatrix} 0 & -2.5 \\ -2 & -6.5 \end{bmatrix}$$

It then follows that

$$A' = [T]_{B'} = P^{-1}AP = \begin{bmatrix} -15.5 & 4.5 \\ -37.5 & 12.5 \end{bmatrix}$$

3. Since $T(\mathbf{u}_1) = (1/\sqrt{2}, 1/\sqrt{2})$ and $T(\mathbf{u}_2) = (-1/\sqrt{2}, 1/\sqrt{2})$, then the matrix of T with respect to B is (cf. Example 3, Section 7.1)

$$A = [T]_B = \begin{bmatrix} 1/\sqrt{2} & -1/\sqrt{2} \\ 1/\sqrt{2} & 1/\sqrt{2} \end{bmatrix}$$

From Exercise 1, we know that

$$P = \begin{bmatrix} 2 & -3 \\ 1 & 4 \end{bmatrix} \quad \text{and} \quad P^{-1} = \frac{1}{11}\begin{bmatrix} 4 & 3 \\ -1 & 2 \end{bmatrix}$$

Thus

$$A' = [T]_{B'} = P^{-1}AP = \frac{1}{11\sqrt{2}}\begin{bmatrix} 13 & -25 \\ 5 & 9 \end{bmatrix}$$

5. Since $T(\mathbf{e}_1) = (1,0,0)$, $T(\mathbf{e}_2) = (0,1,0)$, and $T(\mathbf{e}_3) = (0,0,0)$, we have

$$A = [T]_B = \begin{bmatrix} 1 & 0 & 0 \\ 0 & 1 & 0 \\ 0 & 0 & 0 \end{bmatrix}$$

In order to compute P, we note that $\mathbf{v}_1 = \mathbf{e}_1$, $\mathbf{v}_2 = \mathbf{e}_1 + \mathbf{e}_2$, and $\mathbf{v}_3 = \mathbf{e}_1 + \mathbf{e}_2 + \mathbf{e}_3$. Hence,

$$P = \begin{bmatrix} 1 & 1 & 1 \\ 0 & 1 & 1 \\ 0 & 0 & 1 \end{bmatrix}$$

and

$$P^{-1} = \begin{bmatrix} 1 & -1 & 0 \\ 0 & 1 & -1 \\ 0 & 0 & 1 \end{bmatrix}$$

Thus

$$[T]_{B'} = \begin{bmatrix} 1 & -1 & 0 \\ 0 & 1 & -1 \\ 0 & 0 & 1 \end{bmatrix} \begin{bmatrix} 1 & 0 & 0 \\ 0 & 1 & 0 \\ 0 & 0 & 0 \end{bmatrix} \begin{bmatrix} 1 & 1 & 1 \\ 0 & 1 & 1 \\ 0 & 0 & 1 \end{bmatrix} = \begin{bmatrix} 1 & 0 & 0 \\ 0 & 1 & 1 \\ 0 & 0 & 0 \end{bmatrix}$$

7. Since

$$T(\mathbf{p}_1) = 9 + 3x = \frac{2}{3}\,\mathbf{p}_1 + \frac{1}{2}\,\mathbf{p}_2$$

and

$$T(\mathbf{p}_2) = 12 + 2x = -\frac{2}{9}\,\mathbf{p}_1 + \frac{4}{3}\,\mathbf{p}_2$$

we have

$$[T]_B = \begin{bmatrix} \frac{2}{3} & -\frac{2}{9} \\ \frac{1}{2} & \frac{4}{3} \end{bmatrix}$$

We note that $\mathbf{q}_1 = -\frac{2}{9}\,\mathbf{p}_1 + \frac{1}{3}\,\mathbf{p}_2$ and $\mathbf{q}_2 = \frac{7}{9}\,\mathbf{p}_1 - \frac{1}{6}\,\mathbf{p}_2$. Hence

$$P = \begin{bmatrix} -\frac{2}{9} & \frac{7}{9} \\ \frac{1}{3} & -\frac{1}{6} \end{bmatrix}$$

and

$$P^{-1} = \begin{bmatrix} \frac{3}{4} & \frac{7}{2} \\ \frac{3}{2} & 1 \end{bmatrix}$$

Therefore

$$[T]_{B'} = \begin{bmatrix} \frac{3}{4} & \frac{7}{2} \\ \frac{3}{2} & 1 \end{bmatrix} \begin{bmatrix} \frac{2}{3} & -\frac{2}{9} \\ \frac{1}{2} & \frac{4}{3} \end{bmatrix} \begin{bmatrix} -\frac{2}{9} & \frac{7}{9} \\ \frac{1}{3} & -\frac{1}{6} \end{bmatrix} = \begin{bmatrix} 1 & 1 \\ 0 & 1 \end{bmatrix}$$

8. **(a)** Since $T(1,0) = (3,-1)$ and $T(0,1) = (-4,7)$, we have

$$\det(T) = \begin{vmatrix} 3 & -4 \\ -1 & 7 \end{vmatrix} = 17$$

(c) Since $T(1) = 1$, $T(x) = x - 1$, and $T(x^2) = (x - 1)^2$, we have

$$\det(T) = \begin{vmatrix} 1 & -1 & 1 \\ 0 & 1 & -2 \\ 0 & 0 & 1 \end{vmatrix} = 1$$

9. **(a)** If A and C are similar $n \times n$ matrices, then there exists an invertible $n \times n$ matrix P such that $A = P^{-1}CP$. We can interpret P as being the transition matrix from a basis B' for R^n to a basis B. Moreover, C induces a linear transformation $T:R^n \to R^n$ where $C = [T]_B$. Hence $A = [T]_{B'}$. Thus A and C are matrices for the same transformation with respect to different bases. But from Theorem 7.2.3, we know that the rank of T is equal to the rank of C and hence to the rank of A.

Alternate Solution: We observe that if P is an invertible $n \times n$ matrix, then P represents a linear transformation of R^n onto R^n. Thus the rank of the transformation represented by the matrix CP is the same as that of C. Since P^{-1} is also invertible, its null space contains only the zero vector, and hence the rank of the transformation represented by the matrix $P^{-1}CP$ is also the same as that of C. Thus the ranks of A and C are equal. Again we use the result of Theorem 7.2.3 to equate the rank of a linear transformation with the rank of a matrix which represents it.

Second Alternative: Since the assertion that similar matrices have the same rank deals only with matrices and not with transformations, we outline a proof which involves only matrices. If $A = P^{-1}CP$, then P^{-1} and P can be expressed as products of elementary matrices. But multiplication of the matrix C by an elementary matrix is equivalent to performing an elementary row or column operation on C. From Section 4.6, we know that such operations do not change the rank of C. Thus A and C must have the same rank.

10. **(a)** We use the standard basis for P_4. Since $T(1) = 1$, $T(x) = 2x + 1$, $T(x^2) = (2x + 1)^2$, $T(x^3) = (2x + 1)^3$, and $T(x^4) = (2x + 1)^4$, we have

$$\det(T) = \begin{vmatrix} 1 & 1 & 1 & 1 & 1 \\ 0 & 2 & 4 & 6 & 8 \\ 0 & 0 & 4 & 12 & 24 \\ 0 & 0 & 0 & 8 & 32 \\ 0 & 0 & 0 & 0 & 16 \end{vmatrix} = 1024$$

(b) Thus $\text{rank}(T) = 5$, or $[T]$ is invertible, so, by Theorem 7.5.3, T is one-to-one.

11. **(a)** The matrix for T relative to the standard basis B is

$$[T]_B = \begin{bmatrix} 1 & -1 \\ 2 & 4 \end{bmatrix}$$

The eigenvalues of $[T]_B$ are $\lambda = 2$ and $\lambda = 3$, while corresponding eigenvectors are (1,-1) and (1,-2), respectively. If we let

$$P = \begin{bmatrix} 1 & 1 \\ -1 & -2 \end{bmatrix} \quad \text{then} \quad P^{-1} = \begin{bmatrix} 2 & 1 \\ -1 & -1 \end{bmatrix}$$

and

$$P^{-1}[T]_B P = \begin{bmatrix} 2 & 0 \\ 0 & 3 \end{bmatrix}$$

is diagonal. Since P represents the transition matrix from the basis B' to the standard basis B, we have

$$B' = \left\{\begin{bmatrix} 1 \\ -1 \end{bmatrix}, \begin{bmatrix} 1 \\ -2 \end{bmatrix}\right\}$$

as a basis which produces a diagonal matrix for $[T]_{B'}$.

12. **(b)** The matrix for T relative to the standard basis B is

$$[T]_B = \begin{bmatrix} 0 & -1 & 1 \\ -1 & 0 & 1 \\ 1 & 1 & 0 \end{bmatrix}$$

The eigenvalues of $[T]_B$ are $\lambda = 1$ and $\lambda = -2$. The eigenspace corresponding to $\lambda = 1$ is spanned by the vectors $\begin{bmatrix} -1 \\ 1 \\ 0 \end{bmatrix}$ and $\begin{bmatrix} 1 \\ 0 \\ 1 \end{bmatrix}$ and the eigenspace corresponding to $\lambda = -2$ is spanned by the vector $\begin{bmatrix} 1 \\ 1 \\ -1 \end{bmatrix}$. If we let

$$P = \begin{bmatrix} -1 & 1 & 1 \\ 1 & 0 & 1 \\ 0 & 1 & -1 \end{bmatrix}$$

then $P^{-1}[T]_B P$ will be diagonal. Hence the basis

$$B' = \left\{\begin{bmatrix} -1 \\ 1 \\ 0 \end{bmatrix}, \begin{bmatrix} 1 \\ 0 \\ 1 \end{bmatrix}, \begin{bmatrix} 1 \\ 1 \\ -1 \end{bmatrix}\right\}$$

will produce a diagonal matrix for $[T]_{B'}$.

13. **(a)** The matrix of T with respect to the standard basis for P_2 is

$$A = \begin{bmatrix} 5 & 6 & 2 \\ 0 & -1 & -8 \\ 1 & 0 & -2 \end{bmatrix}$$

The characteristic equation of A is

$$\lambda^3 - 2\lambda^2 - 15\lambda + 36 = (\lambda - 3)^2(\lambda + 4) = 0$$

and the eigenvalues are therefore $\lambda = -4$ and $\lambda = 3$.

(b) If we set $\lambda = -4$, then $(\lambda I - A)\mathbf{x} = \mathbf{0}$ becomes

$$\begin{bmatrix} -9 & -6 & -2 \\ 0 & -3 & 8 \\ -1 & 0 & -2 \end{bmatrix} \begin{bmatrix} x_1 \\ x_2 \\ x_3 \end{bmatrix} = \begin{bmatrix} 0 \\ 0 \\ 0 \end{bmatrix}$$

The augmented matrix reduces to

$$\begin{bmatrix} 1 & 0 & 2 & 0 \\ 0 & 1 & -8/3 & 0 \\ 0 & 0 & 0 & 0 \end{bmatrix}$$

and hence $x_1 = -2s$, $x_2 = \frac{8}{3}s$, and $x_3 = s$. Therefore the vector

$$\begin{bmatrix} -2 \\ 8/3 \\ 1 \end{bmatrix}$$

is a basis for the eigenspace associated with $\lambda = -4$. In P^2, this vector represents the polynomial $-2 + \frac{8}{3}x + x^2$.

If we set $\lambda = 3$ and carry out the above procedure, we find that $x_1 = 5s$, $x_2 = -2s$, and $x_3 = s$. Thus the polynomial $5 - 2x + x^2$ is a basis for the eigenspace associated with $\lambda = 3$.

14. **(a)** We look for values of λ such that

$$T\left(\begin{bmatrix} a & b \\ c & d \end{bmatrix}\right) = \lambda\begin{bmatrix} a & b \\ c & d \end{bmatrix}$$

or

$$\begin{bmatrix} 2c & a + c \\ b - 2c & d \end{bmatrix} = \begin{bmatrix} \lambda a & \lambda b \\ \lambda c & \lambda d \end{bmatrix}$$

If we equate corresponding entries, we find that

$$(*)\qquad \begin{aligned} \lambda a \qquad\qquad - 2c \qquad &= 0 \\ -a + \lambda b \qquad - c \qquad &= 0 \\ - b + (\lambda + 2)c \qquad &= 0 \\ (\lambda - 1)d &= 0 \end{aligned}$$

This system of equations has a nontrivial solution for a, b, c, and d only if

$$\det\begin{bmatrix} \lambda & 0 & -2 & 0 \\ -1 & \lambda & -1 & 0 \\ 0 & -1 & \lambda + 2 & 0 \\ 0 & 0 & 0 & \lambda - 1 \end{bmatrix} = 0$$

or

$$(\lambda - 1)(\lambda + 2)(\lambda^2 - 1) = 0$$

Therefore the eigenvalues are $\lambda = 1$, $\lambda = -2$, and $\lambda = -1$.

(b) We find a basis only for the eigenspace of T associated with $\lambda = 1$. The bases associated with the other eigenvalues are found in a similar way. If $\lambda = 1$, the equation $T(\mathbf{x}) = \lambda\mathbf{x}$ becomes $T\mathbf{x} = \mathbf{x}$ and the augmented matrix for the system of equations (*) above is

$$\begin{bmatrix} 1 & 0 & -2 & 0 & 0 \\ -1 & 1 & -1 & 0 & 0 \\ 0 & -1 & 3 & 0 & 0 \\ 0 & 0 & 0 & 0 & 0 \end{bmatrix}$$

This reduces to

$$\begin{bmatrix} 1 & 0 & -2 & 0 & 0 \\ 0 & 1 & -3 & 0 & 0 \\ 0 & 0 & 0 & 0 & 0 \\ 0 & 0 & 0 & 0 & 0 \end{bmatrix}$$

and hence $a = 2t$, $b = 3t$, $c = t$, and $d = s$. Therefore the matrices

$$\begin{bmatrix} 2 & 3 \\ 1 & 0 \end{bmatrix} \quad \text{and} \quad \begin{bmatrix} 0 & 0 \\ 0 & 1 \end{bmatrix}$$

form a basis for the eigenspace associated with $\lambda = 1$.

15. If $\mathbf{v}$ is an eigenvector of T corresponding to λ, then $\mathbf{v}$ is a nonzero vector such that $T(\mathbf{v}) = \lambda\mathbf{v}$ or $(\lambda I - T)\mathbf{v} = \mathbf{0}$. Thus $\lambda I - T$ maps $\mathbf{v}$ to $\mathbf{0}$, or $\mathbf{v}$ is in the kernel of $\lambda I - T$.

17. Since $C[\mathbf{x}]_B = D[\mathbf{x}]_B$ for all $\mathbf{x}$ in V, we can, in particular, let $\mathbf{x} = \mathbf{v}_i$ for each of the basis vectors $\mathbf{v}_1,\ldots,\mathbf{v}_n$ of V. Since $[\mathbf{v}_i]_B = \mathbf{e}_i$ for each i where $\{\mathbf{e}_1,\ldots,\mathbf{e}_n\}$ is the standard basis for R^n, this yields $C\mathbf{e}_i = D\mathbf{e}_i$ for $i = 1,\ldots,n$. But $C\mathbf{e}_i$ and $D\mathbf{e}_i$ are just the ith columns of C and D, respectively. Since corresponding columns of C and D are all equal, we have $C = D$.

18. (a) Let B be the standard basis for R^2, and let

$$B' = \{(\cos\theta, \sin\theta), (-\sin\theta, \cos\theta)\} = \{\mathbf{v}_1, \mathbf{v}_2\}$$

be the basis consisting of the unit vector $\mathbf{v}_1$ lying along ℓ and the unit vector $\mathbf{v}_2$ perpendicular to ℓ. Note that to change basis from B to B', we rotate through an angle $-\theta$. Now relative to B', T is just the orthogonal projection on the $\mathbf{v}_1$-axis, which is accomplished by setting the $\mathbf{v}_2$-coordinate equal to zero. Then to return to the basis B, we rotate through an angle θ. Thus

$$T\left(\begin{bmatrix} x \\ y \end{bmatrix}\right) = [I]_{B,B'}[T]_{B'}[I]_{B',B}\begin{bmatrix} x \\ y \end{bmatrix}$$

$$= \begin{bmatrix} \cos\theta & -\sin\theta \\ \sin\theta & \cos\theta \end{bmatrix}\begin{bmatrix} 1 & 0 \\ 0 & 0 \end{bmatrix}\begin{bmatrix} \cos\theta & \sin\theta \\ -\sin\theta & \cos\theta \end{bmatrix}\begin{bmatrix} x \\ y \end{bmatrix}$$

$$= \begin{bmatrix} \cos^2\theta & \sin\theta\cos\theta \\ \sin\theta\cos\theta & \sin^2\theta \end{bmatrix}\begin{bmatrix} x \\ y \end{bmatrix}$$

SUPPLEMENTARY EXERCISES 7

3. By the properties of an inner product, we have

$$\begin{aligned} T(\mathbf{v}+\mathbf{w}) &= \langle \mathbf{v}+\mathbf{w},\mathbf{v}_0\rangle \mathbf{v}_0 \\ &= (\langle \mathbf{v},\mathbf{v}_0\rangle + \langle \mathbf{w},\mathbf{v}_0\rangle)\mathbf{v}_0 \\ &= \langle \mathbf{v},\mathbf{v}_0\rangle \mathbf{v}_0 + \langle \mathbf{w},\mathbf{v}_0\rangle \mathbf{v}_0 \\ &= T(\mathbf{v}) + \mathrm{T}(\mathbf{w}) \end{aligned}$$

and

$$\begin{aligned} T(k\mathbf{v}) &= \langle k\mathbf{v},\mathbf{v}_0\rangle \mathbf{v}_0 \\ &= k\langle \mathbf{v},\mathbf{v}_0\rangle \mathbf{v}_0 \\ &= kT(\mathbf{v}) \end{aligned}$$

Thus T is a linear operator on V.

4. **(a)** By direct computation, we have

$$\begin{aligned} T(\mathbf{x}+\mathbf{y}) &= ((\mathbf{x}+\mathbf{y})\cdot \mathbf{v}_1, \ldots, (\mathbf{x}+\mathbf{y})\cdot \mathbf{v}_m) \\ &= (\mathbf{x}\cdot \mathbf{v}_1 + \mathbf{y}\cdot \mathbf{v}_1, \ldots, \mathbf{x}\cdot \mathbf{v}_m + \mathbf{y}\cdot \mathbf{v}_m) \\ &= (\mathbf{x}\cdot \mathbf{v}_1, \ldots, \mathbf{x}\cdot \mathbf{v}_m) + (\mathbf{y}\cdot \mathbf{v}_1, \ldots, \mathbf{y}\cdot \mathbf{v}_m) \\ &= T(\mathbf{x}) + T(\mathbf{y}) \end{aligned}$$

and

$$\begin{aligned} T(k\mathbf{x}) &= ((k\mathbf{x}) \cdot \mathbf{v}_1, \ldots, (k\mathbf{x}) \cdot \mathbf{v}_m) \\ &= (k(\mathbf{x} \cdot \mathbf{v}_1), \ldots, k(\mathbf{x} \cdot \mathbf{v}_m)) \\ &= kT(\mathbf{x}) \end{aligned}$$

4. **(b)** If $\{\mathbf{e}_1, \ldots, \mathbf{e}_n\}$ is the standard basis for R^n, and if $\mathbf{v}_i = (a_{1i}, a_{2i}, \ldots, a_{ni})$ for $i = 1, \ldots, m$, then

$$T(\mathbf{e}_j) = (a_{j1}, a_{j2}, \ldots, a_{jm})$$

But $T(\mathbf{e}_j)$, interpreted as a column vector, is just the jth column of the standard matrix for T. Thus the ith row of this matrix is $(a_{1i}, a_{2i}, \ldots, a_{ni})$, which is just $\mathbf{v}_i$.

5. **(a)** The matrix for T with respect to the standard basis is

$$A = \begin{bmatrix} 1 & 0 & 1 & 1 \\ 2 & 1 & 3 & 1 \\ 1 & 0 & 0 & 1 \end{bmatrix}$$

We first look for a basis for the range of T; that is, for the space of vectors $\mathbf{b}$ such that $A\mathbf{x} = \mathbf{b}$. If we solve the system of equations

$$\begin{aligned} x \qquad + z + w &= b_1 \\ 2x + y + 3z + w &= b_2 \\ x \qquad\qquad + w &= b_3 \end{aligned}$$

we find that $z = b_1 - b_3$ and that any one of x, y, or w will determine the other two. Thus, $T(\mathbf{e}_3)$ and any two of the remaining three columns of A is a basis for $R(T)$.

Alternate Solution: We can use the method of Section 4.6 to find a basis for the column space of A by reducing A^t to row-echelon form. This yields

$$\begin{bmatrix} 1 & 2 & 1 \\ 0 & 1 & 0 \\ 0 & 0 & 1 \\ 0 & 0 & 0 \end{bmatrix}$$

so that the three vectors

$$\begin{bmatrix} 1 \\ 2 \\ 1 \end{bmatrix} \quad \begin{bmatrix} 0 \\ 1 \\ 0 \end{bmatrix} \quad \begin{bmatrix} 0 \\ 0 \\ 1 \end{bmatrix}$$

form a basis for the column space of T and hence for its range space.

Second Alternative: Note that since rank(A) = 3, then $R(T)$ is a 3-dimensional subspace of R^3 and hence is all of R^3. Thus the standard basis for R^3 is also a basis for $R(T)$.

(b) To find a basis for the kernel of T, we consider the solution space of $A\mathbf{x} = \mathbf{0}$. If we set $b_1 = b_2 = b_3 = 0$ in the above system of equations, we find that $z = 0$, $x = -w$, and $y = w$. Thus the vector $(-1,1,0,1)$ forms a basis for the kernel.

7. **(a)** We know that T can be thought of as multiplication by the matrix

$$[T]_B = \begin{bmatrix} 1 & 1 & 2 & -2 \\ 1 & -1 & -4 & 6 \\ 1 & 2 & 5 & -6 \\ 3 & 2 & 3 & -2 \end{bmatrix}$$

where reduction to row-echelon form easily shows that $\text{rank}([T])_B) = 2$. Therefore the rank of T is 2 and the nullity of T is $4 - 2 = 2$.

(b) Since $[T]_B$ is not invertible, T is not one-to-one.

9. **(a)** If $A = P^{-1}BP$, then

$$\begin{aligned} A^t &= (P^{-1}BP)^t \\ &= P^tB^t(P^{-1})^t \\ &= \left[(P^t)^{-1}\right]^{-1}B^t(P^{-1})^t \\ &= \left[(P^{-1})^t\right]^{-1}B^t(P^{-1})^t \end{aligned}$$

Therefore A^t and B^t are similar. You should verify that if P is invertible, then so is P^t and that $(P^t)^{-1} = (P^{-1})^t$.

10. If statement (i) holds, then the range of T is V and hence rank(T) = n. Therefore, by the Dimension Theorem, the nullity of T is zero, so that statement (ii) cannot hold. Thus (i) and (ii) cannot hold simultaneously.

Now if statement (i) does <u>not</u> hold, then the range of T is a proper subspace of V and rank(T) < n. The Dimension Theorem then implies that the nullity of T is greater than zero. Thus statement (ii) <u>must</u> hold. Hence exactly one of the two statements must always hold.

11. If we let $X = \begin{bmatrix} a & b \\ c & d \end{bmatrix}$, then we have

$$\begin{aligned} T\left(\begin{bmatrix} a & b \\ c & d \end{bmatrix}\right) &= \begin{bmatrix} a + c & b + d \\ 0 & 0 \end{bmatrix} + \begin{bmatrix} b & b \\ d & d \end{bmatrix} \\ &= \begin{bmatrix} a + b + c & 2b + d \\ d & d \end{bmatrix} \end{aligned}$$

The matrix X is in the kernel of T if and only if $T(X) = \mathbf{0}$, i.e., if and only if

$$\begin{aligned} a + b + c \qquad &= 0 \\ 2b \qquad + d &= 0 \\ d &= 0 \end{aligned}$$

Hence

$$X = \begin{bmatrix} a & 0 \\ -a & 0 \end{bmatrix}$$

The space of all such matrices X is spanned by the matrix $\begin{bmatrix} 1 & 0 \\ -1 & 0 \end{bmatrix}$, and therefore has dimension 1. Thus the nullity is 1. Since the dimension of M_{22} is 4, the rank of T must be 3.

<u>Alternate Solution</u>. Using the computations done above, we have that the matrix for this transformation with respect to the standard basis in M_{22} is

$$\begin{bmatrix} 1 & 1 & 1 & 0 \\ 0 & 2 & 0 & 1 \\ 0 & 0 & 0 & 1 \\ 0 & 0 & 0 & 1 \end{bmatrix}$$

Since this matrix has rank 3, the rank of T is 3, and therefore the nullity must be 1.

12. We are given that there exist invertible matrices P and Q such that $A = P^{-1}BP$ and $B = Q^{-1}CQ$. Therefore

$$A = P^{-1}(Q^{-1}CQ)P = (QP)^{-1}C(QP)$$

That is, A and C are similar.

13. The standard basis for M_{22} is the set of matrices

$$\begin{bmatrix} 1 & 0 \\ 0 & 0 \end{bmatrix}, \quad \begin{bmatrix} 0 & 1 \\ 0 & 0 \end{bmatrix}, \quad \begin{bmatrix} 0 & 0 \\ 1 & 0 \end{bmatrix}, \quad \begin{bmatrix} 0 & 0 \\ 0 & 1 \end{bmatrix}$$

If we think of the above matrices as the vectors

$$[1\ 0\ 0\ 0]^t, \quad [0\ 1\ 0\ 0]^t, \quad [0\ 0\ 1\ 0]^t, \quad [0\ 0\ 0\ 1]^t$$

then T takes these vectors to

$$[1\ 0\ 0\ 0]^t, \quad [0\ 0\ 1\ 0]^t, \quad [0\ 1\ 0\ 0]^t, \quad [0\ 0\ 0\ 1]^t$$

Therefore the desired matrix for T is

$$\begin{bmatrix} 1 & 0 & 0 & 0 \\ 0 & 0 & 1 & 0 \\ 0 & 1 & 0 & 0 \\ 0 & 0 & 0 & 1 \end{bmatrix}$$

14. **(a)** Reading directly from P, we have

$$\begin{aligned} \mathbf{v}_1 &= 2\mathbf{u}_1 + \mathbf{u}_2 \\ \mathbf{v}_2 &= -\mathbf{u}_1 + \mathbf{u}_2 + \mathbf{u}_3 \\ \mathbf{v}_3 &= 3\mathbf{u}_1 + 4\mathbf{u}_2 + 2\mathbf{u}_3 \end{aligned}$$

(b) Since

$$P^{-1} = \begin{bmatrix} -2 & 5 & -7 \\ -2 & 4 & -5 \\ 1 & -2 & 3 \end{bmatrix}$$

by direct calculation we have

$$\begin{aligned} \mathbf{u}_1 &= -2\mathbf{v}_1 - 2\mathbf{v}_2 + \mathbf{v}_3 \\ \mathbf{u}_2 &= 5\mathbf{v}_1 + 4\mathbf{v}_2 - 2\mathbf{v}_3 \\ \mathbf{u}_3 &= -7\mathbf{v}_1 - 5\mathbf{v}_2 + 3\mathbf{v}_3 \end{aligned}$$

15. The transition matrix P from B' to B is

$$P = \begin{bmatrix} 1 & 1 & 1 \\ 0 & 1 & 1 \\ 0 & 0 & 1 \end{bmatrix}$$

Therefore, by Theorem 7.6.2, we have

$$[T]_{B'} = P^{-1}[T]_B P = \begin{bmatrix} -4 & 0 & 9 \\ 1 & 0 & -2 \\ 0 & 1 & 1 \end{bmatrix}$$

<u>Alternate Solution</u>: We compute the above result more directly. It is easy to show that $\mathbf{u}_1 = \mathbf{v}_1$, $\mathbf{u}_2 = -\mathbf{v}_1 + \mathbf{v}_2$, and $\mathbf{u}_3 = -\mathbf{v}_2 + \mathbf{v}_3$. So

$$T(\mathbf{v}_1) = T(\mathbf{u}_1) = -3\mathbf{u}_1 + \mathbf{u}_2 = -4\mathbf{v}_1 + \mathbf{v}_2$$

$$T(\mathbf{v}_2) = T(\mathbf{u}_1 + \mathbf{u}_2) = T(\mathbf{u}_1) + T(\mathbf{u}_2)$$

$$= \mathbf{u}_1 + \mathbf{u}_2 + \mathbf{u}_3 = \mathbf{v}_3$$

$$T(\mathbf{v}_3) = T(\mathbf{u}_1 + \mathbf{u}_2 + \mathbf{u}_3) = T(\mathbf{u}_1) + T(\mathbf{u}_2) + T(\mathbf{u}_3)$$

$$= 8\mathbf{u}_1 - \mathbf{u}_2 + \mathbf{u}_3$$

$$= 9\mathbf{v}_1 - 2\mathbf{v}_2 + \mathbf{v}_3$$

16. Let $P = \begin{bmatrix} a & b \\ c & d \end{bmatrix}$ and solve for a, b, c, and d so that

$$\begin{bmatrix} 1 & 1 \\ -1 & 4 \end{bmatrix} = P^{-1}\begin{bmatrix} 2 & 1 \\ 1 & 3 \end{bmatrix}P \quad \text{or} \quad P\begin{bmatrix} 1 & 1 \\ -1 & 4 \end{bmatrix} = \begin{bmatrix} 2 & 1 \\ 1 & 3 \end{bmatrix}P$$

This yields the system of equations

$$\begin{aligned} a - b &= 2a + c \\ a + 4b &= 2b + d \\ c - d &= a + 3c \\ c + 4d &= b + 3c \end{aligned}$$

which has solution $b = -a$, $c = 0$, and $d = -a$. Thus, for instance,

$$P = \begin{bmatrix} 1 & -1 \\ 0 & -1 \end{bmatrix} \quad \text{and} \quad P^{-1} = \begin{bmatrix} 1 & -1 \\ 0 & -1 \end{bmatrix}$$

will work.

However, if we try the same procedure on the other pair of matrices, we find that $a = 3d$, $b = d$, and $c = 3d$, so that $\det(P) = 3d^2 - 3d^2 = 0$. Thus P is not invertible.

Alternatively, if there exists an invertible matrix P such that

$$\begin{bmatrix} 3 & 1 \\ -6 & -2 \end{bmatrix} = P^{-1}\begin{bmatrix} -1 & 2 \\ 1 & 0 \end{bmatrix}P$$

then we have a matrix with zero determinant on the left, and one with nonzero determinant on the right. Why? Thus the two matrices cannot be similar.

17. Since

$$T\left(\begin{bmatrix} 1 \\ 0 \\ 0 \end{bmatrix}\right) = \begin{bmatrix} 1 \\ 0 \\ 1 \end{bmatrix}, \quad T\left(\begin{bmatrix} 0 \\ 1 \\ 0 \end{bmatrix}\right) = \begin{bmatrix} -1 \\ 1 \\ 0 \end{bmatrix}, \quad \text{and} \quad T\left(\begin{bmatrix} 0 \\ 0 \\ 1 \end{bmatrix}\right) = \begin{bmatrix} 1 \\ 0 \\ -1 \end{bmatrix}$$

we have

$$[T]_B = \begin{bmatrix} 1 & -1 & 1 \\ 0 & 1 & 0 \\ 1 & 0 & -1 \end{bmatrix}$$

In fact, this result can be read directly from $[T(X)]_B$.

18. We know that $\det(T) \neq 0$ if and only if the matrix of T relative to <u>any</u> basis B has nonzero determinant; that is, if and only if <u>every</u> such matrix is invertible. Choose a particular basis B for V and let V have dimension n. Then the matrix $[T]_B$ represents T as a matrix transformation from R^n to R^n. By Exercise 15 of Section 7.4, such a transformation is one-to-one if and only if $[T]_B$ is invertible. Thus T is one-to-one if and only if $\det(T) \neq 0$.

19. **(b)** Recall that, by the Fundamental Theorem of Calculus,

$$\frac{d}{dx}\int_a^x p(t)\,dt = p(x) \text{ for all polynomials } p(x).$$

21. **(a)** We have

$$T(p(x) + q(x)) = \begin{bmatrix} p(x_1) + q(x_1) \\ p(x_2) + q(x_2) \\ p(x_3) + q(x_3) \end{bmatrix} = \begin{bmatrix} p(x_1) \\ p(x_2) \\ p(x_3) \end{bmatrix} + \begin{bmatrix} q(x_1) \\ q(x_2) \\ q(x_3) \end{bmatrix}$$

$$= T(p(x)) + T(q(x))$$

and

$$T(kp(x) = \begin{bmatrix} kp(x_1) \\ kp(x_2) \\ kp(x_3) \end{bmatrix} = k\begin{bmatrix} p(x_1) \\ p(x_2) \\ p(x_3) \end{bmatrix} kT(p(x))$$

(b) Since T is defined for quadratic polynomials only, and the numbers x_1, x_2, and x_3 are distinct, we can have $p(x_1) = p(x_2) = p(x_3) = 0$ if and only if p is the zero polynomial. (Why?) Thus $\ker(T) = \{\mathbf{0}\}$, so T is one-to-one.

21. **(c)** We have

$$T(a_1P_1(x) + a_2P_2(x) + a_3P_3(x)) = a_1T(P_1(x)) + a_2T(P_2(x)) + a_3T(P_3(x))$$

$$= a_1\begin{bmatrix}1\\0\\0\end{bmatrix} + a_2\begin{bmatrix}0\\1\\0\end{bmatrix} + a_3\begin{bmatrix}0\\0\\1\end{bmatrix}$$

$$= \begin{bmatrix}a_1\\a_2\\a_3\end{bmatrix}$$

(d) From the above calculations, we see that the points must lie on the curve.

23. Since

$$D(x^k) = \begin{cases} 0 & \text{if } k = 0 \\ kx^{k-1} & \text{if } k = 1,2,\ldots,n \end{cases}$$

then

$$[D(x^k)]_B = \begin{cases} (0,\ldots,0) & \text{if } k = 0 \\ (0,\ldots,\underset{\substack{\uparrow \\ k^{\text{th}} \\ \text{component}}}{k},\ldots,0) & \text{if } k = 1,2,\ldots,n \end{cases}$$

where the above vectors all have $n + 1$ components. Thus the matrix of D with respect to B is

$$\begin{bmatrix} 0 & 1 & 0 & 0 & \cdots & 0 \\ 0 & 0 & 2 & 0 & \cdots & 0 \\ 0 & 0 & 0 & 3 & \cdots & 0 \\ \vdots & \vdots & \vdots & \vdots & & \vdots \\ 0 & 0 & 0 & 0 & \cdots & n \\ 0 & 0 & 0 & 0 & \cdots & 0 \end{bmatrix}$$

24. Call the basis B and the vectors $\mathbf{v}_0, \mathbf{v}_1, \ldots, \mathbf{v}_n$. Notice that $D(\mathbf{v}_i) = \mathbf{v}_{i-1}$ for $i = 1, \ldots, n$, while $D(\mathbf{v}_0) = \mathbf{0}$. That is,

$$[D(\mathbf{v}_k)]_B = \begin{cases} (0,\ldots,0) & \text{if } k = 0 \\ (0,\ldots,\underset{\substack{\uparrow \\ k^{\text{th}} \text{ component}}}{1},\ldots,0) & \text{if } k = 1,2,\cdots,n \end{cases}$$

where the above vectors all have $n + 1$ components. Thus the matrix of D with respect to B is

$$\begin{bmatrix} 0 & 1 & 0 & 0 & \cdots & 0 \\ 0 & 0 & 1 & 0 & \cdots & 0 \\ 0 & 0 & 0 & 1 & \cdots & 0 \\ \vdots & \vdots & \vdots & \vdots & & \vdots \\ 0 & 0 & 0 & 0 & \cdots & 1 \\ 0 & 0 & 0 & 0 & \cdots & 0 \end{bmatrix}$$

In fact, the differentiation operator maps P_n to P_{n-1}.

25. Let B_n and B_{n+1} denote the bases for P_n and P_{n+1}, respectively. Since

$$J(x^k) = \frac{x^{k+1}}{k + 1} \quad \text{for } k = 0,\ldots,n$$

we have

$$[J(x^k)]_{B_{n+1}} = (0,\ldots,\underset{\substack{\uparrow \\ k\,+\,2^{\text{nd}} \text{ component}}}{\frac{1}{k + 1}},\ldots,0) \qquad (n + 2 \text{ components})$$

where $[x^k]_{B_n} = [0,\ldots,1,\ldots,0]^t$ with the entry 1 as the $k + 1^{\text{st}}$

component out of a total of $n + 1$ components. Thus the matrix of J with respect to B_{n+1} is

$$\begin{bmatrix} 0 & 0 & 0 & \cdots & 0 \\ 1 & 0 & 0 & \cdots & 0 \\ 0 & 1/2 & 0 & \cdots & 0 \\ 0 & 0 & 1/3 & \cdots & 0 \\ \vdots & \vdots & \vdots & & \vdots \\ 0 & 0 & 0 & \cdots & 1/(n+1) \end{bmatrix}$$

with $n + 2$ rows and $n + 1$ columns.

EXERCISE SET 8.1

1. **(a)** The system is of the form $Y' = AY$ where

$$A = \begin{bmatrix} 1 & 4 \\ 2 & 3 \end{bmatrix}$$

The eigenvalues of A are $\lambda = 5$ and $\lambda = -1$ and the corresponding eigenspaces are spanned by the vectors

$$\begin{bmatrix} 1 \\ 1 \end{bmatrix} \quad \text{and} \quad \begin{bmatrix} -2 \\ 1 \end{bmatrix}$$

respectively. Thus if we let

$$P = \begin{bmatrix} 1 & -2 \\ 1 & 1 \end{bmatrix}$$

we have

$$D = P^{-1}AP = \begin{bmatrix} 5 & 0 \\ 0 & -1 \end{bmatrix}$$

Let $Y = PU$ and hence $Y' = PU'$. Then

$$U' = \begin{bmatrix} 5 & 0 \\ 0 & -1 \end{bmatrix} U$$

or

$$u_1' = 5u_1$$

$$u_2' = -u_2$$

Therefore

$$u_1 = c_1 e^{5x}$$

$$u_2 = c_2 e^{-x}$$

Thus the equation $Y = PU$ is

$$\begin{bmatrix} y_1 \\ y_2 \end{bmatrix} = \begin{bmatrix} 1 & -2 \\ 1 & 1 \end{bmatrix} \begin{bmatrix} c_1 e^{5x} \\ c_2 e^{-x} \end{bmatrix} = \begin{bmatrix} c_1 e^{5x} - 2c_2 e^{-x} \\ c_1 e^{5x} + c_2 e^{-x} \end{bmatrix}$$

or

$$y_1 = c_1 e^{5x} - 2c_2 e^{-x}$$

$$y_2 = c_1 e^{5x} + c_2 e^{-x}$$

1. **(b)** If $y_1(0) = y_2(0) = 0$, then

$$c_1 - 2c_2 = 0$$

$$c_1 + c_2 = 0$$

so that $c_1 = c_2 = 0$. Thus $y_1 = 0$ and $y_2 = 0$.

3. **(a)** The system is of the form $Y' = AY$ where

$$A = \begin{bmatrix} 4 & 0 & 1 \\ -2 & 1 & 0 \\ -2 & 0 & 1 \end{bmatrix}$$

The eigenvalues of A are $\lambda = 1$, $\lambda = 2$. and $\lambda = 3$ and the corresponding eigenspaces are spanned by the vectors

$$\begin{bmatrix} 0 \\ 1 \\ 0 \end{bmatrix} \qquad \begin{bmatrix} -1/2 \\ 1 \\ 1 \end{bmatrix} \qquad \begin{bmatrix} -1 \\ 1 \\ 1 \end{bmatrix}$$

respectively. Thus, if we let

$$P = \begin{bmatrix} 0 & -1/2 & -1 \\ 1 & 1 & 1 \\ 0 & 1 & 1 \end{bmatrix}$$

then

$$D = P^{-1}AP = \begin{bmatrix} 1 & 0 & 0 \\ 0 & 2 & 0 \\ 0 & 0 & 3 \end{bmatrix}$$

Let $Y = PU$ and hence $Y' = PU'$. Then

$$U' = \begin{bmatrix} 1 & 0 & 0 \\ 0 & 2 & 0 \\ 0 & 0 & 3 \end{bmatrix} U$$

so that

$$u_1' = u_1$$

$$u_2' = 2u_2$$

$$u_3' = 3u_3$$

Therefore

$$u_1 = c_1e^{x}$$

$$u_2 = c_2e^{2x}$$

$$u_3 = c_3e^{3x}$$

Thus the equation $Y = PU$ is

$$\begin{bmatrix} y_1 \\ y_2 \\ y_3 \end{bmatrix} = \begin{bmatrix} 0 & -1/2 & -1 \\ 1 & 1 & 1 \\ 0 & 1 & 1 \end{bmatrix} \begin{bmatrix} c_1e^{x} \\ c_2e^{2x} \\ c_3e^{3x} \end{bmatrix}$$

or

$$y_1 = -\frac{1}{2}c_2e^{2x} - c_3e^{3x}$$

$$y_2 = c_1e^{x} + c_2e^{2x} + c_3e^{3x}$$

$$y_3 = c_2e^{2x} + c_3e^{3x}$$

Note: If we use

$$\begin{bmatrix} 0 \\ 1 \\ 0 \end{bmatrix} \quad \begin{bmatrix} -1 \\ 2 \\ 2 \end{bmatrix} \quad \begin{bmatrix} 1 \\ -1 \\ -1 \end{bmatrix}$$

as basis vectors for the eigenspaces, then

$$P = \begin{bmatrix} 0 & -1 & 1 \\ 1 & 2 & -1 \\ 0 & 2 & -1 \end{bmatrix}$$

and

$$y_1 = -c_2e^{2x} + c_3e^{3x}$$

$$y_2 = c_1e^{x} + 2c_2e^{2x} - c_3e^{3x}$$

$$y_3 = 2c_2e^{2x} - c_3e^{3x}$$

There are, of course, infinitely many other ways of writing the answer, depending upon what bases you choose for the eigenspaces.

Since the numbers c_1, c_2, and c_3 are arbitrary, the "different" answers do, in fact, represent the same functions.

3. (b) If we set $x = 0$, then the initial conditions imply that

$$\begin{aligned} -\frac{1}{2}c_2 - c_3 &= -1 \\ c_1 + c_2 + c_3 &= 1 \\ c_2 + c_3 &= 0 \end{aligned}$$

or, equivalently, that $c_1 = 1$, $c_2 = -2$, and $c_3 = 2$. If we had used the "different" solution we found in Part (a), then we would have obtained $c_1 = 1$, $c_2 = -1$, and $c_3 = -2$. In either case, when we substitute these values into the appropriate equations, we find that

$$\begin{aligned} y_1 &= e^{2x} - 2e^{3x} \\ y_2 &= e^{x} - 2e^{2x} + 2e^{3x} \\ y_3 &= -2e^{2x} + 2e^{3x} \end{aligned}$$

5. Following the hint, let $y_1 = y$ and $y_2 = y'$. Thus $y_1' = y_2$ and $y_2' = y'' = y' + 6y = y_2 + 6y_1$. That is,

$$\begin{aligned} y_1' &= \qquad y_2 \\ y_2' &= 6y_1 + y_2 \end{aligned}$$

or $Y' = AY$ where

$$A = \begin{bmatrix} 0 & 1 \\ 6 & 1 \end{bmatrix}$$

The eigenvalues of A and $\lambda = -2$ and $\lambda = 3$ and the corresponding eigenspaces are spanned by the vectors

$$\begin{bmatrix} -1 \\ 2 \end{bmatrix} \quad \text{and} \quad \begin{bmatrix} 1 \\ 3 \end{bmatrix}$$

respectively. Thus, if we let

$$P = \begin{bmatrix} -1 & 1 \\ 2 & 3 \end{bmatrix}$$

then

$$P^{-1}AP = \begin{bmatrix} -2 & 0 \\ 0 & 3 \end{bmatrix}$$

Let $Y = PU$ and hence $Y' = PU'$. Then

$$U' = \begin{bmatrix} -2 & 0 \\ 0 & 3 \end{bmatrix} U$$

or

$$y_1 = -c_1 e^{-2x} + c_2 e^{3x}$$

$$y_2 = 2c_1 e^{-2x} + 3c_2 e^{3x}$$

Therefore

$$u_1 = c_1 e^{-2x}$$

$$u_2 = c_2 e^{3x}$$

Thus the equation $Y = PU$ is

$$\begin{bmatrix} y_1 \\ y_2 \end{bmatrix} = \begin{bmatrix} -1 & 1 \\ 2 & 3 \end{bmatrix} \begin{bmatrix} c_1 e^{-2x} \\ c_2 e^{3x} \end{bmatrix}$$

or

$$\begin{aligned} y_1 &= -c_1e^{-2x} + c_2e^{3x} \\ y_2 &= 2c_1e^{-2x} + 3c_2e^{3x} \end{aligned}$$

Note that $y_1' = y_2$, as required, and, since $y_1 = y$, then

$$y = -c_1e^{-2x} + c_2e^{3x}$$

Since c_1 and c_2 are arbitrary, any answer of the form $y = ae^{-2x} + be^{3x}$ is correct.

7. Following the hint, let $y = f(x)$ be a solution to $y' = ay$, so that $f'(x) = af(x)$. Now consider the function $g(x) = f(x)e^{-ax}$. Observe that

$$\begin{aligned} g'(x) &= f'(x)e^{-ax} - af(x)e^{-ax} \\ &= af(x)e^{-ax} - af(x)e^{-ax} \\ &= 0 \end{aligned}$$

Thus $g(x)$ must be a constant; say $g(x) = c$. Therefore,

$$f(x)e^{-ax} = c$$

or

$$f(x) = ce^{ax}$$

That is, every solution of $y' = ay$ has the form $y = ce^{ax}$

8. Suppose that $Y' = AY$ where A is diagonalizable. Then there is an invertible matrix P such that $P^{-1}AP = D$ where D is a diagonal matrix whose diagonal entries are the eigenvalues of A. If we let $Y = PU$ so that $Y' = PU'$, then $Y' = AY$ becomes $PU' = APU$ or

$$\begin{aligned} U' &= P^{-1}APU \\ &= DU \end{aligned}$$

That is,

$$\begin{aligned} u_1' &= \lambda_1 u_1 \\ &\vdots \\ u_n' &= \lambda_n u_n \end{aligned}$$

and hence

$$\begin{aligned} u_1 &= c_1 e^{\lambda 1 x} \\ &\vdots \\ u_n &= c_n e^{\lambda n x} \end{aligned}$$

Therefore $Y = PU$ can be written

$$\begin{aligned} y_1 &= p_{11}c_1e^{\lambda_1 x} + p_{12}c_2e^{\lambda_2 x} + \cdots + p_{1n}c_ne^{\lambda_n x} \\ &\vdots \\ y_n &= p_{n1}c_1e^{\lambda_1 x} + p_{n2}c_2e^{\lambda_2 x} + \cdots + p_{nn}c_ne^{\lambda_n x} \end{aligned}$$

where p_{ij} is the ij^{th} entry in the matrix P. That is, each of the functions y_i is a linear combination of $e^{\lambda_1 x}, \ldots, e^{\lambda_n x}$.

EXERCISE SET 8.2

1. **(a)** Since $T(x,y) = (-y,-x)$, the standard matrix is

$$\begin{bmatrix} 0 & -1 \\ -1 & 0 \end{bmatrix}$$

(c) Since $T(x,y) = (x,0)$, the standard matrix is

$$\begin{bmatrix} 1 & 0 \\ 0 & 0 \end{bmatrix}$$

3. **(b)** Since $T(x,y,z) = (x,-y,z)$, the standard matrix is

$$\begin{bmatrix} 1 & 0 & 0 \\ 0 & -1 & 0 \\ 0 & 0 & 1 \end{bmatrix}$$

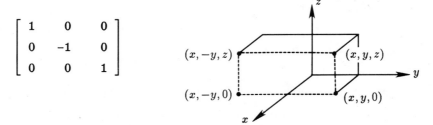

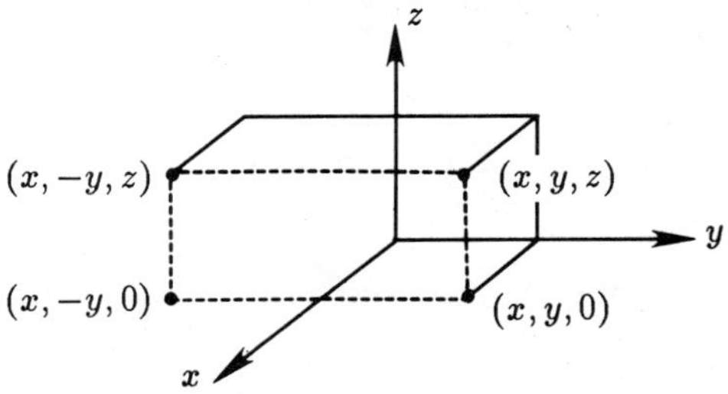

5. **(a)** This transformation leaves the z-coordinate of every point fixed. However it sends (1,0,0) to (0,1,0) and (0,1,0) to (-1,0,0). The standard matrix is therefore

$$\begin{bmatrix} 0 & -1 & 0 \\ 1 & 0 & 0 \\ 0 & 0 & 1 \end{bmatrix}$$

5. **(c)** This transformation leaves the y-coordinate of every point fixed. However it sends (1,0,0) to (0,0,-1) and (0,0,1) to (1,0,0). The standard matrix is therefore

$$\begin{bmatrix} 0 & 0 & 1 \\ 0 & 1 & 0 \\ -1 & 0 & 0 \end{bmatrix}$$

12. **(b)** To reduce this matrix to the identity matrix, we add -2 times Row 1 to Row 2 and then add -4 times Row 2 to Row 1. These operations when performed on I yield

$$\begin{bmatrix} 1 & 0 \\ -2 & 1 \end{bmatrix} \quad \text{and} \quad \begin{bmatrix} 1 & -4 \\ 0 & 1 \end{bmatrix}$$

respectively. The inverses of the above two matrices are

$$\begin{bmatrix} 1 & 0 \\ 2 & 1 \end{bmatrix} \quad \text{and} \quad \begin{bmatrix} 1 & 4 \\ 0 & 1 \end{bmatrix}$$

respectively. Thus,

$$\begin{bmatrix} 1 & 4 \\ 2 & 9 \end{bmatrix} = \begin{bmatrix} 1 & 0 \\ 2 & 1 \end{bmatrix} \begin{bmatrix} 1 & 4 \\ 0 & 1 \end{bmatrix}$$

and therefore the transformation represents a shear by a factor of 4 in the x-direction followed by a shear by a factor of 2 in the y-direction.

(c) To reduce this matrix to the identity matrix, we multiply Row 1 by -1/2, multiply Row 2 by 1/4, and interchange Rows 1 and 2. These operations when performed on I yield

$$\begin{bmatrix} -1/2 & 0 \\ 0 & 1 \end{bmatrix} \quad \begin{bmatrix} 1 & 0 \\ 0 & 1/4 \end{bmatrix} \quad \begin{bmatrix} 0 & 1 \\ 1 & 0 \end{bmatrix}$$

respectively. The inverses of the above matrices are

$$\begin{bmatrix} -2 & 0 \\ 0 & 1 \end{bmatrix} \quad \begin{bmatrix} 1 & 0 \\ 0 & 4 \end{bmatrix} \quad \begin{bmatrix} 0 & 1 \\ 1 & 0 \end{bmatrix}$$

respectively. Thus,

$$\begin{bmatrix} 0 & -2 \\ 4 & 0 \end{bmatrix} = \begin{bmatrix} -2 & 0 \\ 0 & 1 \end{bmatrix}\begin{bmatrix} 1 & 0 \\ 0 & 4 \end{bmatrix}\begin{bmatrix} 0 & 1 \\ 1 & 0 \end{bmatrix}$$

Therefore, the transformation represents a reflection about the line $y = x$ followed by expansions by factors of 4 and -2 in the y- and x-directions, respectively. Note that the order in which the two expansion matrices occur is immaterial. However the position of the reflection matrix cannot be changed.

We return to the original matrix and note that we could, of course, interchange Rows 1 and 2 first and then multiply Row 1 by 1/4 and Row 2 by $-1/2$. This yields the factorization

$$\begin{bmatrix} 0 & -2 \\ 4 & 0 \end{bmatrix} = \begin{bmatrix} 0 & 1 \\ 1 & 0 \end{bmatrix}\begin{bmatrix} 4 & 0 \\ 0 & 1 \end{bmatrix}\begin{bmatrix} 1 & 0 \\ 0 & -2 \end{bmatrix}$$

Here the order of the two expansion matrices can also be interchanged.

Note that there are two other factorizations, for a grand total of six.

13. **(a)** $$\begin{bmatrix} 1 & 0 \\ 0 & 5 \end{bmatrix}\begin{bmatrix} 1/2 & 0 \\ 0 & 1 \end{bmatrix} = \begin{bmatrix} 1/2 & 0 \\ 0 & 5 \end{bmatrix}$$

(c) $$\begin{bmatrix} -1 & 0 \\ 0 & -1 \end{bmatrix}\begin{bmatrix} 0 & 1 \\ 1 & 0 \end{bmatrix} = \begin{bmatrix} 0 & -1 \\ -1 & 0 \end{bmatrix}$$

14. **(a)** $$\begin{bmatrix} 0 & 1 \\ 1 & 0 \end{bmatrix}\begin{bmatrix} 5 & 0 \\ 0 & 1 \end{bmatrix}\begin{bmatrix} -1 & 0 \\ 0 & 1 \end{bmatrix} = \begin{bmatrix} 0 & 1 \\ -5 & 0 \end{bmatrix}$$

15. **(b)** The matrices which represent compressions along the x- and y-axes are $\begin{bmatrix} k & 0 \\ 0 & 1 \end{bmatrix}$ and $\begin{bmatrix} 1 & 0 \\ 0 & k \end{bmatrix}$, respectively, where $0 < k < 1$. But

$$\begin{bmatrix} k & 0 \\ 0 & 1 \end{bmatrix}^{-1} = \begin{bmatrix} 1/k & 0 \\ 0 & 1 \end{bmatrix}$$

and

$$\begin{bmatrix} 1 & 0 \\ 0 & k \end{bmatrix}^{-1} = \begin{bmatrix} 1 & 0 \\ 0 & 1/k \end{bmatrix}$$

Since $0 < k < 1$ implies that $1/k > 1$, the result follows.

(c) The matrices which represent reflections about the x- and y-axes are $\begin{bmatrix} 1 & 0 \\ 0 & -1 \end{bmatrix}$ and $\begin{bmatrix} -1 & 0 \\ 0 & 1 \end{bmatrix}$, respectively. Since these matrices are their own inverses, the result follows.

16. Since $A^{-1} = \begin{bmatrix} -2 & 3 \\ -3 & 4 \end{bmatrix}$, it follows that points (x',y') on the image line must satisfy the equations

$$\begin{aligned} x &= -2x' + 3y' \\ y &= -3x' + 4y' \end{aligned}$$

where $y = -4x + 3$. Hence $-3x' + 4y' = 8x' - 12y' + 3$, or $11x' - 16y' + 3 = 0$. That is, the image of $y = -4x + 3$ has the equation $11x - 16y + 3 = 0$.

17. **(a)** The matrix which represents this shear is $\begin{bmatrix} 1 & 3 \\ 0 & 1 \end{bmatrix}$; its inverse is $\begin{bmatrix} 1 & -3 \\ 0 & 1 \end{bmatrix}$. Thus, points (x',y') on the image line must satisfy the equations

$$\begin{aligned} x &= x' - 3y' \\ y &= y' \end{aligned}$$

where $y = 2x$. Hence $y' = 2x' - 6y'$, or $2x' - 7y' = 0$. That is, the equation of the image line is $2x - 7y = 0$.

Alternatively, we could note that the transformation leaves (0,0) fixed and sends (1,2) to (7,2). Thus (0,0) and (7,2) determine the image line which has the equation $2x - 7y = 0$.

(c) The reflection and its inverse are both represented by the matrix $\begin{bmatrix} 0 & 1 \\ 1 & 0 \end{bmatrix}$. Thus the point (x',y') on the image line must satisfy the equations

$$\begin{aligned} x &= y' \\ y &= x' \end{aligned}$$

where $y = 2x$. Hence $x' = 2y'$, so the image line has the equation $x - 2y = 0$.

(e) The rotation can be represented by the matrix $\begin{bmatrix} 1/2 & -\sqrt{3}/2 \\ \sqrt{3}/2 & 1/2 \end{bmatrix}$.

This sends the origin to itself and the point (1,2) to the point $((1 - 2\sqrt{3})/2,\ (2 + \sqrt{3})/2)$. Since both (0,0) and (1,2) lie on the line $y = 2x$, their images determine the image of the line under the required rotation. Thus, the image line has the equation $(2 + \sqrt{3})x + (2\sqrt{3} - 1)y = 0$.

Alternatively, we could find the inverse of the matrix,

$\begin{bmatrix} 1/2 & \sqrt{3}/2 \\ -\sqrt{3}/2 & 1/2 \end{bmatrix}$, and proceed as we did in Parts (a) and (c).

18. Since the shear is represented by the matrix $\begin{bmatrix} 1 & k \\ 0 & 1 \end{bmatrix}$, it will send (2,1) to $(2+k,k)$ and (3,0) to (3,0). The origin, of course, remains fixed. Thus we must find a value of k for which the vectors $(2+k,k)$ and (3,0) are orthogonal; that is,

$$(2+k,k) \cdot (3,0) = 3(2+k) = 0$$

Clearly $k = -2$.

20. The equation of a line in the plane is $Ax + By + C = 0$ where not both A and B are zero. Moreover

$$\begin{bmatrix} a & b \\ c & d \end{bmatrix}^{-1} = \frac{1}{ad - bc}\begin{bmatrix} d & -b \\ -c & a \end{bmatrix}$$

where $ad - bc \neq 0$. Thus (x,y) is transformed into (x',y') where

$$x = \frac{dx' - by'}{ad - bc}$$

$$y = \frac{-cx' + ay'}{ad - bc}$$

Therefore, the line $Ax + By + C = 0$ is transformed into

$$A\,\frac{dx - by}{ad - bc} + B\,\frac{-cx + ay}{ad - bc} + C = 0$$

or

$$\frac{dA - cB}{ad - bc}\,x + \frac{-bA + aB}{ad - bc}\,y + C = 0$$

This is the equation of a line provided $dA - cB$ and $-bA + aB$ are not both zero. But if both numbers are zero, then we have

$$bdA - bcB = 0$$
$$-bdA + adB = 0$$

or

$$(ad - bc)B = 0$$

This implies that $B = 0$ since $ad - bc \neq 0$. However, $B = 0$ implies that $dA = bA = 0$. Thus either $A = 0$ or $d = b = 0$. But since $ad - bc \neq 0$, b and d cannot both be zero. Hence A must equal zero. Finally, since not both A and B can equal zero, then not both $dA - cB$ and $-bA + aB$ can equal zero.

21. We use the notation and the calculations of Exercise 20. If the line $Ax + By + C = 0$ passes through the origin, then $C = 0$, and the image line has the equation $(dA - cB)x + (-bA + aB)y = 0$. Thus it also must pass through the origin.

The two lines $A_1x + B_1y + C_1 = 0$ and $A_2x + B_2y + C_2 = 0$ are parallel if and only if $A_1B_2 = A_2B_1$. Their image lines are parallel if and only if

$$(dA_1 - cB_1)(-bA_2 + aB_2) = (dA_2 - cB_2)(-bA_1 + aB_1)$$

or

$$bcA_2B_1 + adA_1B_2 = bcA_1B_2 + adA_2B_1$$

or

$$(ad - bc)(A_1B_2 - A_2B_1) = 0$$

or

$$A_1B_2 - A_2B_1 = 0$$

Thus the image lines are parallel if and only if the given lines are parallel.

22. **(a)** Since x is fixed, but y and z are interchanged, we have

$$\begin{bmatrix} 1 & 0 & 0 \\ 0 & 0 & 1 \\ 0 & 1 & 0 \end{bmatrix}$$

23. **(a)** The matrix which transforms (x,y,z) to $(x + kz,\ y + kz,\ z)$ is

$$\begin{bmatrix} 1 & 0 & k \\ 0 & 1 & k \\ 0 & 0 & 1 \end{bmatrix}$$

24. **(a)** A reflection about the x-axis will not change the length of any vector. Thus the only possible eigenvalues are 1 and −1. The only two unit vectors which are transformed to vectors with the same or opposite direction are the standard basis vectors, $\mathbf{e}_1$ and $\mathbf{e}_2$. The vector $\mathbf{e}_1$ goes to itself and $\mathbf{e}_2$ goes to $-\mathbf{e}_2$. Therefore $\mathbf{e}_1$ and $\mathbf{e}_2$ are linearly independent eigenvectors with corresponding eigenvalues 1 and −1.

This is easily checked using the matrix

$$A = \begin{bmatrix} 1 & 0 \\ 0 & -1 \end{bmatrix}$$

with characteristic equation $(\lambda - 1)(\lambda + 1) = 0$. The eigenspaces will be the x and y axes.

(c) This is similar to Part (a) above only the two lines through the origin which remain fixed will be $y = x$ and $y = -x$. Thus 2 linearly independent eigenvectors will be (1,1) and (−1,1). The corresponding eigenvalues will be 1 and −1.

Again, this is easily checked using the matrix

$$A = \begin{bmatrix} 0 & 1 \\ 1 & 0 \end{bmatrix}$$

with characteristic equation $\lambda^2 - 1 = 0$.

(e) The only line through the origin which remains fixed when $k \neq 0$ is the y-axis. Since distance along the y-axis is not changed, the only possible eigenvalue is $\lambda = 1$. A corresponding eigenvector is $\mathbf{e}_2 = (0,1)$.

(f) The only angles through which we can rotate R^2 and keep a line through the origin fixed are integer multiples of 0, π, and 2π. Since rotations by integer multiples of 0 and 2π leave the plane unchanged, we consider only rotations by odd multiples of π. These all leave the x-axis fixed but reverse its direction. Thus such rotations have $\mathbf{e}_1 = (1,0)$ as an eigenvector and -1 as an eigenvalue.

EXERCISE SET 8.3

1. We have

$$\begin{bmatrix} a \\ b \end{bmatrix} = \left[\begin{bmatrix} 1 & 0 \\ 1 & 1 \\ 1 & 2 \end{bmatrix}^t \begin{bmatrix} 1 & 0 \\ 1 & 1 \\ 1 & 2 \end{bmatrix}\right]^{-1} \begin{bmatrix} 1 & 0 \\ 1 & 1 \\ 1 & 2 \end{bmatrix}^t \begin{bmatrix} 0 \\ 2 \\ 7 \end{bmatrix}$$

$$= \begin{bmatrix} 3 & 3 \\ 3 & 5 \end{bmatrix}^{-1} \begin{bmatrix} 9 \\ 16 \end{bmatrix} = \begin{bmatrix} \frac{5}{6} & -\frac{1}{2} \\ -\frac{1}{2} & \frac{1}{2} \end{bmatrix} \begin{bmatrix} 9 \\ 16 \end{bmatrix}$$

$$= \begin{bmatrix} -1/2 \\ 7/2 \end{bmatrix}$$

Thus the desired line is $y = -1/2 + (7/2)x$.

3. Here

$$M = \begin{bmatrix} 1 & 2 & 4 \\ 1 & 3 & 9 \\ 1 & 5 & 25 \\ 1 & 6 & 36 \end{bmatrix}$$

and

$$\begin{bmatrix} a_1 \\ a_2 \\ a_3 \end{bmatrix} = (M^t M)^{-1} M^t \begin{bmatrix} 0 \\ -10 \\ -48 \\ -76 \end{bmatrix}$$

$$= \begin{bmatrix} 4 & 16 & 74 \\ 16 & 74 & 376 \\ 74 & 376 & 2018 \end{bmatrix}^{-1} \begin{bmatrix} -134 \\ -726 \\ -4026 \end{bmatrix}$$

$$= \begin{bmatrix} \frac{221}{10} & -\frac{62}{5} & \frac{3}{2} \\ -\frac{62}{5} & \frac{649}{90} & -\frac{8}{9} \\ \frac{3}{2} & -\frac{8}{9} & \frac{1}{9} \end{bmatrix} \begin{bmatrix} -134 \\ -726 \\ -4026 \end{bmatrix}$$

$$= \begin{bmatrix} 2 \\ 5 \\ -3 \end{bmatrix}$$

Thus the desired quadratic is $y = 2 + 5x - 3x^2$.

5. Since M is an $m \times n$ matrix, M^t is $n \times m$ and hence M^tM is $n \times n$. We can show that the matrix M^tM is invertible if we can show that its rows are linearly independent, which implies that its rank is n. We are given that M has n linearly independent column vectors. Write

$$M = \left[\ \mathbf{c}_1 \ \vdots\ \mathbf{c}_2 \ \vdots\ \cdots \ \vdots\ \mathbf{c}_n \right] \quad \text{and} \quad M^t = \begin{bmatrix} \mathbf{c}_1 \\ \mathbf{c}_2 \\ \vdots \\ \mathbf{c}_n \end{bmatrix}$$

Thus

$$M^tM = \begin{bmatrix} \mathbf{c}_1\cdot\mathbf{c}_1 & \mathbf{c}_1\cdot\mathbf{c}_2 & \cdots & \mathbf{c}_1\cdot\mathbf{c}_n \\ \vdots & \vdots & & \vdots \\ \mathbf{c}_n\cdot\mathbf{c}_1 & \mathbf{c}_n\cdot\mathbf{c}_2 & \cdots & \mathbf{c}_n\cdot\mathbf{c}_n \end{bmatrix}$$

If we suppose that some linear combination of the rows of this matrix equals **0**, then we have constants $a_1, \cdots, a_n$ such that

$$a_1\begin{bmatrix} \mathbf{c}_1\cdot\mathbf{c}_1 & \mathbf{c}_1\cdot\mathbf{c}_2 & \cdots & \mathbf{c}_1\cdot\mathbf{c}_n \end{bmatrix} + \cdots$$
$$+ a_n\begin{bmatrix} \mathbf{c}_n\cdot\mathbf{c}_1 & \mathbf{c}_n\cdot\mathbf{c}_2 & \cdots & \mathbf{c}_n\cdot\mathbf{c}_n \end{bmatrix} = \begin{bmatrix} 0 & 0 & \cdots & 0 \end{bmatrix}$$

or

$$\begin{aligned} a_1\mathbf{c}_1\cdot\mathbf{c}_1 + \cdots + a_n\mathbf{c}_n\cdot\mathbf{c}_1 &= 0 \\ a_1\mathbf{c}_1\cdot\mathbf{c}_2 + \cdots + a_n\mathbf{c}_n\cdot\mathbf{c}_2 &= 0 \\ \vdots \qquad\qquad & \vdots \\ a_1\mathbf{c}_1\cdot\mathbf{c}_n + \cdots + a_n\mathbf{c}_n\cdot\mathbf{c}_n &= 0 \end{aligned}$$

That is,

$$\begin{aligned} \left[a_1\mathbf{c}_1 + \cdots + a_n\mathbf{c}_n\right]\cdot\mathbf{c}_1 &= 0 \\ \left[a_1\mathbf{c}_1 + \cdots + a_n\mathbf{c}_n\right]\cdot\mathbf{c}_2 &= 0 \\ \vdots \qquad\qquad & \vdots \\ \left[a_1\mathbf{c}_1 + \cdots + a_n\mathbf{c}_n\right]\cdot\mathbf{c}_n &= 0 \end{aligned}$$

Since the vectors $\mathbf{c}_1, \cdots, \mathbf{c}_n$ are linearly independent, the vector $\mathbf{v} = a_1\mathbf{c}_1 + \cdots + a_n\mathbf{c}_n$ is the zero vector if and only if $a_1 = a_2 = \cdots = a_n = 0$. Now $\mathbf{v}$ lies in the space S spanned by the vectors $\mathbf{c}_1, \cdots, \mathbf{c}_n$ and is, by the above equations, orthogonal to each of these n vectors. That is, **v** is in S, but is orthogonal to every vector in S and hence to itself. Thus **v** must be the zero vector, so that all the numbers $a_i = 0$. That is, if a linear combination of the rows of M^tM is **0**, then all of the coefficients in that combination must be 0, so that the rows are linearly independent. This completes the proof.

6. The two column vectors of M are linearly independent if and only if neither is a nonzero multiple of the other. Since all of the entries in the first column are equal, the columns are linearly independent if and only if the second column has at least two different entries, or if and only if at least two of the numbers x_i are distinct, or if and only if not all of the points (x_i, y_i) lie on the same vertical line.

7. Since we have

$$M = \begin{bmatrix} 1 & x_1 & x_1^2 & \cdots & x_1^m \\ \vdots & \vdots & \vdots & & \vdots \\ 1 & x_n & x_n^2 & \cdots & x_n^m \end{bmatrix}$$

the columns of M will be linearly independent if and only if the only solution to

$$a_0 \begin{bmatrix} 1 & \cdots & 1 \end{bmatrix}^t + a_1 \begin{bmatrix} x_1 & \cdots & x_n \end{bmatrix}^t + a_2 \begin{bmatrix} x_1^2 & \cdots & x_n^2 \end{bmatrix}^t + \cdots a_m \begin{bmatrix} x_1^m & \cdots & x_n^m \end{bmatrix}^t = \begin{bmatrix} 0 & \cdots & 0 \end{bmatrix}^t$$

is $a_0 = a_1 = \cdots = a_m = 0$. That is, if and only if the system of equations

$$\begin{array}{ccccccccc} a_0 & + & a_1x_1 & + & a_2x_1^2 & + \cdots + & a_mx_1^m & = & 0 \\ \vdots & & \vdots & & \vdots & & \vdots & & \vdots \\ a_0 & + & a_1x_n & + & a_2x_n^2 & + \cdots + & a_mx_n^m & = & 0 \end{array}$$

has only the trivial solution for the numbers $a_0, \cdots, a_m$. But if any of the numbers $a_i \neq 0$, then there are at least $m + 1$ distinct numbers x_i which must satisfy the equation

$$a_0 + a_1x + a_2x^2 + \cdots a_mx^m = 0$$

This is a polynomial of degree at most m and can therefore have at most m distinct real roots. Thus, indeed, $a_0 = \cdots = a_m = 0$.

9. Let x_i denote the month, starting with January = 1, and y_i denote the sales for that month. Then we have

$$\begin{bmatrix} a_0 \\ a_1 \\ a_2 \end{bmatrix} = \left[M^tM \right]^{-1} M^t \begin{bmatrix} 4.0 \\ 4.4 \\ 5.2 \\ 6.4 \\ 8.0 \end{bmatrix}$$

where

$$M = \begin{bmatrix} 1 & 1 & 1^2 \\ 1 & 2 & 2^2 \\ 1 & 3 & 3^2 \\ 1 & 4 & 4^2 \\ 1 & 5 & 5^2 \end{bmatrix} \qquad M^tM = \begin{bmatrix} 5 & 15 & 55 \\ 15 & 55 & 225 \\ 55 & 225 & 979 \end{bmatrix}$$

$$\left[M^t M \right]^{-1} = \begin{bmatrix} \frac{23}{5} & -\frac{33}{10} & \frac{1}{2} \\ -\frac{33}{10} & \frac{187}{70} & -\frac{3}{7} \\ \frac{1}{2} & -\frac{3}{7} & \frac{1}{14} \end{bmatrix} \quad \text{and} \quad M^t y = \begin{bmatrix} 28.0 \\ 94.0 \\ 370.8 \end{bmatrix}$$

so that the desired quadratic is $y = 4 - .2x + .2x^2$. When $x = 12$, $y = 30.4$, so the projected sales for December are \$30,400.

EXERCISE SET 8.4

1. **(a)** Since $f(x) = 1 + x$, we have

$$a_0 = \frac{1}{\pi}\int_0^{2\pi}(1 + x)\,dx = 2 + 2\pi$$

Using Example 1 and some simple integration, we obtain

$$a_k = \frac{1}{\pi}\int_0^{2\pi}(1 + x)\cos(kx)\,dx = 0$$

$$k = 1,2,\ldots$$

$$b_k = \frac{1}{\pi}\int_0^{2\pi}(1 + x)\sin(kx)\,dx = -\frac{2}{k}$$

Thus, the least squares approximation to $1 + x$ on $[0,2\pi]$ by a trigonometric polynomial of order ≤ 2 is

$$1 + x \simeq (1 + \pi) - 2\sin x - \sin 2x$$

2. **(a)** Since $f(x) = x^2$,

$$a_0 = \frac{1}{\pi}\int_0^{2\pi}x^2\,dx = \frac{8}{3}\pi^2$$

Using Example 1 and integration by parts or a table of integrals, we find that

$$a_k = \frac{1}{\pi}\int_0^{2\pi} x^2\cos(kx)\,dx = -\frac{2}{k\pi}\int_0^{2\pi} x\sin(kx)\,dx = \frac{4}{k^2}$$

$$b_k = \frac{1}{\pi}\int_0^{2\pi} x^2\sin(kx)\,dx = -\frac{(2\pi)^2}{k\pi} + \frac{2}{k\pi}\int_0^{2\pi} x\cos(kx)\,dx = -\frac{4\pi}{k}$$

$$k = 1,2,\ldots$$

Thus, the least squares approximation to x^2 on $[0,2\pi]$ by a trigonometric polynomial of order at most 3 is

$$x^2 \simeq \frac{4}{3}\pi^2 + 4\cos x + \cos 2x + \frac{4}{9}\cos 3x - 4\pi\sin x$$
$$-2\pi\sin 2x - \frac{4\pi}{3}\sin 3x$$

3. (a) The space W of continuous functions of the form $a + be^x$ over $[0,1]$ is spanned by the functions $\mathbf{u}_1 = 1$ and $\mathbf{u}_2 = e^x$. First we use the Gram-Schmidt process to find an orthonormal basis $\{\mathbf{v}_1, \mathbf{v}_2\}$ for W.

Since $\langle \mathbf{f},\mathbf{g}\rangle = \int_0^1 f(x)g(x)\,dx$, then $\|\mathbf{u}_1\| = 1$ and hence

$$\mathbf{v}_1 = 1$$

Thus

$$\mathbf{v}_2 = \frac{e^x - \langle e^x,1\rangle 1}{\|e^x - \langle e^x,1\rangle 1\|} = \frac{e^x - e + 1}{a}$$

where a is the constant

$$a = \|e^x - e + 1\| = \left[\int_0^1 \left[e^x - e + 1\right]^2 dx\right]^{1/2}$$

$$= \left[\frac{(3 - e)(e - 1)}{2}\right]^{1/2}$$

Therefore the orthogonal projection of x on W is

$$\begin{aligned}
\text{proj}_W x &= \langle x,1\rangle 1 + \left\langle x, \frac{e^x - e + 1}{a}\right\rangle \frac{e^x - e + 1}{a} \\
&= \int_0^1 x\,dx + \frac{e^x - e + 1}{a}\int_0^1 \frac{x(e^x - e + 1)}{a}\,dx \\
&= \frac{1}{2} + \frac{e^x - e + 1}{a}\left[\frac{3 - e}{2a}\right] \\
&= \frac{1}{2} + \left[\frac{1}{e - 1}\right](e^x - e + 1) \\
&= -\frac{1}{2} + \left[\frac{1}{e - 1}\right]e^x
\end{aligned}$$

(b) The mean square error is

$$\begin{aligned}
\int_0^1 \left[x - \left(-\frac{1}{2} + \left[\frac{1}{e - 1}\right]e^x\right)\right]^2 dx &= \frac{13}{12} + \frac{1 + e}{2(1 - e)} \\
&= \frac{1}{12} + \frac{3 - e}{2(1 - e)}
\end{aligned}$$

The answer above is deceptively short since a great many calculations are involved.

To shortcut some of the work, we derive a different expression for the mean square error (m.s.e.). By definition,

$$\begin{aligned}
\text{m.s.e.} &= \int_a^b [f(x) - g(x)]^2 dx \\
&= \|\mathbf{f} - \mathbf{g}\|^2 \\
&= \langle \mathbf{f} - \mathbf{g}, \mathbf{f} - \mathbf{g}\rangle \\
&= \langle \mathbf{f}, \mathbf{f}-\mathbf{g}\rangle - \langle \mathbf{g}, \mathbf{f} - \mathbf{g}\rangle
\end{aligned}$$

Recall that $\mathbf{g} = \text{proj}_W\mathbf{f}$, so that $\mathbf{g}$ and $\mathbf{f} - \mathbf{g}$ are orthogonal (see Figure 3). Therefore,

$$\begin{aligned}
\text{m.s.e.} &= \langle \mathbf{f}, \mathbf{f} - \mathbf{g}\rangle \\
&= \langle \mathbf{f},\mathbf{f}\rangle - \langle \mathbf{f},\mathbf{g}\rangle
\end{aligned}$$

But $\mathbf{g} = \langle \mathbf{f},\mathbf{v}_1\rangle\mathbf{v}_1 + \langle \mathbf{f},\mathbf{v}_2\rangle\mathbf{v}_2$, so that

$$(*) \qquad \text{m.s.e.} = \langle \mathbf{f},\mathbf{f}\rangle - \langle \mathbf{f},\mathbf{v}_1\rangle^2 - \langle \mathbf{f},\mathbf{v}_2\rangle^2$$

Now back to the problem at hand. We know $\langle \mathbf{f},\mathbf{v}_1\rangle$ and $\langle \mathbf{f},\mathbf{v}_2\rangle$ from Part (a). Thus, in this case,

$$\text{m.s.e.} = \int_0^1 x^2 dx - \left[\frac{1}{2}\right]^2 - \left[\frac{3 - e}{2a}\right]^2$$

$$= \frac{1}{12} - \frac{3 - e}{2(e - 1)} \approx .0014$$

Clearly the formula (*) above can be generalized. If W is an n-dimensional space with orthonormal basis $\{\mathbf{v}_1,\mathbf{v}_2,\cdots,\mathbf{v}_n\}$, then

$$(**) \qquad \text{m.s.e.} = \|\mathbf{f}\|^2 - \langle \mathbf{f},\mathbf{v}_1\rangle^2 - \cdots - \langle \mathbf{f},\mathbf{v}_n\rangle^2$$

4. **(a)** The space W of polynomials of the form $a_0 + a_1x$ over $[0,1]$ has the basis $\{1,x\}$. The Gram-Schmidt process yields the orthonormal basis $\{1, \sqrt{3}(2x - 1)\}$. Therefore, the orthogonal projection of e^x on W is

$$\text{proj}_W e^x = \langle e^x,1\rangle 1 + \langle e^x, \sqrt{3}(2x - 1)\rangle\sqrt{3}(2x - 1)$$

$$= e - 1 + \sqrt{3}(3 - e)[\sqrt{3}(2x - 1)]$$

$$= (4e - 10) + (18 - 6e)x$$

(b) From Part (a) and (*) in the solution to 3.(b), we have

$$\text{m.s.e.} = \int_0^1 e^{2x}dx - (e - 1)^2 - [\sqrt{3}(3 - e)]^2$$

$$= \frac{(3 - e)(7e - 19)}{2} \approx .004$$

5. (a) The space V of polynomials of the form $a_0 + a_1x + a_2x^2$ over $[-1,1]$ has the basis $\{1,x,x^2\}$. Using the inner product $\langle\mathbf{u},\mathbf{v}\rangle = \int_{-1}^{1} u(x)v(x)\,dx$ and the Gram-Schmidt process, we obtain the orthonormal basis

$$\left\{\frac{1}{\sqrt{2}},\ \sqrt{\frac{3}{2}}\,x,\ \frac{1}{2}\sqrt{\frac{5}{2}}\,(3x^2 - 1)\right\}$$

(See Exercise 27, Section 5.3.) Thus

$$\langle\sin \pi x,\mathbf{v}_1\rangle = \frac{1}{\sqrt{2}}\int_{-1}^{1} \sin(\pi x)\,dx = 0$$

$$\langle\sin \pi x,\mathbf{v}_2\rangle = \sqrt{\frac{3}{2}}\int_{-1}^{1} x\sin(\pi x)\,dx = \frac{2}{\pi}\sqrt{\frac{3}{2}}$$

$$\langle\sin \pi x,\mathbf{v}_3\rangle = \frac{1}{2}\sqrt{\frac{5}{2}}\int_{-1}^{1} (3x^2 - 1)\sin(\pi x)\,dx = 0$$

Therefore,

$$\sin \pi x \simeq \frac{3}{\pi}\,x$$

5. (b) From Part (a) and (**) in the solution to 3.(b), we have

$$\text{m.s.e.} = \int_{-1}^{1} \sin^2(\pi x)\,dx - \frac{6}{\pi^2} = 1 - \frac{6}{\pi^2} \approx .39$$

8. The Fourier series of $\pi - x$ is

$$\frac{a_0}{2} + \sum_{k=1}^{\infty} \left[a_k\cos kx + b_k\sin kx\right]$$

where

$$a_0 = \frac{1}{\pi}\int_0^{2\pi} (\pi - x)\,dx = 0$$

and

$$a_k = \frac{1}{\pi}\int_0^{2\pi}(\pi - x)\cos(kx)\,dx = 0$$

$$k = 1,2,\ldots$$

$$b_k = \frac{1}{\pi}\int_0^{2\pi}(\pi - x)\sin(kx)\,dx = \frac{2}{k}$$

Thus the required Fourier series is $\sum_{k=1}^{\infty} \frac{2}{k} \sin kx$. Notice that this is consistent with the result of Example 1, Part (b).

EXERCISE SET 8.5

1. **(d)** Since one of the terms in this expression is the product of 3 rather than 2 variables, the expression is not a quadratic form.

5. **(b)** The quadratic form can be written as

$$\begin{bmatrix} x_1 & x_2 \end{bmatrix} \begin{bmatrix} 7 & 1/2 \\ 1/2 & 4 \end{bmatrix} \begin{bmatrix} x_1 \\ x_2 \end{bmatrix} = \mathbf{x}^t A\mathbf{x}$$

The characteristic equation of A is $(\lambda - 7)(\lambda - 4) - 1/4 = 0$ or

$$4\lambda^2 - 44\lambda + 111 = 0$$

which gives us

$$\lambda = \frac{11 \pm \sqrt{10}}{2}$$

If we solve for the eigenvectors, we find that for $\lambda = \dfrac{11 + \sqrt{10}}{2}$

$$x_1 = (3 + \sqrt{10})x_2$$

and for $\lambda = \dfrac{11 + \sqrt{10}}{2}$

$$x_1 = (3 - \sqrt{10})x_2$$

Therefore the normalized eigenvectors are

$$\begin{bmatrix} \dfrac{3+\sqrt{10}}{\sqrt{20+6\sqrt{10}}} \\ \dfrac{1}{\sqrt{20+6\sqrt{10}}} \end{bmatrix} \quad \text{and} \quad \begin{bmatrix} \dfrac{3-\sqrt{10}}{\sqrt{20-6\sqrt{10}}} \\ \dfrac{1}{\sqrt{20-6\sqrt{10}}} \end{bmatrix}$$

or, if we simplify,

$$\begin{bmatrix} \dfrac{1}{\sqrt{20-6\sqrt{10}}} \\ \dfrac{1}{\sqrt{20+6\sqrt{10}}} \end{bmatrix} \quad \text{and} \quad \begin{bmatrix} \dfrac{-1}{\sqrt{20+6\sqrt{10}}} \\ \dfrac{1}{\sqrt{20-6\sqrt{10}}} \end{bmatrix}$$

Thus the maximum value of the given form with its constraint is

$$\frac{11+\sqrt{10}}{2} \quad \text{at} \quad x_1 = \frac{1}{\sqrt{20-6\sqrt{10}}} \quad \text{and} \quad x_2 = \frac{1}{\sqrt{20+6\sqrt{10}}}$$

The minimum value is

$$\frac{11-\sqrt{10}}{2} \quad \text{at} \quad x_1 = \frac{-1}{\sqrt{20+6\sqrt{10}}} \quad \text{and} \quad x_2 = \frac{1}{\sqrt{20-6\sqrt{10}}}$$

6. **(b)** We write the quadratic form as

$$\begin{bmatrix} x_1 & x_2 & x_3 \end{bmatrix} \begin{bmatrix} 2 & 1 & 1 \\ 1 & 1 & 0 \\ 1 & 0 & 1 \end{bmatrix} \begin{bmatrix} x_1 \\ x_2 \\ x_3 \end{bmatrix} = \mathbf{x}^t A\mathbf{x}$$

The characteristic equation of A is

$$\lambda^3 - 4\lambda^2 + 3\lambda = \lambda(\lambda - 3)(\lambda - 1) = 0$$

with roots $\lambda = 0, 1, 3$. The corresponding normalized eigenvectors are

$$\begin{bmatrix} 1/\sqrt{3} \\ -1/\sqrt{3} \\ -1/\sqrt{3} \end{bmatrix} \quad \begin{bmatrix} 0 \\ 1/\sqrt{2} \\ -1/\sqrt{2} \end{bmatrix} \quad \begin{bmatrix} 2/\sqrt{6} \\ 1/\sqrt{6} \\ 1/\sqrt{6} \end{bmatrix}$$

respectively. Thus the maximum value of the given form with its constraint is

$$3 \text{ at } x_1 = 2/\sqrt{6},\ x_2 = 1/\sqrt{6},\ x_3 = 1/\sqrt{6}$$

The minimum value is

$$0 \text{ at } x_1 = 1/\sqrt{3},\ x_2 = -1/\sqrt{3},\ x_3 = -1/\sqrt{3}$$

7. **(b)** The eigenvalues of this matrix are the roots of the equation $\lambda^2 - 10\lambda + 24 = 0$. They are $\lambda = 6$ and $\lambda = 4$ which are both positive. Therefore the matrix is positive definite.

8. **(b)** The principal submatrices are

$$[5] \quad \text{and} \quad \begin{bmatrix} 5 & -1 \\ -1 & 5 \end{bmatrix}$$

Their determinants are 5 and 24, so the matrix is positive definite.

9. **(b)** The characteristic equation of this matrix is $\lambda^3 - 3\lambda + 2 = (\lambda + 1)^2(\lambda - 2)$. Since two of the eigenvalues are negative, the matrix is not positive definite.

11. **(a)** Since $x_1^2 + x_2^2 > 0$ unless $x_1 = x_2 = 0$, the form is positive definite.

(c) Since $(x_1 - x_2)^2 \geq 0$ the form is positive semidefinite. It is not positive definite because it can equal zero whenever $x_1 = x_2$ even when $x_1 = x_2 \neq 0$.

11. **(e)** If $|x_1| > |x_2|$, then the form has a positive value, but if $|x_1| < |x_2|$, then the form has a negative value. Thus it is indefinite.

12. **(a)** The related quadratic form is $3x_1^2 - 2x_2^2 + x_3^2$ which is indefinite since its values can be both positive and negative. Hence the matrix is indefinite.

(c) The related quadratic form is $6x_1^2 + 9x_2^2 + x_3^2 + 7x_1x_2 + x_1x_3 + 2x_2x_3$. The possible signs of the values which this form can assume are not obvious and its characteristic equation is rather messy. However, if we convert the matrix to the symmetric one

$$\begin{bmatrix} 6 & 7/2 & 1/2 \\ 7/2 & 9 & 1 \\ 1/2 & 1 & 1 \end{bmatrix}$$

The determinant of every principal submatrix is positive. Hence the matrix is positive definite.

13. **(a)** By definition,

$$\begin{aligned} T(\mathbf{x} + \mathbf{y}) &= (\mathbf{x} + \mathbf{y})^t A(\mathbf{x} + \mathbf{y}) \\ &= (\mathbf{x}^t + \mathbf{y}^t)A(\mathbf{x} + \mathbf{y}) \\ &= \mathbf{x}^t A\mathbf{x} + \mathbf{x}^t A\mathbf{y} + \mathbf{y}^t A\mathbf{x} + \mathbf{y}^t A\mathbf{y} \\ &= T(\mathbf{x}) + \mathbf{x}^t A\mathbf{y} + \left[\mathbf{x}^t A^t \mathbf{y}\right]^t + T(\mathbf{y}) \\ &= T(\mathbf{x}) + \mathbf{x}^t A\mathbf{y} + \mathbf{x}^t A^t \mathbf{y} + T(\mathbf{y}) \end{aligned}$$

(The transpose of a 1×1 matrix is itself.)

$$= T(\mathbf{x}) + 2\mathbf{x}^t A\mathbf{y} + T(\mathbf{y})$$

(Assuming that A is symmetric, $A^t = A$.)

(b) We have

$$\begin{aligned} T(k\mathbf{x}) &= (k\mathbf{x})^t A(k\mathbf{x}) \\ &= k^2\mathbf{x}^t A\mathbf{x} \qquad \text{(Every term has a factor of } k^2.\text{)} \\ &= k^2 T(\mathbf{x}) \end{aligned}$$

(c) The transformation is not linear because $T(k\mathbf{x}) \neq kT(\mathbf{x})$ unless $k = 1$ by Part (b).

14. **(b)** The symmetric matrix associated with the quadratic form is

$$\begin{bmatrix} 5 & 2 & -1 \\ 2 & 1 & -1 \\ -1 & -1 & k \end{bmatrix}$$

The determinants of the principal submatrices are 5, 1, and $k - 2$. These are positive provided $k > 2$. Thus the quadratic form is positive definite if and only if $k > 2$.

15. If we expand the quadratic form, it becomes

$$c_1^2x_1^2 + c_2^2x_2^2 + \cdots + c_n^2x_n^2 + 2c_1c_2x_1x_2 + 2c_1c_3x_1x_3 + \cdots + 2c_1c_nx_1x_n + 2c_2c_3x_2x_3 + \cdots + 2c_{n-1}c_nx_{n-1}x_n$$

Thus we have

$$A = \begin{bmatrix} c_1^2 & c_1c_2 & c_1c_3 & \cdots & c_1c_n \\ c_1c_2 & c_2^2 & c_2c_3 & \cdots & c_2c_n \\ c_1c_3 & c_2c_3 & c_3^2 & \cdots & c_3c_n \\ \vdots & \vdots & \vdots & & \vdots \\ c_1c_n & c_2c_n & c_3c_n & \cdots & c_n^2 \end{bmatrix}$$

and the quadratic form is given by $\mathbf{x}^t A\mathbf{x}$ where $\mathbf{x} = \begin{bmatrix} x_1 & x_2 & \cdots & x_n \end{bmatrix}^t$.

16. **(a)** We have

$$s_{\mathbf{x}}^2 = \frac{1}{n-1}\left[x_1^2 + x_2^2 + \cdots + x_n^2 - 2\bar{x}\left[x_1 + \cdots + x_n\right] + n\bar{x}^2\right]$$

$$= \frac{1}{n-1}\left[x_1^2+\cdots+x_n^2 - \frac{2}{n}\left[x_1+\cdots+x_n\right]^2 + \frac{1}{n}\left[x_1+\cdots+x_n\right]^2\right]$$

$$= \frac{1}{n-1}\left[x_1^2+\cdots+x_n^2 - \frac{1}{n}\left[x_1^2+\cdots+x_n^2 + 2\left[x_1x_2+\cdots+x_{n-1}x_n\right]\right]\right]$$

$$= \frac{1}{n-1}\left[\frac{n-1}{n}\left[x_1^2 + \cdots + x_n^2\right] - \frac{2}{n}\left[x_1x_2 + \cdots + x_{n-1}x_n\right]\right]$$

Thus $s_{\mathbf{x}}^2 = \mathbf{x}^t A\mathbf{x}$ where

$$A = \begin{bmatrix} \frac{1}{n} & \frac{-1}{n(n-1)} & \frac{-1}{n(n-1)} & \cdots & \frac{-1}{n(n-1)} \\ \frac{-1}{n(n-1)} & \frac{1}{n} & \frac{-1}{n(n-1)} & \cdots & \frac{-1}{n(n-1)} \\ \vdots & \vdots & \vdots & & \vdots \\ \frac{-1}{n(n-1)} & \frac{-1}{n(n-1)} & \frac{-1}{n(n-1)} & \cdots & \frac{1}{n} \end{bmatrix}$$

(b) Since $s_{\mathbf{x}}^2$ is $1/(n-1)$ times a sum of squares, it cannot be negative. In fact, it can only be zero if $(x_i - \bar{x}) = 0$ for $i = 1, \cdots, n$, which can only be true in case $x_1 = x_2 = \cdots = x_n$. If we let all the values of x_i be equal but not zero, we have that $s_{\mathbf{x}}^2 = 0$ while $\mathbf{x} \neq \mathbf{0}$. Thus the form is positive semidefinite.

17. To show that $\lambda_n \leq \mathbf{x}^t A\mathbf{x}$ if $\|\mathbf{x}\| = 1$, we use the equation from the proof dealing with λ_1 and the fact that λ_n is the smallest eigenvalue to obtain

$$\begin{aligned}
\mathbf{x}^t A\mathbf{x} = \langle\mathbf{x}, A\mathbf{x}\rangle &= \lambda_1\langle\mathbf{x},\mathbf{v}_1\rangle^2 + \lambda_2\langle\mathbf{x},\mathbf{v}_2\rangle^2 + \cdots + \lambda_n\langle\mathbf{x},\mathbf{v}_n\rangle^2 \\
&\geq \lambda_n\langle\mathbf{x},\mathbf{v}_1\rangle^2 + \lambda_n\langle\mathbf{x},\mathbf{v}_2\rangle^2 + \cdots + \lambda_n\langle\mathbf{x},\mathbf{v}_n\rangle^2 \\
&= \lambda_n\left[\langle\mathbf{x},\mathbf{v}_1\rangle^2 + \cdots + \langle\mathbf{x},\mathbf{v}_n\rangle^2\right] \\
&= \lambda_n
\end{aligned}$$

Now suppose that $\mathbf{x}$ is an eigenvector of A corresponding to λ_n. As in the proof dealing with λ_1, we have

$$\mathbf{x}^t A\mathbf{x} = \langle\mathbf{x}, A\mathbf{x}\rangle = \langle\mathbf{x}, \lambda_n\mathbf{x}\rangle = \lambda_n\langle\mathbf{x},\mathbf{x}\rangle = \lambda_n\|\mathbf{x}\|^2 = \lambda_n$$

EXERCISE SET 8.6

1. **(a)** The quadratic form $\mathbf{x}^t A\mathbf{x}$ can be written as

$$\begin{bmatrix} x_1 & x_2 \end{bmatrix} \begin{bmatrix} 2 & -1 \\ -1 & 2 \end{bmatrix} \begin{bmatrix} x_1 \\ x_2 \end{bmatrix}$$

The characteristic equation of A is $\lambda^2 - 4\lambda + 3 = 0$. The eigenvalues are $\lambda = 3$ and $\lambda = 1$. The corresponding eigenspaces are spanned by the vectors

$$\begin{bmatrix} 1 \\ -1 \end{bmatrix} \quad \text{and} \quad \begin{bmatrix} 1 \\ 1 \end{bmatrix}$$

respectively. These vectors are orthogonal. If we normalize them, we can use the result to obtain a matrix P such that the substitution $\mathbf{x} = P\mathbf{y}$ or

$$\begin{bmatrix} x_1 \\ x_2 \end{bmatrix} = \begin{bmatrix} 1/\sqrt{2} & 1/\sqrt{2} \\ -1/\sqrt{2} & 1/\sqrt{2} \end{bmatrix} \begin{bmatrix} y_1 \\ y_2 \end{bmatrix}$$

will eliminate the cross-product term. This yields the new quadratic form

$$\begin{bmatrix} y_1 & y_2 \end{bmatrix} \begin{bmatrix} 3 & 0 \\ 0 & 1 \end{bmatrix} \begin{bmatrix} y_1 \\ y_2 \end{bmatrix}$$

or $3y_1^2 + y_2^2$.

2. **(a)** The quadratic form can be written as $\mathbf{x}^t A\mathbf{x}$ where

$$\mathbf{x}^t = \begin{bmatrix} x_1 & x_2 & x_3 \end{bmatrix}$$

and

$$A = \begin{bmatrix} 3 & 2 & 0 \\ 2 & 4 & -2 \\ 0 & -2 & 5 \end{bmatrix}$$

The characteristic equation of A is

$$\lambda^3 - 12\lambda^2 + 39\lambda - 28 = (\lambda - 7)(\lambda - 4)(\lambda - 1) = 0$$

The eigenspaces corresponding to $\lambda = 7$, $\lambda = 4$, and $\lambda = 1$ are spanned by

$$\begin{bmatrix} 1 \\ 2 \\ -2 \end{bmatrix} \quad \begin{bmatrix} 2 \\ 1 \\ 2 \end{bmatrix} \quad \text{and} \quad \begin{bmatrix} 2 \\ -2 \\ -1 \end{bmatrix}$$

respectively. If we normalize these vectors, we obtain the matrix

$$P = \begin{bmatrix} 1/3 & 2/3 & 2/3 \\ 2/3 & 1/3 & -2/3 \\ -2/3 & 2/3 & -1/3 \end{bmatrix}$$

By direct calculation, we have

$$P^t AP = \begin{bmatrix} 7 & 0 & 0 \\ 0 & 4 & 0 \\ 0 & 0 & 1 \end{bmatrix} = B$$

so the substitution $\mathbf{x} = P\mathbf{y}$ will yield the quadratic form

$$\mathbf{y}^t B\mathbf{y} \quad \text{or} \quad 7y_1^2 + 4y_2^2 + y_3^2$$

7. **(a)** If we complete the squares, then the equation $9x^2 + 4y^2 - 36x - 24y + 36 = 0$ becomes

$$9(x^2 - 4x + 4) + 4(y^2 - 6y + 9) = -36 + 9(4) + 4(9)$$

or

$$9(x - 2)^2 + 4(y - 3)^2 = 36$$

or

$$\frac{(x')^2}{4} + \frac{(y')^2}{9} = 1$$

This is an ellipse.

(c) If we complete the square, then $y^2 - 8x - 14y + 49 = 0$ becomes

$$y^2 - 14y + 49 = 8x$$

or

$$(y - 7)^2 = 8x$$

This is the parabola $(y')^2 = 8x'$.

(e) If we complete the squares, then $2x^2 - 3y^2 + 6x + 20y = -41$ becomes

$$2\left[x^2 + 3x + \frac{9}{4}\right] - 3\left[y^2 - \frac{20}{3}y + \frac{100}{9}\right] = -41 + \frac{9}{2} - \frac{100}{3}$$

or

$$2\left[x + \frac{3}{2}\right]^2 - 3\left[y - \frac{10}{3}\right]^2 = -\frac{419}{6}$$

or

$$12(x')^2 - 18(y')^2 = -419$$

This is the hyperbola

$$\frac{(y')^2}{\frac{419}{18}} - \frac{(x')^2}{\frac{419}{12}} = 1$$

8. **(a)** The matrix form for the conic $2x^2 - 4xy - y^2 + 8 = 0$ is

(*) $$\mathbf{x}^t A\mathbf{x} + 8 = 0$$

where

$$A = \begin{bmatrix} 2 & -2 \\ -2 & -1 \end{bmatrix}$$

The eigenvalues of A are $\lambda_1 = 3$ and $\lambda_2 = -2$ and the eigenspaces are spanned by the vectors

$$\begin{bmatrix} -2 \\ 1 \end{bmatrix} \quad \text{and} \quad \begin{bmatrix} 1 \\ 2 \end{bmatrix}$$

respectively. Normalizing these vectors yields

$$\begin{bmatrix} \frac{-2}{\sqrt{5}} \\ \frac{1}{\sqrt{5}} \end{bmatrix} \quad \text{and} \quad \begin{bmatrix} \frac{1}{\sqrt{5}} \\ \frac{2}{\sqrt{5}} \end{bmatrix}$$

We choose

$$P = \begin{bmatrix} \frac{1}{\sqrt{5}} & \frac{-2}{\sqrt{5}} \\ \frac{2}{\sqrt{5}} & \frac{1}{\sqrt{5}} \end{bmatrix}$$

Note that $\det(P) = 1$. (Interchanging columns yields a matrix with determinant -1.) If we substitute $\mathbf{x} = P\mathbf{x}'$ into (*), we have

$$(P\mathbf{x}')^t A(P\mathbf{x}') + 8 = 0$$

or

$$(\mathbf{x}')^t (P^t AP)\mathbf{x}' + 8 = 0$$

Since

$$P^t AP = \begin{bmatrix} -2 & 0 \\ 0 & 3 \end{bmatrix}$$

this yields the hyperbola

$$-2(x')^2 + 3(y')^2 + 8 = 0$$

Note that if we had let

$$P = \begin{bmatrix} \frac{2}{\sqrt{5}} & \frac{1}{\sqrt{5}} \\ \frac{-1}{\sqrt{5}} & \frac{2}{\sqrt{5}} \end{bmatrix}$$

then $\det(P) = 1$ and the rotation would have produced the diagonal matrix

$$P^t AP = \begin{bmatrix} 3 & 0 \\ 0 & -2 \end{bmatrix}$$

and hence the hyperbola

$$3(x')^2 - 2(y')^2 + 8 = 0$$

That is, there are two different rotations which will put the hyperbola into the two different standard positions.

8. **(b)** The matrix form for the conic $x^2 + 2xy + y^2 + 8x + y = 0$ is

$$\mathbf{x}^t A\mathbf{x} + K\mathbf{x} = 0$$

where

$$A = \begin{bmatrix} 1 & 1 \\ 1 & 1 \end{bmatrix} \quad \text{and} \quad K = [8 \quad 1]$$

The eigenvalues of A are $\lambda_1 = 0$ and $\lambda_2 = 2$ and the eigenspaces are spanned by the vectors

$$\begin{bmatrix} -1 \\ 1 \end{bmatrix} \quad \text{and} \quad \begin{bmatrix} 1 \\ 1 \end{bmatrix}$$

respectively. Thus, one possibility is

$$P_1 = \begin{bmatrix} \frac{1}{\sqrt{2}} & \frac{-1}{\sqrt{2}} \\ \frac{1}{\sqrt{2}} & \frac{1}{\sqrt{2}} \end{bmatrix}$$

Note that $\det(P_1) = 1$. Hence

$$P_1^t A P_1 = \begin{bmatrix} 2 & 0 \\ 0 & 0 \end{bmatrix} \quad \text{and} \quad KP_1 = \begin{bmatrix} \frac{9}{\sqrt{2}} & \frac{-7}{\sqrt{2}} \end{bmatrix}$$

Another possibility is

$$P_2 = \begin{bmatrix} \frac{1}{\sqrt{2}} & \frac{1}{\sqrt{2}} \\ \frac{-1}{\sqrt{2}} & \frac{1}{\sqrt{2}} \end{bmatrix}$$

so that

$$P_2^t A P_2 = \begin{bmatrix} 0 & 0 \\ 0 & 2 \end{bmatrix} \quad \text{and} \quad KP_2 = \begin{bmatrix} \frac{7}{\sqrt{2}} & \frac{9}{\sqrt{2}} \end{bmatrix}$$

A third possibility is

$$P_3 = \begin{bmatrix} \frac{-1}{\sqrt{2}} & \frac{-1}{\sqrt{2}} \\ \frac{1}{\sqrt{2}} & \frac{-1}{\sqrt{2}} \end{bmatrix}$$

so that

$$P_3^tAP_3 = \begin{bmatrix} 0 & 0 \\ 0 & 2 \end{bmatrix} \quad \text{and} \quad KP_3 = \begin{bmatrix} \frac{-7}{\sqrt{2}} & \frac{-9}{\sqrt{2}} \end{bmatrix}$$

Finally, we could let

$$P_4 = \begin{bmatrix} \frac{-1}{\sqrt{2}} & \frac{1}{\sqrt{2}} \\ \frac{-1}{\sqrt{2}} & \frac{-1}{\sqrt{2}} \end{bmatrix}$$

so that

$$P_4^tAP_4 = \begin{bmatrix} 2 & 0 \\ 0 & 0 \end{bmatrix} \quad \text{and} \quad KP_4 = \begin{bmatrix} \frac{-9}{\sqrt{2}} & \frac{7}{\sqrt{2}} \end{bmatrix}$$

Thus if we let $\mathbf{x} = P_i\mathbf{x}'$, so that

$$(\mathbf{x}')^t(P_i^tAP_i)\mathbf{x}' + KP_i\mathbf{x}' = 0 \qquad \text{for } i = 1,\ldots,4$$

we obtain the equations

$$2(x')^2 + \frac{9}{\sqrt{2}}x' - \frac{7}{\sqrt{2}}y' = 0 \qquad (i = 1)$$

$$2(y')^2 + \frac{7}{\sqrt{2}}x' + \frac{9}{\sqrt{2}}y' = 0 \qquad (i = 2)$$

$$2(y')^2 - \frac{7}{\sqrt{2}}x' - \frac{9}{\sqrt{2}}y' = 0 \qquad (i = 3)$$

$$2(x')^2 - \frac{9}{\sqrt{2}}x' + \frac{7}{\sqrt{2}}y' = 0 \qquad (i = 4)$$

These represent the parabola $(x'')^2 = \frac{7}{2\sqrt{2}}y''$ in each of its four standard positions.

8. (c) The matrix form for the conic $5x^2 + 4xy + 5y^2 = 9$ is

$$\mathbf{x}^t A\mathbf{x} = 9$$

where

$$A = \begin{bmatrix} 5 & 2 \\ 2 & 5 \end{bmatrix}$$

The eigenvalues of A are $\lambda_1 = 3$ and $\lambda_2 = 7$ and the eigenspaces are spanned by the vectors

$$\begin{bmatrix} -1 \\ 1 \end{bmatrix} \quad \text{and} \quad \begin{bmatrix} 1 \\ 1 \end{bmatrix}$$

If we normalize these vectors, we obtain

$$\begin{bmatrix} \frac{-1}{\sqrt{2}} \\ \frac{1}{\sqrt{2}} \end{bmatrix} \quad \text{and} \quad \begin{bmatrix} \frac{1}{\sqrt{2}} \\ \frac{1}{\sqrt{2}} \end{bmatrix}$$

Hence we can let

$$P = \begin{bmatrix} \frac{1}{\sqrt{2}} & \frac{-1}{\sqrt{2}} \\ \frac{1}{\sqrt{2}} & \frac{1}{\sqrt{2}} \end{bmatrix}$$

Note that $\det(P) = 1$. Thus if we let $\mathbf{x} = P\mathbf{x}'$, then we have

$$(\mathbf{x}')^t(P^tAP)\mathbf{x}' = 9$$

where

$$P^tAP = \begin{bmatrix} 7 & 0 \\ 0 & 3 \end{bmatrix}$$

This yields the ellipse,

$$7(x')^2 + 3(y')^2 = 9$$

Had we let

$$P = \begin{bmatrix} \frac{1}{\sqrt{2}} & \frac{1}{\sqrt{2}} \\ \frac{-1}{\sqrt{2}} & \frac{1}{\sqrt{2}} \end{bmatrix}$$

then we would have obtained

$$3(x')^2 + 7(y')^2 = 9$$

which is the same ellipse rotated 90°.

9. The matrix form for the conic $9x^2 - 4xy + 6y^2 - 10x - 20y = 5$ is

$$\mathbf{x}^t A\mathbf{x} + K\mathbf{x} = 5$$

where

$$A = \begin{bmatrix} 9 & -2 \\ -2 & 6 \end{bmatrix} \quad \text{and} \quad K = [-10 \quad -20]$$

The eigenvalues of A are $\lambda_1 = 5$ and $\lambda_2 = 10$ and the eigenspaces are spanned by the vectors

$$\begin{bmatrix} 1 \\ 2 \end{bmatrix} \quad \text{and} \quad \begin{bmatrix} -2 \\ 1 \end{bmatrix}$$

Thus we can let

$$P = \begin{bmatrix} \frac{1}{\sqrt{5}} & \frac{-2}{\sqrt{5}} \\ \frac{2}{\sqrt{5}} & \frac{1}{\sqrt{5}} \end{bmatrix}$$

Note that $\det(P) = 1$. If we let $\mathbf{x} = P\mathbf{x}'$, then

$$(\mathbf{x}')^t(P^tAP)\mathbf{x}' + KP\mathbf{x}' = 5$$

where

$$P^tAP = \begin{bmatrix} 5 & 0 \\ 0 & 10 \end{bmatrix} \quad \text{and} \quad KP = [-10\sqrt{5} \quad 0]$$

Thus we have the equation

$$5(x')^2 + 10(y')^2 - 10\sqrt{5}\, x' = 5$$

If we complete the square, we obtain the equation

$$5\left[(x')^2 - 2\sqrt{5}\, x' + 5\right] + 10(y')^2 = 5 + 25$$

or

$$(x'')^2 + 2(y'')^2 = 6$$

where $x'' = x' - \sqrt{5}$ and $y'' = y'$. This is the ellipse

$$\frac{(x'')^2}{6} + \frac{(y'')^2}{3} = 1$$

Of course we could also rotate to obtain the same ellipse in the form $2(x'')^2 + (y'')^2 = 6$, which is just the other standard position.

11. The matrix form for the conic $2x^2 - 4xy - y^2 - 4x - 8y = -14$ is

$$\mathbf{x}^tA\mathbf{x} + K\mathbf{x} = -14$$

where

$$A = \begin{bmatrix} 2 & -2 \\ -2 & -1 \end{bmatrix} \quad \text{and} \quad K = [-4 \quad -8]$$

The eigenvalues of A are $\lambda_1 = 3$, $\lambda_2 = -2$ and the eigenspaces are spanned by the vectors

$$\begin{bmatrix} -2 \\ 1 \end{bmatrix} \quad \text{and} \quad \begin{bmatrix} 1 \\ 2 \end{bmatrix}$$

Thus we can let

$$P = \begin{bmatrix} \frac{2}{\sqrt{5}} & \frac{1}{\sqrt{5}} \\ \frac{-1}{\sqrt{5}} & \frac{2}{\sqrt{5}} \end{bmatrix}$$

Note that $\det(P) = 1$. If we let $\mathbf{x} = P\mathbf{x}'$, then

$$(\mathbf{x}')^t(P^tAP)\mathbf{x}' + KP\mathbf{x}' = -14$$

where

$$P^tAP = \begin{bmatrix} 3 & 0 \\ 0 & -2 \end{bmatrix} \quad \text{and} \quad KP = [0 \quad -4\sqrt{5}]$$

Thus we have the equation

$$3(x')^2 - 2(y')^2 - 4\sqrt{5}\, y' = -14$$

If we complete the square, then we obtain

$$3(x')^2 - 2\left[(y')^2 + 2\sqrt{5}\, y' + 5\right] = -14 - 10$$

or

$$3(x'')^2 - 2(y'')^2 = -24$$

where $x'' = x'$ and $y'' = y' + \sqrt{5}$. This is the hyperbola

$$\frac{(y'')^2}{12} - \frac{(x'')^2}{8} = 1$$

Of course, we could also rotate to obtain the same hyperbola in the form $2(x'')^2 - 3(y'')^2 = 24$.

14. The matrix form for the conic $4x^2 - 20xy + 25y^2 - 15x - 6y = 0$ is

$$\mathbf{x}^t A\mathbf{x} + K\mathbf{x} = 0$$

where

$$A = \begin{bmatrix} 4 & -10 \\ -10 & 25 \end{bmatrix} \quad \text{and} \quad K = \begin{bmatrix} -15 & -6 \end{bmatrix}$$

The eigenvalues of A are $\lambda_1 = 29$ and $\lambda_2 = 0$ and the eigenspaces are spanned by the vectors

$$\begin{bmatrix} -2 \\ 5 \end{bmatrix} \quad \text{and} \quad \begin{bmatrix} 5 \\ 2 \end{bmatrix}$$

Thus we can let

$$P = \begin{bmatrix} \frac{2}{\sqrt{29}} & \frac{5}{\sqrt{29}} \\ \frac{-5}{\sqrt{29}} & \frac{2}{\sqrt{29}} \end{bmatrix}$$

Note that $\det(P) = 1$. If we let $\mathbf{x} = P\mathbf{x}'$, then

$$(\mathbf{x}')^t(P^t AP)\mathbf{x}' + KP\mathbf{x}' = 0$$

where

$$P^t AP = \begin{bmatrix} 29 & 0 \\ 0 & 0 \end{bmatrix} \quad \text{and} \quad KP = \begin{bmatrix} 0 & -87/\sqrt{29} \end{bmatrix}$$

Thus we have the equation

$$29(x')^2 - \frac{87}{\sqrt{29}}\, y' = 0$$

which is the parabola $(x')^2 = \frac{3}{\sqrt{29}} y'$. The other standard positions for this parabola are represented by the equations

$$(x')^2 = -\frac{3}{\sqrt{29}} y' \qquad \text{and} \qquad (y')^2 = \pm \frac{3}{\sqrt{29}} x'$$

15. (a) The equation $x^2 - y^2 = 0$ can be written as $(x - y)(x + y) = 0$. Thus it represents the two intersecting lines $x \pm y = 0$.

(b) The equation $x^2 + 3y^2 + 7 = 0$ can be written as $x^2 + 3y^2 = -7$. Since the left side of this equation cannot be negative, then there are no points (x,y) which satisfy the equation.

(c) If $8x^2 + 7y^2 = 0$, then $x = y = 0$. Thus the graph consists of the single point $(0,0)$.

(d) This equation can be rewritten as $(x - y)^2 = 0$. Thus it represents the single line $y = x$.

(e) The equation $9x^2 + 12xy + 4y^2 - 52 = 0$ can be written as $(3x + 2y)^2 = 52$ or $3x + 2y = \pm\sqrt{52}$. Thus its graph is the two parallel lines $3x + 2y \pm 2\sqrt{13} = 0$.

(f) The equation $x^2 + y^2 - 2x - 4y = -5$ can be written as $x^2 - 2x + 1 + y^2 - 4y + 4 = 0$ or $(x - 1)^2 + (y - 2)^2 = 0$. Thus it represents the point $(1,2)$.

EXERCISE SET 8.7

5. **(a)** If we complete the squares, the quadric becomes

$$9(x^2 - 2x + 1) + 36(y^2 - 4y + 4) + 4(z^2 - 6z + 9) = -153 + 9 + 144 + 36$$

or

$$9(x - 1)^2 + 36(y - 2)^2 + 4(z - 3)^2 = 36$$

or

$$\frac{(x')^2}{4} + \frac{(y')^2}{1} + \frac{(z')^2}{9} = 1$$

This is an ellipsoid.

(c) If we complete the square, the quadric becomes

$$3(x^2 + 14x + 49) - 3y^2 - z^2 = -144 + 147$$

or

$$3(x + 7)^2 - 3y^2 - z^2 = 3$$

or

$$\frac{(x')^2}{1} - \frac{(y')^2}{1} - \frac{(z')^2}{3} = 1$$

This is a hyperboloid of two sheets.

5. **(e)** If we complete the squares, the quadric becomes

$$(x^2 + 2x + 1) + 16(y^2 - 2y + 1) - 16z = 15 + 1 + 16$$

or

$$(x + 1)^2 + 16(y - 1)^2 - 16(z + 2) = 0$$

or

$$\frac{(x')^2}{16} + \frac{(y')^2}{1} - z' = 0$$

This is an elliptic paraboloid.

(g) If we complete the squares, the quadric becomes

$$(x^2 - 2x + 1) + (y^2 + 4y + 4) + (z^2 - 6z + 9) = 11 + 1 + 4 + 9$$

or

$$(x - 1)^2 + (y + 2)^2 + (z - 3)^2 = 25$$

or

$$\frac{(x')^2}{25} + \frac{(y')^2}{25} + \frac{(z')^2}{25} = 1$$

This is a sphere.

6. **(a)** The matrix form for the quadric is $\mathbf{x}^t A\mathbf{x} + 150 = 0$ where

$$A = \begin{bmatrix} 2 & 0 & 36 \\ 0 & 3 & 0 \\ 36 & 0 & 23 \end{bmatrix}$$

The eigenvalues of A are $\lambda_1 = -25$, $\lambda_2 = 3$, and $\lambda_3 = 50$. The corresponding eigenspaces are spanned by the orthogonal vectors

$$\begin{bmatrix} 4 \\ 0 \\ -3 \end{bmatrix} \quad \begin{bmatrix} 0 \\ 1 \\ 0 \end{bmatrix} \quad \begin{bmatrix} 3 \\ 0 \\ 4 \end{bmatrix}$$

Thus we can let $\mathbf{x} = P\mathbf{x}'$ where

$$P = \begin{bmatrix} 4/5 & 0 & 3/5 \\ 0 & 1 & 0 \\ -3/5 & 0 & 4/5 \end{bmatrix} \quad \text{and} \quad P^tAP = \begin{bmatrix} -25 & 0 & 0 \\ 0 & 3 & 0 \\ 0 & 0 & 50 \end{bmatrix}$$

Note that $\det(P) = 1$. (This is the only reason for actually calculating P in this problem.) Using the above transformation, we obtain the quadric

$$-25(x')^2 + 3(y')^2 + 50(z')^2 + 150 = 0$$

or

$$25(x')^2 - 3(y')^2 - 50(z')^2 - 150 = 0$$

This is a hyperboloid of two sheets.

(c) The matrix form for the quadric is $\mathbf{x}^tA\mathbf{x} + K\mathbf{x} = 0$ where

$$A = \begin{bmatrix} 144 & 0 & -108 \\ 0 & 100 & 0 \\ -108 & 0 & 81 \end{bmatrix} \quad \text{and} \quad K = \begin{bmatrix} -540 & 0 & -720 \end{bmatrix}$$

The eigenvalues of A are $\lambda_1 = 225$, $\lambda_2 = 100$, and $\lambda_3 = 0$. The eigenspaces are spanned by the vectors

$$\begin{bmatrix} 4 \\ 0 \\ -3 \end{bmatrix} \quad \begin{bmatrix} 0 \\ 1 \\ 0 \end{bmatrix} \quad \begin{bmatrix} 3 \\ 0 \\ 4 \end{bmatrix}$$

Thus we can let $\mathbf{x} = P\mathbf{x}'$ where

$$P = \begin{bmatrix} 4/5 & 0 & 3/5 \\ 0 & 1 & 0 \\ -3/5 & 0 & 4/5 \end{bmatrix} \quad \text{and} \quad P^tAP = \begin{bmatrix} 225 & 0 & 0 \\ 0 & 100 & 0 \\ 0 & 0 & 0 \end{bmatrix}$$

Note that $\det(P) = 1$. Hence the quadric becomes

$$\mathbf{x}^t(P^tAP)\mathbf{x} + KP\mathbf{x}' = 0$$

or

$$225(x')^2 + 100(y')^2 - 900z' = 0$$

This is the elliptic paraboloid

$$9(x')^2 + 4(y')^2 - 36z' = 0$$

7. The matrix form for the quadric is $\mathbf{x}^t A\mathbf{x} + K\mathbf{x} = -9$ where

$$A = \begin{bmatrix} 0 & 1 & 1 \\ 1 & 0 & 1 \\ 1 & 1 & 0 \end{bmatrix} \quad \text{and} \quad K = \begin{bmatrix} -6 & -6 & -4 \end{bmatrix}$$

The eigenvalues of A are $\lambda_1 = \lambda_2 = -1$ and $\lambda_3 = 2$. and the vectors

$$\mathbf{u}_1 = \begin{bmatrix} -1 \\ 0 \\ 1 \end{bmatrix} \quad \mathbf{u}_2 = \begin{bmatrix} -1 \\ 1 \\ 0 \end{bmatrix} \quad \text{and} \quad \mathbf{u}_3 = \begin{bmatrix} 1 \\ 1 \\ 1 \end{bmatrix}$$

span the corresponding eigenspaces. Note that $\mathbf{u}_1 \cdot \mathbf{u}_3 = \mathbf{u}_2 \cdot \mathbf{u}_3 = 0$ but that $\mathbf{u}_1 \cdot \mathbf{u}_2 \neq 0$. Hence, we must apply the Gram-Schmidt process to $\{\mathbf{u}_1,\mathbf{u}_2\}$. We must also normalize $\mathbf{u}_3$. This gives the orthonormal set

$$\begin{bmatrix} \frac{-1}{\sqrt{2}} \\ 0 \\ \frac{1}{\sqrt{2}} \end{bmatrix} \quad \begin{bmatrix} \frac{-1}{\sqrt{6}} \\ \frac{2}{\sqrt{6}} \\ \frac{-1}{\sqrt{6}} \end{bmatrix} \quad \begin{bmatrix} \frac{1}{\sqrt{3}} \\ \frac{1}{\sqrt{3}} \\ \frac{-1}{\sqrt{3}} \end{bmatrix}$$

Thus we can let

$$P = \begin{bmatrix} \frac{1}{\sqrt{2}} & \frac{-1}{\sqrt{6}} & \frac{1}{\sqrt{3}} \\ 0 & \frac{2}{\sqrt{6}} & \frac{1}{\sqrt{3}} \\ \frac{-1}{\sqrt{2}} & \frac{-1}{\sqrt{6}} & \frac{1}{\sqrt{3}} \end{bmatrix}$$

Note that $\det(P) = 1$,

$$P^tAP = \begin{bmatrix} -1 & 0 & 0 \\ 0 & -1 & 0 \\ 0 & 0 & 2 \end{bmatrix} \quad \text{and} \quad KP = \begin{bmatrix} -\sqrt{2} & \frac{-2}{\sqrt{6}} & \frac{-16}{\sqrt{3}} \end{bmatrix}$$

Therefore the transformation $\mathbf{x} = P\mathbf{x}'$ reduces the quadric to

$$-(x')^2 - (y')^2 + 2(z')^2 - \sqrt{2}\,x' - \frac{2}{\sqrt{6}}\,y' - \frac{16}{\sqrt{3}}\,z' = -9$$

If we complete the squares, this becomes

$$\left[(x')^2 + \sqrt{2}\,x' + \frac{1}{2}\right] + \left[(y')^2 + \frac{2}{\sqrt{6}}\,y' + \frac{1}{6}\right]$$

$$- 2\left[(z')^2 - \frac{8}{\sqrt{3}}\,z' + \frac{16}{3}\right] = 9 + \frac{1}{2} + \frac{1}{6} - \frac{32}{3}$$

Letting $x'' = x' + \frac{1}{\sqrt{2}}$, $y'' = y' + \frac{1}{\sqrt{6}}$, $z'' = z' - \frac{4}{\sqrt{3}}$ yields

$$(x'')^2 + (y'')^2 - 2(z'')^2 = -1$$

This is the hyperboloid of two sheets

$$\frac{(z'')^2}{1/2} - \frac{(x'')^2}{1} - \frac{(y'')^2}{1} = 1$$

9. The matrix form for the quadric is $\mathbf{x}^t A\mathbf{x} + K\mathbf{x} - 31 = 0$ where

$$A = \begin{bmatrix} 0 & 1 & 0 \\ 1 & 0 & 0 \\ 0 & 0 & 0 \end{bmatrix} \quad \text{and} \quad K = [-6 \quad 10 \quad 1]$$

The eigenvalues of A are $\lambda_1 = 1$, $\lambda_2 = -1$, and $\lambda_3 = 0$, and the corresponding eigenspaces are spanned by the orthogonal vectors

$$\begin{bmatrix} 1 \\ 1 \\ 0 \end{bmatrix} \quad \begin{bmatrix} -1 \\ 1 \\ 0 \end{bmatrix} \quad \begin{bmatrix} 0 \\ 0 \\ 1 \end{bmatrix}$$

Thus, we let $\mathbf{x} = P\mathbf{x}'$ where

$$P = \begin{bmatrix} \frac{1}{\sqrt{2}} & \frac{-1}{\sqrt{2}} & 0 \\ \frac{1}{\sqrt{2}} & \frac{1}{\sqrt{2}} & 0 \\ 0 & 0 & 1 \end{bmatrix}$$

Note that $\det(P) = 1$,

$$P^t AP = \begin{bmatrix} 1 & 0 & 0 \\ 0 & -1 & 0 \\ 0 & 0 & 0 \end{bmatrix} \quad \text{and} \quad KP = \begin{bmatrix} 2\sqrt{2} & 8\sqrt{2} & 1 \end{bmatrix}$$

Therefore, the equation of the quadric is reduced to

$$(x')^2 - (y')^2 + 2\sqrt{2}\, x' + 8\sqrt{2}\, y' + z' - 31 = 0$$

If we complete the squares, this becomes

$$\left[(x')^2 + 2\sqrt{2}x' + 2\right] - \left[(y')^2 - 8\sqrt{2}y' + 32\right] + z' = 31 + 2 - 32$$

Letting $x'' = x' + \sqrt{2}$, $y'' = y' - 4\sqrt{2}$, and $z'' = z' - 1$ yields

$$(x'')^2 - (y'')^2 + z'' = 0$$

This is a hyperbolic paraboloid.

11. We know that the equation of a general quadric Q can be put into the matrix form $\mathbf{x}^t A\mathbf{x} + K\mathbf{x} + j = 0$ where

$$A = \begin{bmatrix} a & d & e \\ d & b & f \\ e & f & c \end{bmatrix} \qquad \text{and} \qquad K = \begin{bmatrix} g & h & i \end{bmatrix}$$

Since A is a symmetric matrix, then A is orthogonally diagonalizable by Theorem 6.3.1. Thus, by Theorem 6.2.1, A has 3 linearly independent eigenvectors. Now let T be the matrix whose column vectors are the 3 linearly independent eigenvectors of A. It follows from the proof of Theorem 6.2.1 and the discussion immediately following that theorem, that $T^{-1}AT$ will be a diagonal matrix whose diagonal entries are the eigenvalues λ_1, λ_2, and λ_3 of A. Theorem 6.3.3 guarantees that these eigenvalues are real.

As noted immediately after Theorem 6.3.2, we can, if necessary, transform the matrix T to a matrix S whose column vectors form an orthonormal set. To do this, orthonormalize the basis of each eigenspace before using its elements as column vectors of S.

Furthermore, by Theorem 5.4.3, we know that S is orthogonal; that is, $S^{-1} = S^t$. Hence $\det(S^{-1}) = \det(S^t)$, or, since $\det(S) \neq 0$,

$$\frac{1}{\det(S)} = \det(S)$$

This implies that $[\det(S)]^2 = 1$, or $\det(S) = \pm 1$.

In case $\det(S) = -1$, we interchange two columns in S to obtain a matrix P such that $\det(P) = 1$. If $\det(S) = 1$, we let $P = S$. Thus, by the assertion before Example 8 of Section 5.4, P represents a rotation. Note that P is orthogonal, so that $P^{-1} = P^t$, and also, that P orthogonally diagonalizes A. In fact,

$$P^tAP = \begin{bmatrix} \lambda_1 & 0 & 0 \\ 0 & \lambda_2 & 0 \\ 0 & 0 & \lambda_3 \end{bmatrix}$$

Hence, if we let $\mathbf{x} = P\mathbf{x}'$, then the equation of the quadric Q becomes

$$(\mathbf{x}')^t \left[P^t A P\right] \mathbf{x}' + KP\mathbf{x}' + j = 0$$

or

$$\lambda_1 (x')^2 + \lambda_2 (y')^2 + \lambda_3 (z')^2 + g'x' + h'y' + i'z' + j = 0$$

where

$$\begin{bmatrix} g' & h' & i' \end{bmatrix} = KP$$

Thus we have proved Theorem 8.7.1.

EXERCISE SET 9.1

1. If $AB = C$ where A is $m \times n$, B is $n \times p$, and C is $m \times p$, then C has $m \cdot p$ entries, each of the form

$$c_{ij} = a_{i1}b_{1j} + a_{i2}b_{2j} + \cdots + a_{in}b_{nj}$$

Thus we need n multiplications and $n - 1$ additions to compute each of the numbers c_{ij}. Therefore we need $m \cdot n \cdot p$ multiplications and $m \cdot (n - 1) \cdot p$ additions to compute C.

2. If A is an $n \times n$ matrix, then for any other $n \times n$ matrix, B, it will take (by Exercise 1) n^3 multiplications and $n^2(n - 1)$ additions to compute AB. To compute A^k, we must carry out $k - 1$ of these matrix multiplications, one to compute A^2, a second to compute $A^3 = A(A^2)$, and in all, $k - 1$ to compute $A^k = A(A^{k-1})$. Therefore we must carry out a total of $n^3(k - 1)$ multiplications and $n^2(n - 1)(k - 1)$ additions to compute A^k.

5. Following the hint, we have

$$S_n = 1 + 2 + 3 + \cdots + n$$

$$S_n = n + (n - 1) + (n - 2) + \cdots + 1$$

or

$$2S_n = (n + 1) + (n + 1) + (n + 1) + \cdots + (n + 1)$$

Thus

$$S_n = \frac{n(n+1)}{2}$$

7. **(a)** By direct computation,

$$(k+1)^3 - k^3 = k^3 + 3k^2 + 3k + 1 - k^3 = 3k^2 + 3k + 1$$

(b) The sum "telescopes". That is,

$$\begin{aligned}&[2^3 - 1^3] + [3^3 - 2^3] + [4^3 - 3^3] + \cdots + [(n+1)^3 - n^3]\\ &= 2^3 - 1^3 + 3^3 - 2^3 + 4^3 - 3^3 + \cdots + (n+1)^3 - n^3\\ &= (n+1)^3 - 1\end{aligned}$$

(c) By Parts (a) and (b), we have

$$\begin{aligned}&3(1)^2 + 3(1) + 1 + 3(2)^2 + 3(2) + 1 + 3(3)^2 + 3(3) + 1 + \cdots + 3n^2 + 3n + 1\\ &= 3(1^2 + 2^2 + 3^2 + \cdots + n^2) + 3(1 + 2 + 3 + \cdots + n) + n\\ &= (n+1)^3 - 1\end{aligned}$$

(d) Thus, by Part (c) and Exercise 6, we have

$$\begin{aligned}1^2 + 2^2 + 3^2 + \cdots + n^2 &= \frac{1}{3}\left[(n+1)^3 - 1 - 3\,\frac{n(n+1)}{2} - n\right]\\ &= \frac{(n+1)^3}{3} - \frac{1}{3} - \frac{n(n+1)}{2} - \frac{n}{3}\\ &= \frac{2(n+1)^3 - 2 - 3n(n+1) - 2n}{6}\\ &= \frac{(n+1)[2(n+1)^2 - 3n - 2]}{6}\\ &= \frac{(n+1)(2n^2 + 4n + 2 - 3n - 2)}{6}\end{aligned}$$

$$= \frac{(n + 1)(2n^2 + n)}{6}$$

$$= \frac{n(n + 1)(2n + 1)}{6}$$

9. Since R is a row-echelon form of an invertible $n \times n$ matrix, it has ones down the main diagonal and nothing but zeros below. If, as usual, $\mathbf{x} = [x_1 \quad x_2 \quad \cdots \quad x_n]^t$ and $\mathbf{b} = [b_1 \quad b_2 \quad \cdots \quad b_n]^t$, then we have $x_n = b_n$ with no computations. However $x_{n-1} = b_{n-1} - cx_n$ for some number c, so that it will require one multiplication and one addition to find x_{n-1}. In general,

$$x_i = b_i - \text{some linear combination of } x_{i+1}, x_{i+2}, \cdots, x_n$$

Therefore it will require two multiplications and two additions to find x_{n-2}, three of each to find x_{n-3}, and finally, $n - 1$ of each to find x_1. That is, it will require

$$1 + 2 + 3 + \cdots + (n - 1) = \frac{(n - 1)n}{2}$$

multiplications and the same number of additions to solve the system by back substitution.

10. To reduce an invertible $n \times n$ matrix to I_n, we first divide Row 1 by a_{11} (which we are assuming is not zero since no row interchanges are required). This requires $n - 1$ multiplications because we can ignore the first row, first column position which must be one. We then subtract a_{i1} times Row 1 from Row i for $i = 2, 3, \ldots, n$ to reduce the first column to that of I_n. This requires $(n - 1)^2$ multiplications and the same number of additions because we can ignore those entries which we know must be zero. Thus we have used $n - 1 + (n - 1)^2 = n(n - 1)$ multiplications and $(n - 1)^2$ additions.

We then repeat the process on the $(n - 1) \times (n - 1)$ submatrix obtained from the original by deleting the first row and the first column. This will require $(n - 1)(n - 2)$ multiplicaitons and $(n - 2)^2$ additions.

We continue in this fashion until we have reduced every diagonal entry to one and every entry below the diagonal to zero. The number of multiplications required so far is

$$\begin{aligned}
&n(n-1)+(n-1)(n-2)+(n-2)(n-3) + \cdots + 3 \cdot 2 + 2 \cdot 1 \\
&\quad = n^2 - n + (n-1)^2 - (n-1) + (n-2)^2 - (n-2) + \cdots + 3^2 - 3 + 2^2 - 2 \\
&\quad = [n^2 + (n-1)^2 + \cdots + 3^2 + 2^2] - [n + (n-1) + \cdots + 3 + 2] \\
&\quad = \frac{n(n+1)(2n+1)}{6} - 1 - \left[\frac{n(n+1)}{2} - 1\right] \\
&\quad = \frac{n^3 - n}{3}
\end{aligned}$$

and the number of additions is

$$\begin{aligned}
(n-1)^2 + (n-2)^2 + \cdots + 2^2 + 1^2 &= \frac{(n-1)(n)2n-1)}{6} \\
&= \frac{n^3}{3} - \frac{n^2}{2} + \frac{n}{6}
\end{aligned}$$

Finally, we must reduce the entries above the main diagonal to zero. We start with the nth column. We multiply Row n by a_{in} for $i = 1, 2, \ldots, n - 1$ and subtract the result from Row i. This would appear to involve $n - 1$ multiplications and the same number of additions since we can ignore everything but the entries in the last column. However, the result is a foregone conclusion. In fact, once we have reduced the matrix to row-echelon form, the final reduction is automatic. Thus the above formulas give all the multiplications and additions required.

11. To solve a linear system whose coefficient matrix is an invertible $n \times n$ matrix, A, we form the $n \times (n + 1)$ matrix $[A \,|\, \mathbf{b}]$ and reduce A to I_n. Thus we first divide Row 1 by a_{11}, using n multiplications (ignoring the multiplication whose result must be one and assuming that $a_{11} \neq 0$ since no row interchanges are required). We then subtract a_{i1} times Row 1 from Row i for i = 2, ..., n to reduce the first column to that of I_n. This requires $n(n - 1)$ multiplications and the same number of additions (again ignoring the operations whose results we already know). The total number of multiplications so far is n^2 and the total number of additions is $n(n - 1)$.

To reduce the second column to that of I_n, we repeat the procedure, starting with Row 2 and ignoring Column 1. Thus $n - 1$ multiplications assure us that there is a one on the main diagonal, and $(n - 1)^2$ multiplications and additions will make all $n - 1$ of the remaining column entries zero. This requires $n(n - 1)$ new multiplications and $(n - 1)^2$ new additions.

In general, to reduce Column i to the ith column of I_n, we require $n + 1 - i$ multiplications followed by $(n + 1 - i)(n - 1)$ multiplications and additions, for a total of $n(n + 1 - i)$ multiplications and $(n + 1 - i)(n - 1)$ additions.

If we add up all these numbers, we find that we need

$$\begin{aligned} n^2 + n(n-1) + n(n-2) + \cdots + n(2) + n(1) &= n\Big[n + (n-1) + \cdots + 2 + 1\Big] \\ &= \frac{n^2(n+1)}{2} \\ &= \frac{n^3}{2} + \frac{n^2}{2} \end{aligned}$$

multiplications and

$$n(n-1)+(n-1)^2+(n-2)(n-1)+\cdots+2(n-1)+(n-1)$$
$$=(n-1)\left[n+(n-1)+\cdots+2+1\right]$$
$$=\frac{(n-1)(n)(n+1)}{2}$$
$$=\frac{n^3}{2}-\frac{n}{2}$$

additions to compute the reduction.

12. Since $\lim_{x\to+\infty}(1/x)=0$, we have

$$\lim_{x\to+\infty}\frac{P(x)}{a_kx^k}=\lim_{x\to+\infty}\frac{a_0+a_1x+\cdots+a_kx^k}{a_kx^k}$$
$$=\lim_{x\to+\infty}\left[\frac{a_0}{a_kx^k}+\frac{a_1}{a_kx^{k-1}}+\cdots+\frac{a_{k-1}}{a_kx}+1\right]$$
$$=0+0+\cdots+0+1$$
$$=1$$

EXERCISE SET 9.2

1. The system in matrix form is

$$\begin{bmatrix} 3 & -6 \\ -2 & 5 \end{bmatrix} \begin{bmatrix} x_1 \\ x_2 \end{bmatrix} = \begin{bmatrix} 3 & 0 \\ -2 & 1 \end{bmatrix} \begin{bmatrix} 1 & -2 \\ 0 & 1 \end{bmatrix} \begin{bmatrix} x_1 \\ x_2 \end{bmatrix} = \begin{bmatrix} 0 \\ 1 \end{bmatrix}$$

This reduces to two matrix equations

$$\begin{bmatrix} 1 & -2 \\ 0 & 1 \end{bmatrix} \begin{bmatrix} x_1 \\ x_2 \end{bmatrix} = \begin{bmatrix} y_1 \\ y_2 \end{bmatrix}$$

and

$$\begin{bmatrix} 3 & 0 \\ -2 & 1 \end{bmatrix} \begin{bmatrix} y_1 \\ y_2 \end{bmatrix} = \begin{bmatrix} 0 \\ 1 \end{bmatrix}$$

The second matrix equation yields the system

$$\begin{aligned} 3y_1 \qquad &= 0 \\ -2y_1 + y_2 &= 1 \end{aligned}$$

which has $y_1 = 0$, $y_2 = 1$ as its solution. If we substitute these values into the first matrix equation, we obtain the system

$$\begin{aligned} x_1 - 2x_2 &= 0 \\ x_2 &= 1 \end{aligned}$$

This yields the final solution $x_1 = 2$, $x_2 = 1$.

3. To reduce the matrix of coefficients to a suitable upper triangular matrix, we carry out the following operations:

$$\begin{bmatrix} 2 & 8 \\ -1 & -1 \end{bmatrix} \rightarrow \begin{bmatrix} 1 & 4 \\ -1 & -1 \end{bmatrix} \rightarrow \begin{bmatrix} 1 & 4 \\ 0 & 3 \end{bmatrix} \rightarrow \begin{bmatrix} 1 & 4 \\ 0 & 1 \end{bmatrix} = U$$

These operations involve multipliers 1/2, 1, and 1/3. Thus the corresponding lower triangular matrix is

$$L = \begin{bmatrix} 2 & 0 \\ -1 & 3 \end{bmatrix}$$

We therefore have the two matrix equations

$$\begin{bmatrix} 1 & 4 \\ 0 & 1 \end{bmatrix} \begin{bmatrix} x_1 \\ x_2 \end{bmatrix} = \begin{bmatrix} y_1 \\ y_2 \end{bmatrix}$$

and

$$\begin{bmatrix} 2 & 0 \\ -1 & 3 \end{bmatrix} \begin{bmatrix} y_1 \\ y_2 \end{bmatrix} = \begin{bmatrix} -2 \\ -2 \end{bmatrix}$$

The second matrix equation yields the system

$$\begin{aligned} 2y_1 \phantom{{}+3y_2} &= -2 \\ -y_1 + 3y_2 &= -2 \end{aligned}$$

which has $y_1 = -1$, $y_2 = -1$ as its solution. If we substitute these values into the first matrix equation, we obtain the system

$$\begin{aligned} x_1 + 4x_2 &= -1 \\ x_2 &= -1 \end{aligned}$$

This yields the final solution $x_1 = 3$, $x_2 = -1$.

5. To reduce the matrix of coefficients to a suitable upper triangular matrix, we carry out the following operations:

$$\begin{bmatrix} 2 & -2 & -2 \\ 0 & -2 & 2 \\ -1 & 5 & 2 \end{bmatrix} \rightarrow \begin{bmatrix} 1 & -1 & -1 \\ 0 & -2 & 2 \\ -1 & 5 & 2 \end{bmatrix} \rightarrow \begin{bmatrix} 1 & -1 & -1 \\ 0 & -2 & 2 \\ 0 & 4 & 1 \end{bmatrix} \rightarrow$$

$$\begin{bmatrix} 1 & -1 & -1 \\ 0 & 1 & -1 \\ 0 & 4 & 1 \end{bmatrix} \rightarrow \begin{bmatrix} 1 & -1 & -1 \\ 0 & 1 & -1 \\ 0 & 0 & 5 \end{bmatrix} \rightarrow \begin{bmatrix} 1 & -1 & -1 \\ 0 & 1 & -1 \\ 0 & 0 & 1 \end{bmatrix} = U$$

These operations involve multipliers of 1/2, 0 (for the 2nd row), 1, -1/2, -4, and 1/5. Thus the corresponding lower triangular matrix is

$$L = \begin{bmatrix} 2 & 0 & 0 \\ 0 & -2 & 0 \\ -1 & 4 & 5 \end{bmatrix}$$

We therefore have the matrix equations

$$\begin{bmatrix} 1 & -1 & -1 \\ 0 & 1 & -1 \\ 0 & 0 & 1 \end{bmatrix} \begin{bmatrix} x_1 \\ x_2 \\ x_3 \end{bmatrix} = \begin{bmatrix} y_1 \\ y_2 \\ y_3 \end{bmatrix}$$

and

$$\begin{bmatrix} 2 & 0 & 0 \\ 0 & -2 & 0 \\ -1 & 4 & 5 \end{bmatrix} \begin{bmatrix} y_1 \\ y_2 \\ y_3 \end{bmatrix} = \begin{bmatrix} -4 \\ -2 \\ 6 \end{bmatrix}$$

The second matrix equation yields the system

$$\begin{aligned} 2y_1 \qquad\qquad\qquad &= -4 \\ -\ 2y_2 \qquad\quad &= -2 \\ -y_1 + 4y_2 + 5y_3 &= \ \ 6 \end{aligned}$$

which has $y_1 = -2$, $y_2 = 1$ and $y_3 = 0$ as its solution. If we substitute these values into the first matrix equation, we obtain the system

$$\begin{aligned} x_1 - x_2 - x_3 &= -2 \\ x_2 - x_3 &= 1 \\ x_3 &= 0 \end{aligned}$$

This yields the final solution $x_1 = -1$, $x_2 = 1$, $x_3 = 0$.

11. **(a)** To reduce A to row-echelon form, we carry out the following operations:

$$\begin{bmatrix} 2 & 1 & -1 \\ -2 & -1 & 2 \\ 2 & 1 & 0 \end{bmatrix} \longrightarrow \begin{bmatrix} 1 & 1/2 & -1/2 \\ -2 & -1 & 2 \\ 2 & 1 & 0 \end{bmatrix} \longrightarrow \begin{bmatrix} 1 & 1/2 & -1/2 \\ 0 & 0 & 1 \\ 2 & 1 & 0 \end{bmatrix}$$

$$\longrightarrow \begin{bmatrix} 1 & 1/2 & -1/2 \\ 0 & 0 & 1 \\ 0 & 0 & 1 \end{bmatrix} \longrightarrow \begin{bmatrix} 1 & 1/2 & -1/2 \\ 0 & 0 & 1 \\ 0 & 0 & 0 \end{bmatrix} = U$$

This involves multipliers 1/2, 2, −2, 1 (for the 2nd diagonal entry), and −1. Where no multiplier is needed in the second entry of the last row, we use the multiplier 1, thus obtaining the lower triangular matrix

$$L = \begin{bmatrix} 2 & 0 & 0 \\ -2 & 1 & 0 \\ 2 & 1 & 1 \end{bmatrix}$$

In fact, if we compute LU, we see that it will equal A no matter what entry we choose for the lower right-hand corner of L.

If we stop just before we reach row-echelon form, we obtain the matrices

$$U = \begin{bmatrix} 1 & 1/2 & -1/2 \\ 0 & 0 & 1 \\ 0 & 0 & 1 \end{bmatrix} \qquad L = \begin{bmatrix} 2 & 0 & 0 \\ -2 & 1 & 0 \\ 2 & 0 & 1 \end{bmatrix}$$

which will also serve.

(b) We have that $A = LU$ where

$$L = \begin{bmatrix} 2 & 0 & 0 \\ -2 & 1 & 0 \\ 2 & 1 & 1 \end{bmatrix} \qquad U = \begin{bmatrix} 1 & 1/2 & -1/2 \\ 0 & 0 & 1 \\ 0 & 0 & 0 \end{bmatrix}$$

If we let

$$L_1 = \begin{bmatrix} 1 & 0 & 0 \\ -1 & 1 & 0 \\ 1 & 1 & 1 \end{bmatrix} \quad \text{and} \quad D = \begin{bmatrix} 2 & 0 & 0 \\ 0 & 1 & 0 \\ 0 & 0 & 1 \end{bmatrix}$$

then $A = L_1DU$ as desired. (See the matrices at the very end of Section 9.2.)

(c) Let $U_2 = DU$ and note that this matrix is upper triangular. Then $A = L_1U_2$ is of the required form.

12. Suppose that

$$\begin{bmatrix} 0 & 1 \\ 1 & 0 \end{bmatrix} = \begin{bmatrix} a & 0 \\ b & c \end{bmatrix} \begin{bmatrix} d & e \\ 0 & f \end{bmatrix} = \begin{bmatrix} ad & ae \\ bd & be + cf \end{bmatrix}$$

This implies that $ad = 0$ but that $ae = bd = 1$. That is, if the matrix has an LU-decomposition, then we must have 4 numbers which satisfy the above equations. Now if $ad = 0$, then either a or d must be 0. However, this cannot be true if $ae = bd = 1$. Thus the matrix has no such decomposition.

13. **(a)** If A has such an LU-decomposition, we can write it as

$$\begin{bmatrix} a & b \\ c & d \end{bmatrix} = \begin{bmatrix} 1 & 0 \\ w & 1 \end{bmatrix} \begin{bmatrix} x & y \\ 0 & z \end{bmatrix} = \begin{bmatrix} x & y \\ wx & yw + z \end{bmatrix}$$

This yields the system of equations

$$x = a \qquad y = b \qquad wx = c \qquad yw + z = d$$

Since $a \neq 0$, this has the unique solution

$$x = a \qquad y = b \qquad w = c/a \qquad z = (ad - bc)/a$$

The uniqueness of the solution guarantees the uniqueness of the LU-decomposition.

(b) By the above,

$$\begin{bmatrix} a & b \\ c & d \end{bmatrix} = \begin{bmatrix} 1 & 0 \\ c/a & 1 \end{bmatrix} \begin{bmatrix} a & b \\ 0 & (ad - bc)/a \end{bmatrix}$$

14. To solve the system $A\mathbf{x} = \mathbf{b}$ by the method of Example 1, we must first reduce A to row-echelon form in order to find its LU-decomposition. This reduction will serve to compute both L and U simultaneously without additional multiplications or additions. Using the argument of Exercise 10 in Section 9.1, we have that this will require

$$\frac{n^3 - n}{3}$$

multiplications and

$$\frac{n^3}{3} - \frac{n^2}{2} + \frac{n}{6}$$

additions.

Next, we must use forward substitution to solve the system $L\mathbf{y} = \mathbf{b}$. By virtue of Exercise 9 of Section 9.1, this requires

$$\frac{(n - 1)n}{2}$$

multiplications and the same number of additions.

Finally, we must use back substitution to solve the system $U\mathbf{x} = \mathbf{y}$, which requires the same number of multiplications and additions as the forward substitution above.

Thus the grand total for all the operations is

$$\frac{n^3 - n}{3} + (n - 1)n = \frac{n^3}{3} + n^2 - \frac{4n}{3}$$

multiplications and

$$\frac{n^3}{3} - \frac{n^2}{2} + \frac{n}{6} + (n - 1)n = \frac{n^3}{3} + \frac{n^2}{2} - \frac{5n}{6}$$

additions.

15. **(a)** Let a_{ij}, b_{jk}, and c_{ik} denote the entries of L_1, L_2, and L_1L_2, respectively. Then

$$c_{ik} = a_{i1}b_{1k} + a_{i2}b_{2k} + \cdots + a_{ij}b_{jk} + \cdots + a_{in}b_{nk}$$

Fix i and k, where $i < k$. We must show that $c_{ik} = 0$ for all such i and k.

First we consider the terms $a_{ij}b_{jk}$ of c_{ik} for which $j < k$. Because L_2 is lower triangular, $b_{jk} = 0$ for all such j, and hence $a_{ij}b_{jk} = 0$.

There remain the terms of c_{ik} for which $k \leq j$. For all such j we have $i < k \leq j$ so that $i < j$. Because L_1 is lower triangular, this guarantees that $a_{ij} = 0$ and hence $a_{ij}b_{jk} = 0$. Thus $a_{ij}b_{jk} = 0$ for all j and hence $c_{ik} = 0$ whenever $i < k$. That is, L_1L_2 is lower triangular.

(b) We have that $L = E_1^{-1}E_2^{-1} \cdots E_k^{-1}$ where each of the matrices E_i is an elementary matrix which does not involve interchanging rows. By Exercise 27 of Section 2.4, we know that if A is an invertible upper triangular matrix, then A^{-1} is also upper triangular. Now the matrices E_i are all lower triangular by their construction. Therefore for $i = 1, \ldots, k$ we have that E_i^t is upper triangular. Hence $\left[E_i^t\right]^{-1}$ is also upper triangular. But

$$\left[E_i^t\right]^{-1} = \left[E_i^{-1}\right]^t$$

which implies that E_i^{-1} is lower triangular. Hence L, as the product of lower triangular matrices, must be lower triangular.

17. Let A be any $n \times n$ matrix. We know that A can be reduced to row-echelon form and that this may require row interchanges. If we perform these interchanges (if any) first, we reduce A to the matrix

$$E_k \cdots E_1 A = B$$

where E_i is the elementary matrix corresponding to the ith such interchange. Now we know that B has an LU-decomposition, call it LU where U is a row-echelon form of A. That is,

$$E_k \cdots E_1 A = LU$$

where each of the matrices E_i is elementary and hence invertible. (In fact, $E_i^{-1} = E_i$ for all E_i. Why?) If we let

$$P = (E_k \cdots E_1)^{-1} = E_1^{-1} \cdots E_k^{-1} \qquad \text{if } k > 0$$

and

$$P = I \qquad \text{if no row interchanges are required}$$

then we have $A = PLU$ as desired.

18. First we interchange Row 2 and Row 3 of A to obtain the equation

$$B = \begin{bmatrix} 3 & -1 & 0 \\ 0 & 2 & 1 \\ 3 & -1 & 1 \end{bmatrix} = \begin{bmatrix} 1 & 0 & 0 \\ 0 & 0 & 1 \\ 0 & 1 & 0 \end{bmatrix} \begin{bmatrix} 3 & -1 & 0 \\ 3 & -1 & 1 \\ 0 & 2 & 1 \end{bmatrix} = EA$$

Next we find an LU-decomposition of B as follows

$$B \longrightarrow \begin{bmatrix} 1 & -1/3 & 0 \\ 0 & 2 & 1 \\ 3 & -1 & 1 \end{bmatrix} \longrightarrow \begin{bmatrix} 1 & -1/3 & 0 \\ 0 & 2 & 1 \\ 0 & 0 & 1 \end{bmatrix}$$

$$\longrightarrow \begin{bmatrix} 1 & -1/3 & 0 \\ 0 & 1 & 1/2 \\ 0 & 0 & 1 \end{bmatrix} = U$$

This gives us

$$L = \begin{bmatrix} 3 & 0 & 0 \\ 0 & 2 & 0 \\ 3 & 0 & 1 \end{bmatrix}$$

so that $B = EA = LU$ or $A = E^{-1}LU$. Since

$$E^{-1} = \begin{bmatrix} 1 & 0 & 0 \\ 0 & 0 & 1 \\ 0 & 1 & 0 \end{bmatrix} = E$$

we have $A = ELU$, or

$$A = \begin{bmatrix} 1 & 0 & 0 \\ 0 & 0 & 1 \\ 0 & 1 & 0 \end{bmatrix} \begin{bmatrix} 3 & 0 & 0 \\ 0 & 2 & 0 \\ 3 & 0 & 1 \end{bmatrix} \begin{bmatrix} 1 & -1/3 & 0 \\ 0 & 1 & 1/2 \\ 0 & 0 & 1 \end{bmatrix}$$

EXERCISE SET 9.3

NOTE: Throughout this section, calculations are carried out to three significant digits. This means that only three significant digits are retained at each stage of the calculation. Thus, if you perform the calculation on a pocket calculator which carries more than three significant digits and only round off to obtain the final result, then your answer will probably differ from ours. Moreover, the order in which the calculations are performed can influence the result (see, for example, the solution to Exercise 3). Thus, slight variations in the answers are to be expected.

1. First rewrite the equations as

$$x_1 = -\frac{1}{2}x_2 + \frac{7}{2}$$

$$x_2 = \frac{1}{2}x_1 - \frac{1}{2}$$

If we start with $x_1 = x_2 = 0$, then the first approximation is

$$x_1 = -\frac{1}{2}(0) + \frac{7}{2} = \frac{7}{2} = 3.5$$

$$x_2 = \frac{1}{2}(0) - \frac{1}{2} = -\frac{1}{2} = -.5$$

This yields the second approximation

$$x_1 = -\frac{1}{2}\left[-\frac{1}{2}\right] + \frac{7}{2} = \frac{15}{4} = 3.75$$

$$x_2 = \frac{1}{2}\left[\frac{7}{2}\right] - \frac{1}{2} = \frac{5}{4} = 1.25$$

Thus, the third approximation is

$$x_1 = -\frac{1}{2}(1.25) + \frac{7}{2} \approx 2.88$$

$$x_2 = \frac{1}{2}(3.75) - \frac{1}{2} \approx 1.38$$

correct to 3 significant digits. Hence, the fourth approximation is

$$x_1 = -\frac{1}{2}(1.38) + \frac{7}{2} = 2.81$$

$$x_2 = \frac{1}{2}(2.88) - \frac{1}{2} = .94$$

The exact solution is $x_1 = 3$, $x_2 = 1$.

3. First rewrite the equations as

$$x_1 = \frac{2}{5}x_2 - \frac{13}{5}$$

$$x_2 = -\frac{1}{7}x_1 - \frac{10}{7}$$

If we start with $x_1 = x_2 = 0$, then the first approximation is

$$x_1 = -\frac{13}{5} = -2.60$$

$$x_2 = -\frac{10}{7} \approx -1.43$$

This yields the second approximation

$$x_1 = \frac{2}{5}(-1.43) - \frac{13}{5} \approx -3.17$$

$$x_2 = -\frac{1}{7}(-2.6) - \frac{10}{7} \approx -1.06$$

Thus, the third approximation is

$$x_1 = \frac{2}{5}(-1.06) - \frac{13}{5} \approx -3.02$$

$$x_2 = -\frac{1}{7}(-3.17) - \frac{10}{7} \approx -.977$$

Finally, the fourth approximation is

$$x_1 = \frac{2}{5}(-.977) - \frac{13}{5} \approx -2.99$$

$$x_2 = -\frac{1}{7}(-3.02) - \frac{10}{7} \approx -.998$$

The exact solution is $x_1 = -3$, $x_2 = -1$.

In the final computation for x_2, we can write

$$x_2 \approx -(.143)(-3.02) - 1.43 \approx .432 - 1.43 = -.998$$

or

$$x_2 = \frac{3.02 - 10}{7} = -\frac{6.98}{7} \approx -.997$$

or

$$x_2 \approx .431 - 1.43 \approx -.999$$

Thus the order in which the calculations are performed can, indeed, affect the result.

5. First rewrite the equations as

$$x_1 = -\frac{1}{2}x_2 + \frac{7}{2}$$

$$x_2 = \frac{1}{2}x_1 - \frac{1}{2}$$

If we start with $x_2 = 0$, then the first approximation is

$$x_1 = 0 + \frac{7}{2} = 3.5$$

$$x_2 = \frac{1}{2}\left[\frac{7}{2}\right] - \frac{1}{2} = 1.25$$

This yields the second approximation

$$x_1 = -\frac{1}{2}(1.25) + \frac{7}{2} \approx 2.88$$

$$x_2 = \frac{1}{2}(2.88) - \frac{1}{2} = .940$$

Thus, the third approximation is

$$x_1 = -\frac{1}{2}(.94) + \frac{7}{2} = 3.03$$

$$x_2 = \frac{1}{2}(3.03) - \frac{1}{2} \approx 1.02$$

The exact solution is $x_1 = 3$, $x_2 = 1$.

7. First rewrite the equations as

$$x_1 = \frac{2}{5}x_2 - \frac{13}{5}$$

$$x_2 = -\frac{1}{7}x_1 - \frac{10}{7}$$

If we start with $x_2 = 0$, then the first approximation is

$$x_1 = -\frac{13}{5} = -2.60$$

$$x_2 = -\frac{1}{7}(-2.6) - \frac{10}{7} \approx -1.06$$

The second approximation is

$$x_1 = \frac{2}{5}(-1.06) - \frac{13}{5} \approx -3.02$$

$$x_2 = -\frac{1}{7}(-3.02) - \frac{10}{7} \approx -.998$$

Thus, the third approximation is

$$x_1 = \frac{2}{5}(-.998) - \frac{13}{5} \approx -3.00$$

$$x_2 = -\frac{1}{7}(-3.00) - \frac{10}{7} = -1.00$$

The exact solution is $x_1 = -3$, $x_2 = -1$.

9. First rewrite the equations as

$$x_1 = -\frac{1}{10}x_2 - \frac{2}{10}x_3 + \frac{3}{10}$$

$$x_2 = -\frac{1}{10}x_1 + \frac{1}{10}x_3 + \frac{3}{20}$$

$$x_3 = -\frac{2}{10}x_1 - \frac{1}{10}x_2 - \frac{9}{10}$$

If we start with $x_1 = x_2 = x_3 = 0$, then the first approximation is

$$x_1 = \frac{3}{10} = .3$$

$$x_2 = \frac{3}{20} = .15$$

$$x_3 = -\frac{9}{10} = -.9$$

The second approximation is

$$x_1 = -\frac{1}{10}\left[\frac{3}{20}\right] - \frac{2}{10}\left[-\frac{9}{10}\right] + \frac{3}{10} = .465$$

$$x_2 = -\frac{1}{10}\left[\frac{3}{10}\right] + \frac{1}{10}\left[-\frac{9}{10}\right] + \frac{3}{20} = .03$$

$$x_3 = -\frac{2}{10}\left[\frac{3}{10}\right] - \frac{1}{10}\left[\frac{3}{20}\right] - \frac{9}{10} = -.975$$

Thus, the third approximation is

$$x_1 = -\frac{1}{10}(.03) - \frac{2}{10}(-.975) + \frac{3}{10} = .492$$

$$x_2 = -\frac{1}{10}(.465) + \frac{1}{10}(-.975) + \frac{3}{20} = .006$$

$$x_3 = -\frac{2}{10}(.465) - \frac{1}{10}(.03) - \frac{9}{10} = -.996$$

The exact solution is $x_1 = \frac{1}{2}$, $x_2 = 0$, $x_3 = -1$.

11. First rewrite the equations as

$$x_1 = -\frac{1}{10}x_2 - \frac{2}{10}x_3 + \frac{3}{10}$$

$$x_2 = -\frac{1}{10}x_1 + \frac{1}{10}x_3 + \frac{3}{20}$$

$$x_3 = -\frac{2}{10}x_1 - \frac{1}{10}x_2 - \frac{9}{10}$$

If we start with $x_2 = x_3 = 0$, then the first approximation is

$$x_1 = \frac{3}{10} = .3$$

$$x_2 = -\frac{1}{10}\left[\frac{3}{10}\right] + \frac{1}{10}(0) + \frac{3}{20} = .12$$

$$x_3 = -\frac{2}{10}(.3) - \frac{1}{10}(.12) - \frac{9}{10} = -.972$$

The second approximation is

$$x_1 = -\frac{1}{10}(.12) - \frac{2}{10}(-.972) + \frac{3}{10} \approx .482$$

$$x_2 = -\frac{1}{10}(.482) + \frac{1}{10}(-.972) + \frac{3}{20} = .00460$$

$$x_3 = -\frac{2}{10}(.482) - \frac{1}{10}(.0046) - \frac{9}{10} = -.997$$

Thus the third approximation is

$$x_1 = -\frac{1}{10}(.0046) - \frac{2}{10}(-.997) + \frac{3}{10} \approx .499$$

$$x_2 = -\frac{1}{10}(.499) + \frac{1}{10}(-.997) + \frac{3}{20} = .0004$$

$$x_3 = -\frac{2}{10}(.499) - \frac{1}{10}(.0004) - \frac{9}{10} \approx -1.00$$

The next approximation gives $x_1 = .500$, $x_2 = 0$, $x_3 = -1$, which is the exact solution.

13. **(a)** The matrix is strictly diagonally dominant because $2 > 1$ and $4 > |-1|$.

(c) The matrix is not strictly diagonally dominant because, in the second row, $5 \not> 3 + 3$.

14. **(a)** We apply Jacobi iteration to the equations

$$(*) \qquad \begin{aligned} x_1 &= -3x_2 + 4 \\ x_2 &= x_1 \end{aligned}$$

If, as usual, we start with $x_1 = x_2 = 0$, then we obtain the following approximations:

	0th	1st	2nd	3rd	4th	5th	6th	7th	8th
x_1	0	4	4	-8	-8	28	28	-80	-80
x_2	0	0	4	4	-8	-8	28	28	-80

We now show that the Jacobi iterations diverge. If we let $x_1^{(n)}$ denote the nth approximation to x_1, then, from (*), we have

$$x_1^{(n)} = 4 - 3x_1^{(n-2)} \qquad n = 3,4,\ldots$$

Since $|a - b| \geq |a| - |b|$, then

$$|x_1^{(n)}| = |4 - 3x_1^{(n-2)}| = |3x_1^{(n-2)} - 4|$$
$$\geq 3|x_1^{(n-2)}| - 4$$

But since $|x_1^{(n-2)}| > 2$ for $n = 3,4,\ldots$, then

$$3|x_1^{(n-2)}| - 4 > |x_1^{(n-2)}|$$

Therefore,

$$|x_1^{(n)}| > |x_1^{(n-2)}| \qquad \text{for } n = 3,4,\ldots$$

Thus, the successive approximations for x_1 form a sequence of integers where each is greater in absolute value than the one two stages before. That is,

$$\lim_{n \to \infty} |x_1^{(n)}| = \infty$$

14. **(b)** The coefficient matrix is not strictly diagonally dominant because $1 \not> 3$ and $|1| \not> 1$.

15. Let $A = [a_{ij}]$ denote the coefficient matrix. Since the system has a unique solution, then $\det(A) \neq 0$, so at least one of the elementary products $a_{i_1 1}a_{i_2 2} \cdots a_{i_n n}$ is nonzero. Thus, we can interchange Row 1 and Row i_1, if necessary, to put $a_{i_1 1}$ in the first row and the first column. We also interchange subscripts on x_1 and x_{i_1}. We carry on with this process, interchanging rows and subscripts until we have a new system of equations with coefficient matrix $B = [b_{ij}]$ where $b_{ij} = a_{i_j j}$. Hence B has no zero entries on the diagonal.

EXERCISE SET 9.4

4. The augmented matrix of the system is

$$\begin{bmatrix} 3 & 1 & -2 \\ -5 & 1 & 22 \end{bmatrix}$$

The pivot entry is -5. If we interchange rows to bring it to the top of the column, we obtain

$$\begin{bmatrix} -5 & 1 & 22 \\ 3 & 1 & -2 \end{bmatrix}$$

Next we multiply the top row by $-\frac{1}{5}$ to get

$$\begin{bmatrix} 1 & -\frac{1}{5} & -\frac{22}{5} \\ 3 & 1 & -2 \end{bmatrix}$$

Now we add -3 times Row 1 to Row 2 to obtain

$$\begin{bmatrix} 1 & -\frac{1}{5} & -\frac{22}{5} \\ 0 & \frac{8}{5} & \frac{56}{5} \end{bmatrix}$$

Finally, we multiply Row 2 by $\frac{5}{8}$, which yields

$$\begin{bmatrix} 1 & -\frac{1}{5} & -\frac{22}{5} \\ 0 & 1 & 7 \end{bmatrix}$$

The corresponding system of equations is

$$x_1 - \frac{1}{5}x_2 = -\frac{22}{5}$$

$$x_2 = 7$$

and the solution is $x_1 = -3$, $x_2 = 7$. We leave the check to you.

6. The augmented matrix is

$$\begin{bmatrix} 2 & 3 & -1 & 5 \\ 4 & 4 & -3 & 3 \\ 2 & -3 & 1 & -1 \end{bmatrix}$$

The pivot entry is 4. Thus, we interchange Rows 1 and 2 to get

$$\begin{bmatrix} 4 & 4 & -3 & 3 \\ 2 & 3 & -1 & 5 \\ 2 & -3 & 1 & -1 \end{bmatrix}$$

Next we multiply the top row by $\frac{1}{4}$, which yields

$$\begin{bmatrix} 1 & 1 & -\frac{3}{4} & \frac{3}{4} \\ 2 & 3 & -1 & 5 \\ 2 & -3 & 1 & -1 \end{bmatrix}$$

Now if we add -2 times Row 1 to Rows 2 and 3, we obtain

$$\begin{bmatrix} 1 & 1 & -\frac{3}{4} & \frac{3}{4} \\ 0 & 1 & \frac{1}{2} & \frac{7}{2} \\ 0 & -5 & \frac{5}{2} & -\frac{5}{2} \end{bmatrix}$$

The new pivot entry is −5, so we interchange Rows 2 and 3, which gives

$$\begin{bmatrix} 1 & 1 & -\frac{3}{4} & \frac{3}{4} \\ 0 & -5 & \frac{5}{2} & -\frac{5}{2} \\ 0 & 1 & \frac{1}{2} & \frac{7}{2} \end{bmatrix}$$

Now we multiply Row 2 by $-\frac{1}{5}$ to get

$$\begin{bmatrix} 1 & 1 & -\frac{3}{4} & \frac{3}{4} \\ 0 & 1 & -\frac{1}{2} & \frac{1}{2} \\ 0 & 1 & \frac{1}{2} & \frac{7}{2} \end{bmatrix}$$

Finally, we add −1 times Row 2 to Row 3 to obtain

$$\begin{bmatrix} 1 & 1 & -\frac{3}{4} & \frac{3}{4} \\ 0 & 1 & -\frac{1}{2} & \frac{1}{2} \\ 0 & 0 & 1 & 3 \end{bmatrix}$$

The corresponding system of equations is

$$\begin{aligned} x_1 + x_2 - \frac{3}{4}x_3 &= \frac{3}{4} \\ x_2 - \frac{1}{2}x_3 &= \frac{1}{2} \\ x_3 &= 3 \end{aligned}$$

and the solution is $x_1 = 1$, $x_2 = 2$, $x_3 = 3$. We leave the check to you.

8. The augmented matrix is

$$\begin{bmatrix} .21 & .33 & .54 \\ .70 & .24 & .94 \end{bmatrix}$$

The pivot entry is .70, so we interchange rows to obtain

$$\begin{bmatrix} .70 & .24 & .94 \\ .21 & .33 & .54 \end{bmatrix}$$

Next we multiply the top row by $\frac{1}{.70}$ and round off, which gives

$$\begin{bmatrix} 1 & .343 & 1.34 \\ .21 & .33 & .54 \end{bmatrix}$$

Now we add −.21 times Row 1 to Row 2 and round off to get

$$\begin{bmatrix} 1 & .343 & 1.34 \\ 0 & .258 & .259 \end{bmatrix}$$

Finally, we multiply Row 2 by $\frac{1}{.258}$ and round off to obtain

$$\begin{bmatrix} 1 & .343 & 1.34 \\ 0 & 1 & 1.00 \end{bmatrix}$$

The corresponding system of equations is

$$\begin{aligned} x_1 + .343x_2 &= 1.34 \\ x_2 &= 1.00 \end{aligned}$$

and the solution is $x_1 = .997$, $x_2 = 1.00$.

10. First we solve with pivoting. The augmented matrix is

$$\begin{bmatrix} .0001 & 1 & 1 \\ 1 & 1 & 2 \end{bmatrix}$$

The pivot entry is 1, so we interchange rows to get

$$\begin{bmatrix} 1 & 1 & 2 \\ .0001 & 1 & 1 \end{bmatrix}$$

Now we add −.0001 times Row 1 to Row 2 and round off to obtain

$$\begin{bmatrix} 1 & 1 & 2 \\ 0 & 1.00 & 1.00 \end{bmatrix}$$

The corresponding system of equations is

$$x_1 + x_2 = 2$$
$$x_2 = 1$$

and the solution is $x_1 = 1$, $x_2 = 1$. Notice that the exact solution is $x_1 = 1.\overline{0001}$, $x_2 = .\overline{9998}$.

Now we solve without pivoting. Recall that the augmented matrix is

$$\begin{bmatrix} .0001 & 1 & 1 \\ 1 & 1 & 2 \end{bmatrix}$$

We multiply Row 1 by 10^4, then subtract Row 1 from Row 2, and round off to get

$$\begin{bmatrix} 1 & 10^4 & 10^4 \\ 0 & -10^4 & -10^4 \end{bmatrix}$$

Note that to 3 significant digits, $10^4 - 1 = 10^4 - 2 = 10^4$. Finally, we multiply Row 2 by -10^{-4} to obtain

$$\begin{bmatrix} 1 & 10^4 & 10^4 \\ 0 & 1 & 1 \end{bmatrix}$$

The corresponding system of equations is

$$x_1 + 10^4 x_2 = 10^4$$
$$x_2 = 1$$

and the solution is $x_1 = 0$, $x_2 = 1$.

EXERCISE SET 9.5*

1. **(a)** The characteristic equation is $\lambda^2 + 2\lambda - 3 = 0$, so the eigenvalues are $\lambda_1 = -3$ and $\lambda_2 = 1$. Clearly $\lambda = -3$ is dominant.

   (b) The characteristic equation is $\lambda^2 - 4 = 0$, so the eigenvalues are $\lambda = \pm 2$. Since $|2| = |-2|$, there is no dominant eigenvalue.

   (d) The characteristic equation is $(\lambda - 3)(\lambda^2 - \lambda + 12) = 0$. The only real root is $\lambda = 3$. Hence $\lambda = 3$ is the dominant eigenvalue if we disregard nonreal solutions to the characteristic equation.

2. **(a)** The first approximation is

$$A\mathbf{x}_0 = \begin{bmatrix} 3 & 4 \\ 1 & 3 \end{bmatrix} \begin{bmatrix} 1 \\ 1 \end{bmatrix} = \begin{bmatrix} 7 \\ 4 \end{bmatrix} \Rightarrow \mathbf{x}_1 = \frac{1}{7} \begin{bmatrix} 7 \\ 4 \end{bmatrix} \approx \begin{bmatrix} 1 \\ .571 \end{bmatrix}$$

   The second approximation is

$$A\mathbf{x}_1 = \begin{bmatrix} 3 & 4 \\ 1 & 3 \end{bmatrix} \begin{bmatrix} 1 \\ .571 \end{bmatrix} \approx \begin{bmatrix} 5.28 \\ 2.71 \end{bmatrix} \Rightarrow \mathbf{x}_2 = \frac{1}{5.28} \begin{bmatrix} 5.28 \\ 2.71 \end{bmatrix} \approx \begin{bmatrix} 1 \\ .513 \end{bmatrix}$$

   The third approximation is

$$A\mathbf{x}_2 = \begin{bmatrix} 3 & 4 \\ 1 & 3 \end{bmatrix} \begin{bmatrix} 1 \\ .513 \end{bmatrix} \approx \begin{bmatrix} 5.05 \\ 2.54 \end{bmatrix} \Rightarrow \mathbf{x}_3 = \frac{1}{5.05} \begin{bmatrix} 5.05 \\ 2.54 \end{bmatrix} \approx \begin{bmatrix} 1 \\ .503 \end{bmatrix}$$

*See the note preceding the solutions for Exercise Set 9.3.

In addition,

$$A\mathbf{x}_3 = \begin{bmatrix} 3 & 4 \\ 1 & 3 \end{bmatrix} \begin{bmatrix} 1 \\ .503 \end{bmatrix} = \begin{bmatrix} 5.01 \\ 2.51 \end{bmatrix}$$

2. **(b)** If λ_1 is the dominant eigenvalue, then

$$\lambda_1 \approx \frac{\langle \mathbf{x}_3, A\mathbf{x}_3 \rangle}{\langle \mathbf{x}_3, \mathbf{x}_3 \rangle} \approx \frac{1(5.01) + .503(2.51)}{1 + (.503)^2} \approx 5.02$$

(c) The eigenvalues are $\lambda = 5$ and $\lambda = 1$. Hence $\lambda = 5$ is the dominant eigenvalue. The corresponding eigenspace is spanned by the vector $\begin{bmatrix} 1 \\ 1/2 \end{bmatrix}$, which is the dominant eigenvector.

(d) The percentage error is

$$\left| \frac{5 - 5.02}{5} \right| \times 100\% = .4\%$$

4. **(a)** The first three approximations are

$$A\mathbf{x}_0 = \begin{bmatrix} -3 & 2 \\ 2 & 0 \end{bmatrix} \begin{bmatrix} 1 \\ 1 \end{bmatrix} = \begin{bmatrix} -1 \\ 2 \end{bmatrix} \Rightarrow \mathbf{x}_1 = \begin{bmatrix} -1/2 \\ 1 \end{bmatrix}$$

$$A\mathbf{x}_1 = \begin{bmatrix} -3 & 2 \\ 2 & 0 \end{bmatrix} \begin{bmatrix} -1/2 \\ 1 \end{bmatrix} = \begin{bmatrix} 7/2 \\ -1 \end{bmatrix} \Rightarrow \mathbf{x}_2 \approx \begin{bmatrix} 1 \\ -.286 \end{bmatrix}$$

$$A\mathbf{x}_2 = \begin{bmatrix} -3 & 2 \\ 2 & 0 \end{bmatrix} \begin{bmatrix} 1 \\ -.286 \end{bmatrix} \approx \begin{bmatrix} -3.57 \\ 2 \end{bmatrix} \Rightarrow \mathbf{x}_3 \approx \begin{bmatrix} 1 \\ -.560 \end{bmatrix}$$

Also

$$A\mathbf{x}_3 = \begin{bmatrix} -3 & 2 \\ 2 & 0 \end{bmatrix} \begin{bmatrix} 1 \\ -.560 \end{bmatrix} = \begin{bmatrix} -4.12 \\ 2 \end{bmatrix}$$

(b) If λ_1 is the dominant eigenvalue, then

$$\lambda_1 \approx \frac{\langle \mathbf{x}_3, A\mathbf{x}_3 \rangle}{\langle \mathbf{x}_3, \mathbf{x}_3 \rangle} \approx \frac{-4.12 + 2(-.560)}{1 + (-.560)^2} \approx -4.00$$

(c) The eigenvalues are $\lambda = -4$ and $\lambda = 1$. Thus $\lambda_1 = -4$ is the dominant eigenvalue. The corresponding eigenspace is spanned by the vector $\begin{bmatrix} 1 \\ -1/2 \end{bmatrix}$.

6. (a) The first approximation is

$$A\mathbf{x}_0 = \begin{bmatrix} -5 & 5 \\ 6 & -4 \end{bmatrix}\begin{bmatrix} 1 \\ 1 \end{bmatrix} = \begin{bmatrix} 0 \\ 2 \end{bmatrix} \Rightarrow \mathbf{x}_1 = \begin{bmatrix} 0 \\ 1 \end{bmatrix}$$

The second approximation is

$$A\mathbf{x}_1 = \begin{bmatrix} -5 & 5 \\ 6 & -4 \end{bmatrix}\begin{bmatrix} 0 \\ 1 \end{bmatrix} = \begin{bmatrix} 5 \\ -4 \end{bmatrix} \Rightarrow \mathbf{x}_2 = \begin{bmatrix} 1 \\ -.8 \end{bmatrix}$$

which yields

$$\tilde{\lambda}(1) = \frac{\langle \mathbf{x}_1, A\mathbf{x}_1 \rangle}{\langle \mathbf{x}_1, \mathbf{x}_1 \rangle} = \frac{-4}{1} = -4$$

The third approximation is

$$A\mathbf{x}_2 = \begin{bmatrix} -5 & 5 \\ 6 & -4 \end{bmatrix}\begin{bmatrix} 1 \\ -.8 \end{bmatrix} = \begin{bmatrix} -9.0 \\ 9.2 \end{bmatrix} \Rightarrow \mathbf{x}_3 = \begin{bmatrix} -.978 \\ 1 \end{bmatrix}$$

which yields

$$\tilde{\lambda}(2) = \frac{\langle \mathbf{x}_2, A\mathbf{x}_2 \rangle}{\langle \mathbf{x}_2, \mathbf{x}_2 \rangle} = \frac{-9 - .8(9.2)}{1 + (-.8)^2} \approx -10.0$$

Thus the estimated percentage error is

$$\left|\frac{-10 - (-4)}{-10}\right| \times 100\% = 60\%$$

The fourth approximation is

$$A\mathbf{x}_3 = \begin{bmatrix} -5 & 5 \\ 6 & -4 \end{bmatrix}\begin{bmatrix} -.978 \\ 1 \end{bmatrix} \approx \begin{bmatrix} 9.89 \\ -9.87 \end{bmatrix} \Rightarrow \mathbf{x}_4 \approx \begin{bmatrix} 1 \\ -.998 \end{bmatrix}$$

which yields

$$\tilde{\lambda}(3) = \frac{\langle \mathbf{x}_3, A\mathbf{x}_3 \rangle}{\langle \mathbf{x}_3, \mathbf{x}_3 \rangle} = \frac{-.978(9.89) - 9.87}{(-.978)^2 + 1} \approx -9.95$$

Thus the estimated percentage error is

$$\left| \frac{-9.95 - (-10)}{-9.95} \right| \times 100\% \approx .5\%$$

Since this is less than 2%, we can stop at the third iteration with the estimates $\lambda_1 \approx -9.95$ and $\mathbf{x} \approx \begin{bmatrix} -.978 \\ 1 \end{bmatrix}$ for the dominant eigenvalue and the dominant eigenvector, respectively.

7. **(a)** The first three approximations are

$$A\mathbf{x}_0 = \begin{bmatrix} 2 & 1 & 0 \\ 1 & 2 & 0 \\ 0 & 0 & 10 \end{bmatrix} \begin{bmatrix} 1 \\ 1 \\ 1 \end{bmatrix} = \begin{bmatrix} 3 \\ 3 \\ 10 \end{bmatrix} \Rightarrow \mathbf{x}_1 = \begin{bmatrix} .3 \\ .3 \\ 1 \end{bmatrix}$$

$$A\mathbf{x}_1 = \begin{bmatrix} 2 & 1 & 0 \\ 1 & 2 & 0 \\ 0 & 0 & 10 \end{bmatrix} \begin{bmatrix} .3 \\ .3 \\ 1 \end{bmatrix} = \begin{bmatrix} .9 \\ .9 \\ 10 \end{bmatrix} \Rightarrow \mathbf{x}_2 = \begin{bmatrix} .09 \\ .09 \\ 1 \end{bmatrix}$$

$$A\mathbf{x}_2 = \begin{bmatrix} 2 & 1 & 0 \\ 1 & 2 & 0 \\ 0 & 0 & 10 \end{bmatrix} \begin{bmatrix} .09 \\ .09 \\ 1 \end{bmatrix} = \begin{bmatrix} .27 \\ .27 \\ 10 \end{bmatrix} \Rightarrow \mathbf{x}_3 = \begin{bmatrix} .027 \\ .027 \\ 1 \end{bmatrix}$$

Also

$$A\mathbf{x}_3 = \begin{bmatrix} 2 & 1 & 0 \\ 1 & 2 & 0 \\ 0 & 0 & 10 \end{bmatrix} \begin{bmatrix} .027 \\ .027 \\ 1 \end{bmatrix} = \begin{bmatrix} .081 \\ .081 \\ 10 \end{bmatrix}$$

(b) If λ_1 is the dominant eigenvalue, then

$$\lambda_1 \approx \frac{\langle \mathbf{x}_3, A\mathbf{x}_3 \rangle}{\langle \mathbf{x}_3, \mathbf{x}_3 \rangle} = \frac{.081(.27) + .081(.27) + 10}{(.027)^2 + (.027)^2 + 1} \approx 10.0$$

EXERCISE SET 9.6*

1. **(a)** The first three approximations are

$$A\mathbf{x}_0 = \begin{bmatrix} 6 & 2 \\ 2 & 3 \end{bmatrix}\begin{bmatrix} 1 \\ 1 \end{bmatrix} = \begin{bmatrix} 8 \\ 5 \end{bmatrix} \approx 8\begin{bmatrix} 1 \\ .625 \end{bmatrix}$$

$$A\mathbf{x}_1 = \begin{bmatrix} 6 & 2 \\ 2 & 3 \end{bmatrix}\begin{bmatrix} 8 \\ 5 \end{bmatrix} = \begin{bmatrix} 58 \\ 31 \end{bmatrix} \approx 58\begin{bmatrix} 1 \\ .534 \end{bmatrix}$$

$$A\mathbf{x}_2 = \begin{bmatrix} 6 & 2 \\ 2 & 3 \end{bmatrix}\begin{bmatrix} 58 \\ 31 \end{bmatrix} = \begin{bmatrix} 410 \\ 209 \end{bmatrix} \approx 410\begin{bmatrix} 1 \\ .510 \end{bmatrix}$$

Also

$$A\mathbf{x}_3 = \begin{bmatrix} 6 & 2 \\ 2 & 3 \end{bmatrix}\begin{bmatrix} 410 \\ 209 \end{bmatrix} \approx \begin{bmatrix} 2880 \\ 1450 \end{bmatrix}$$

If we stop after three iterations, we find that the dominant eigenvector can be approximated by either of the vectors

$$\begin{bmatrix} 410 \\ 209 \end{bmatrix} \quad \text{or} \quad \begin{bmatrix} 1 \\ .510 \end{bmatrix}$$

---

*See the note preceding the solutions for Exercise Set 9.3.

If we use the power method with scaling, we have that the first three approximations are

$$A\mathbf{x}_0 = \begin{bmatrix} 6 & 2 \\ 2 & 3 \end{bmatrix} \begin{bmatrix} 1 \\ 1 \end{bmatrix} = \begin{bmatrix} 8 \\ 5 \end{bmatrix} \Rightarrow \mathbf{x}_1 = \begin{bmatrix} 1 \\ .625 \end{bmatrix}$$

$$A\mathbf{x}_1 = \begin{bmatrix} 6 & 2 \\ 2 & 3 \end{bmatrix} \begin{bmatrix} 1 \\ .625 \end{bmatrix} \approx \begin{bmatrix} 7.25 \\ 3.88 \end{bmatrix} \Rightarrow \mathbf{x}_2 \approx \begin{bmatrix} 1 \\ .535 \end{bmatrix}$$

$$A\mathbf{x}_2 = \begin{bmatrix} 6 & 2 \\ 2 & 3 \end{bmatrix} \begin{bmatrix} 1 \\ .535 \end{bmatrix} \approx \begin{bmatrix} 7.07 \\ 3.60 \end{bmatrix} \Rightarrow \mathbf{x}_3 \approx \begin{bmatrix} 1 \\ .509 \end{bmatrix}$$

Also

$$A\mathbf{x}_3 = \begin{bmatrix} 6 & 2 \\ 2 & 3 \end{bmatrix} \begin{bmatrix} 1 \\ .509 \end{bmatrix} \approx \begin{bmatrix} 7.02 \\ 3.53 \end{bmatrix}$$

This yields the approximate eigenvector $\begin{bmatrix} 1 \\ .509 \end{bmatrix}$, and is clearly a saner method for further iterations.

1. **(b)** If λ_1 is the dominant eigenvalue then

$$\lambda_1 \approx \frac{\langle \mathbf{x}_3, A\mathbf{x}_3 \rangle}{\langle \mathbf{x}_3, \mathbf{x}_3 \rangle} \approx \frac{2880(410) + 1450(209)}{(410)^2 + (209)^2} \approx 6.98$$

or

$$\lambda_1 \approx \frac{\langle \mathbf{x}_3, A\mathbf{x}_3 \rangle}{\langle \mathbf{x}_3, \mathbf{x}_3 \rangle} \approx \frac{7.02 + 3.53(.509)}{1 + (.509)^2} \approx 7.00$$

(c) If we normalize the approximate eigenvector $\begin{bmatrix} 1 \\ .510 \end{bmatrix}$ and round off, we obtain the vector $\begin{bmatrix} .893 \\ .455 \end{bmatrix}$. Thus we let

$$B = \begin{bmatrix} 6 & 2 \\ 2 & 3 \end{bmatrix} - 6.98 \begin{bmatrix} .893 \\ .455 \end{bmatrix} [.893 \quad .455] \approx \begin{bmatrix} .44 & -.83 \\ -.83 & 1.55 \end{bmatrix}$$

Then

$$B\mathbf{x}_0 = \begin{bmatrix} .44 & -.83 \\ -.83 & 1.55 \end{bmatrix} \begin{bmatrix} 1 \\ 1 \end{bmatrix} = \begin{bmatrix} -.39 \\ .72 \end{bmatrix}$$

$$B\mathbf{x}_1 = \begin{bmatrix} .44 & -.83 \\ -.83 & 1.55 \end{bmatrix} \begin{bmatrix} -.39 \\ .72 \end{bmatrix} \approx \begin{bmatrix} -.77 \\ 1.44 \end{bmatrix}$$

$$B\mathbf{x}_2 = \begin{bmatrix} .44 & -.83 \\ -.83 & 1.55 \end{bmatrix} \begin{bmatrix} -.77 \\ 1.44 \end{bmatrix} \approx \begin{bmatrix} -1.54 \\ 2.87 \end{bmatrix} \longrightarrow \begin{bmatrix} -.537 \\ 1 \end{bmatrix}$$

Also

$$B\mathbf{x}_3 = \begin{bmatrix} .44 & -.83 \\ -.83 & 1.55 \end{bmatrix} \begin{bmatrix} -1.54 \\ 2.87 \end{bmatrix} \approx \begin{bmatrix} -3.06 \\ 5.73 \end{bmatrix}$$

Thus if λ_2 is the remaining eigenvalue then

$$\lambda_2 = \frac{\langle \mathbf{x}_3, B\mathbf{x}_3 \rangle}{\langle \mathbf{x}_3, \mathbf{x}_3 \rangle} \approx \frac{(-1.54)(-3.06) + (2.87)(5.73)}{(-1.54)^2 + (2.87)^2} \approx 1.99$$

where $\mathbf{x}_3 = \begin{bmatrix} -.537 \\ 1 \end{bmatrix}$ is an approximation to the corresponding eigenvector.

If we use the power method with scaling, then if we normalize the approximate eigenvector $\begin{bmatrix} 1 \\ .509 \end{bmatrix}$ and round off, we obtain the vector $\begin{bmatrix} .893 \\ .454 \end{bmatrix}$. Thus we let

$$B = \begin{bmatrix} 6 & 2 \\ 2 & 3 \end{bmatrix} - 7 \begin{bmatrix} .893 \\ .454 \end{bmatrix} [.893 \quad .454] \approx \begin{bmatrix} .42 & -.84 \\ -.84 & 1.56 \end{bmatrix}$$

Then

$$B\mathbf{x}_0 = \begin{bmatrix} .42 & -.84 \\ -.84 & 1.56 \end{bmatrix} \begin{bmatrix} 1 \\ 1 \end{bmatrix} = \begin{bmatrix} -.42 \\ .72 \end{bmatrix} \Rightarrow \mathbf{x}_1 \approx \begin{bmatrix} -.583 \\ 1 \end{bmatrix}$$

$$Bx_1 = \begin{bmatrix} .42 & -.84 \\ -.84 & 1.56 \end{bmatrix} \begin{bmatrix} -.583 \\ 1 \end{bmatrix} \approx \begin{bmatrix} -1.08 \\ 2.05 \end{bmatrix} \Rightarrow x_2 \approx \begin{bmatrix} -.527 \\ 1 \end{bmatrix}$$

$$Bx_2 = \begin{bmatrix} .42 & -.84 \\ -.84 & 1.56 \end{bmatrix} \begin{bmatrix} -.527 \\ 1 \end{bmatrix} \approx \begin{bmatrix} -1.06 \\ 2.00 \end{bmatrix} \Rightarrow x_3 \approx \begin{bmatrix} -.53 \\ 1 \end{bmatrix}$$

$$Bx_3 = \begin{bmatrix} .42 & -.84 \\ -.84 & 1.56 \end{bmatrix} \begin{bmatrix} -.53 \\ 1 \end{bmatrix} \approx \begin{bmatrix} -1.06 \\ 2.00 \end{bmatrix}$$

Thus if λ_2 is the remaining eigenvalue, then

$$\lambda_2 \approx \frac{\langle x_3, Bx_3 \rangle}{\langle x_3, x_3 \rangle} \approx \frac{-.53(-1.06) + 2}{(-.53)^2 + 1} \approx 2.00$$

2. **(a)** We use only the power method with scaling. The first three approximations are

$$Ax_0 = \begin{bmatrix} 10 & 4 \\ 4 & 4 \end{bmatrix} \begin{bmatrix} 1 \\ 1 \end{bmatrix} = \begin{bmatrix} 14 \\ 8 \end{bmatrix} \Rightarrow x_1 \approx \begin{bmatrix} 1 \\ .571 \end{bmatrix}$$

$$Ax_1 = \begin{bmatrix} 10 & 4 \\ 4 & 4 \end{bmatrix} \begin{bmatrix} 1 \\ .571 \end{bmatrix} \approx \begin{bmatrix} 12.3 \\ 6.28 \end{bmatrix} \Rightarrow x_2 \approx \begin{bmatrix} 1 \\ .511 \end{bmatrix}$$

$$Ax_2 = \begin{bmatrix} 10 & 4 \\ 4 & 4 \end{bmatrix} \begin{bmatrix} 1 \\ .511 \end{bmatrix} \approx \begin{bmatrix} 12.0 \\ 6.04 \end{bmatrix} \Rightarrow x_3 \approx \begin{bmatrix} 1 \\ .503 \end{bmatrix}$$

Also

$$Ax_3 = \begin{bmatrix} 10 & 4 \\ 4 & 4 \end{bmatrix} \begin{bmatrix} 1 \\ .503 \end{bmatrix} \approx \begin{bmatrix} 12.0 \\ 6.01 \end{bmatrix}$$

(b) If λ_1 is the dominant eigenvalue, then

$$\lambda_1 \approx \frac{\langle x_3, Ax_3 \rangle}{\langle x_3, x_3 \rangle} \approx \frac{12 + 6.01(.503)}{1 + (.503)^2} \approx 12.0$$

(c) If we normalize the approximate eigenvector $\begin{bmatrix} 1 \\ .503 \end{bmatrix}$ and round off, we obtain the vector $\begin{bmatrix} .893 \\ .449 \end{bmatrix}$. Thus we let

$$B = \begin{bmatrix} 10 & 4 \\ 4 & 4 \end{bmatrix} - 12 \begin{bmatrix} .893 \\ .449 \end{bmatrix} \begin{bmatrix} .893 & .449 \end{bmatrix} \approx \begin{bmatrix} .44 & -.81 \\ -.81 & 1.58 \end{bmatrix}$$

Then

$$B\mathbf{x}_0 = \begin{bmatrix} .44 & -.81 \\ -.81 & 1.58 \end{bmatrix} \begin{bmatrix} 1 \\ 1 \end{bmatrix} = \begin{bmatrix} -.37 \\ .77 \end{bmatrix} \Rightarrow \mathbf{x}_1 \approx \begin{bmatrix} -.481 \\ 1 \end{bmatrix}$$

$$B\mathbf{x}_1 = \begin{bmatrix} .44 & -.81 \\ -.81 & 1.58 \end{bmatrix} \begin{bmatrix} -.481 \\ 1 \end{bmatrix} \approx \begin{bmatrix} -1.02 \\ 1.97 \end{bmatrix} \Rightarrow \mathbf{x}_2 \approx \begin{bmatrix} -.518 \\ 1 \end{bmatrix}$$

$$B\mathbf{x}_2 = \begin{bmatrix} .44 & -.81 \\ -.81 & 1.58 \end{bmatrix} \begin{bmatrix} -.518 \\ 1 \end{bmatrix} \approx \begin{bmatrix} -1.04 \\ 2.00 \end{bmatrix} \Rightarrow \mathbf{x}_3 \approx \begin{bmatrix} -.52 \\ 1 \end{bmatrix}$$

Also

$$B\mathbf{x}_3 = \begin{bmatrix} .44 & -.81 \\ -.81 & 1.58 \end{bmatrix} \begin{bmatrix} -.52 \\ 1 \end{bmatrix} \approx \begin{bmatrix} -1.04 \\ 2.00 \end{bmatrix}$$

Thus if λ_2 is the remaining eigenvalue, then

$$\lambda_2 \approx \frac{\langle \mathbf{x}_3, B\mathbf{x}_3 \rangle}{\langle \mathbf{x}_3, \mathbf{x}_3 \rangle} \approx \frac{-.52(-1.04) + 2}{(-.52)^2 + 1} \approx 2.00$$

3. First we use the power method on A. The first ten approximations are

$$A\mathbf{x}_0 = \begin{bmatrix} -2 & 1 \\ 4 & 1 \end{bmatrix} \begin{bmatrix} 1 \\ 1 \end{bmatrix} = \begin{bmatrix} -1 \\ 5 \end{bmatrix}$$

$$A\mathbf{x}_1 = \begin{bmatrix} -2 & 1 \\ 4 & 1 \end{bmatrix} \begin{bmatrix} -1 \\ 5 \end{bmatrix} = \begin{bmatrix} 7 \\ 1 \end{bmatrix}$$

$$A\mathbf{x}_2 = \begin{bmatrix} -2 & 1 \\ 4 & 1 \end{bmatrix}\begin{bmatrix} 7 \\ 1 \end{bmatrix} = \begin{bmatrix} -13 \\ 29 \end{bmatrix}$$

$$A\mathbf{x}_3 = \begin{bmatrix} -2 & 1 \\ 4 & 1 \end{bmatrix}\begin{bmatrix} -13 \\ 29 \end{bmatrix} = \begin{bmatrix} 55 \\ -23 \end{bmatrix}$$

$$A\mathbf{x}_4 = \begin{bmatrix} -2 & 1 \\ 4 & 1 \end{bmatrix}\begin{bmatrix} 55 \\ -23 \end{bmatrix} = \begin{bmatrix} -133 \\ 197 \end{bmatrix} \longrightarrow \begin{bmatrix} -.629 \\ 1 \end{bmatrix}$$

$$A\mathbf{x}_5 = \begin{bmatrix} -2 & 1 \\ 4 & 1 \end{bmatrix}\begin{bmatrix} -133 \\ 197 \end{bmatrix} = \begin{bmatrix} 463 \\ -335 \end{bmatrix} \longrightarrow \begin{bmatrix} 1 \\ -.724 \end{bmatrix}$$

$$A\mathbf{x}_6 = \begin{bmatrix} -2 & 1 \\ 4 & 1 \end{bmatrix}\begin{bmatrix} 463 \\ -335 \end{bmatrix} \approx \begin{bmatrix} -1260 \\ 1520 \end{bmatrix} \longrightarrow \begin{bmatrix} -.829 \\ 1 \end{bmatrix}$$

$$A\mathbf{x}_7 = \begin{bmatrix} -2 & 1 \\ 4 & 1 \end{bmatrix}\begin{bmatrix} -1260 \\ 1520 \end{bmatrix} \approx \begin{bmatrix} 4040 \\ -3520 \end{bmatrix} \longrightarrow \begin{bmatrix} 1 \\ -.871 \end{bmatrix}$$

$$A\mathbf{x}_8 = \begin{bmatrix} -2 & 1 \\ 4 & 1 \end{bmatrix}\begin{bmatrix} 4040 \\ -3520 \end{bmatrix} \approx \begin{bmatrix} -11600 \\ 12700 \end{bmatrix} \longrightarrow \begin{bmatrix} -.913 \\ 1 \end{bmatrix}$$

$$A\mathbf{x}_9 = \begin{bmatrix} -2 & 1 \\ 4 & 1 \end{bmatrix}\begin{bmatrix} -11600 \\ 12700 \end{bmatrix} \approx \begin{bmatrix} 35900 \\ -33700 \end{bmatrix} \longrightarrow \begin{bmatrix} 1 \\ -.939 \end{bmatrix}$$

Since the fluctuations seem to be getting less wild and we are running out of ink, we stop and compute the desired approximations.

If λ_1 is the dominant eigenvalue, then

$$\lambda_1 \approx \frac{\langle \mathbf{x}_9, A\mathbf{x}_9 \rangle}{\langle \mathbf{x}_9, \mathbf{x}_9 \rangle} \approx \frac{(-11600)(35900) + (12700)(-33700)}{(-11600)^2 + (12700)^2} \approx -2.85$$

The corresponding eigenvector is approximately $\begin{bmatrix} 1 \\ -.939 \end{bmatrix}$ or $\begin{bmatrix} -.913 \\ 1 \end{bmatrix}$ if you prefer. Since $\lambda_1 = -3$, it's good that we didn't stop sooner. If we had used $\mathbf{x}_3$ in our calculations, we would have gotten $\lambda_1 \approx -1.37$. Since the actual eigenvector is any nonzero multiple of $[1 \quad -1]^t$, we're getting close, but a few more iterations wouldn't hurt.

Next we apply the inverse power method to A. That is, we apply the power method to A^{-1}. We start scaling part way through for sanity's sake. The first ten approximations are

$$A^{-1}\mathbf{x}_0 = \begin{bmatrix} -.167 & .167 \\ .667 & .333 \end{bmatrix} \begin{bmatrix} 1 \\ 1 \end{bmatrix} = \begin{bmatrix} 0 \\ 1 \end{bmatrix}$$

$$A^{-1}\mathbf{x}_1 = \begin{bmatrix} -.167 & .167 \\ .667 & .333 \end{bmatrix} \begin{bmatrix} 0 \\ 1 \end{bmatrix} = \begin{bmatrix} .167 \\ .333 \end{bmatrix}$$

$$A^{-1}\mathbf{x}_2 = \begin{bmatrix} -.167 & .167 \\ .667 & .333 \end{bmatrix} \begin{bmatrix} .167 \\ .333 \end{bmatrix} \approx \begin{bmatrix} .0277 \\ .222 \end{bmatrix}$$

$$A^{-1}\mathbf{x}_3 = \begin{bmatrix} -.167 & .167 \\ .667 & .333 \end{bmatrix} \begin{bmatrix} .0277 \\ .222 \end{bmatrix} \approx \begin{bmatrix} .0325 \\ .0924 \end{bmatrix} \longrightarrow \begin{bmatrix} .352 \\ 1 \end{bmatrix} = \mathbf{x}_4$$

$$A^{-1}\mathbf{x}_4 = \begin{bmatrix} -.167 & .167 \\ .667 & .333 \end{bmatrix} \begin{bmatrix} .352 \\ 1 \end{bmatrix} \approx \begin{bmatrix} .108 \\ .568 \end{bmatrix} \longrightarrow \begin{bmatrix} .190 \\ 1 \end{bmatrix} = \mathbf{x}_5$$

$$A^{-1}\mathbf{x}_5 = \begin{bmatrix} -.167 & .167 \\ .667 & .333 \end{bmatrix} \begin{bmatrix} .190 \\ 1 \end{bmatrix} \approx \begin{bmatrix} .135 \\ .460 \end{bmatrix} \longrightarrow \begin{bmatrix} .293 \\ 1 \end{bmatrix} = \mathbf{x}_6$$

$$A^{-1}\mathbf{x}_6 = \begin{bmatrix} -.167 & .167 \\ .667 & .333 \end{bmatrix} \begin{bmatrix} .293 \\ 1 \end{bmatrix} \approx \begin{bmatrix} .118 \\ .528 \end{bmatrix} \longrightarrow \begin{bmatrix} .223 \\ 1 \end{bmatrix} = \mathbf{x}_7$$

$$A^{-1}\mathbf{x}_7 = \begin{bmatrix} -.167 & .167 \\ .667 & .333 \end{bmatrix} \begin{bmatrix} .223 \\ 1 \end{bmatrix} \approx \begin{bmatrix} .130 \\ .482 \end{bmatrix} \longrightarrow \begin{bmatrix} .270 \\ 1 \end{bmatrix} = \mathbf{x}_8$$

$$A^{-1}\mathbf{x}_8 = \begin{bmatrix} -.167 & .167 \\ .667 & .333 \end{bmatrix} \begin{bmatrix} .270 \\ 1 \end{bmatrix} \approx \begin{bmatrix} .122 \\ .513 \end{bmatrix} \longrightarrow \begin{bmatrix} .238 \\ 1 \end{bmatrix} = \mathbf{x}_9$$

$$A^{-1}\mathbf{x}_9 = \begin{bmatrix} -.167 & .167 \\ .667 & .333 \end{bmatrix} \begin{bmatrix} .238 \\ 1 \end{bmatrix} \approx \begin{bmatrix} .127 \\ .492 \end{bmatrix} \longrightarrow \begin{bmatrix} .258 \\ 1 \end{bmatrix}$$

Since the approximations to the eigenvector seem to be settling down, we stop.

If ρ_1 is the dominant eigenvalue of A^{-1}, then

$$\rho_1 \approx \frac{\langle \mathbf{x}_9, A^{-1}\mathbf{x}_9 \rangle}{\langle \mathbf{x}_9, \mathbf{x}_9 \rangle} \approx \frac{(.238)(.127) + .492}{(.238)^2 + 1^2} \approx .492$$

Thus if λ_2 is the remaining eigenvalue of A, we have

$$\lambda_2 \approx \frac{1}{\rho_1} \approx 2.03$$

Since $\lambda_2 = 2$, this is close. The eigenvector corresponding to λ_2 is actually any nonzero multiple of $[.25 \quad 1]^t$, which is reasonably close to $\mathbf{x}_9$.

5. **(a)** Suppose that λ is an eigenvalue of the invertible matrix A. That is, there is a vector $\mathbf{x}$ such that

$$A\mathbf{x} = \lambda\mathbf{x}$$

This is true if and only if

$$A^{-1}A\mathbf{x} = A^{-1}(\lambda\mathbf{x})$$

or

$$I\mathbf{x} = \lambda A^{-1}\mathbf{x}$$

or, since $\lambda \neq 0$,

$$\frac{1}{\lambda}\,\mathbf{x} = A^{-1}\mathbf{x}$$

That is,

$$A\mathbf{x} = \lambda\mathbf{x} \iff A^{-1}\mathbf{x} = \frac{1}{\lambda}\,\mathbf{x}$$

which is what we were asked to prove.

(b) By Part (a), the vector $\mathbf{x}$ is an eigenvector of A corresponding to λ if and only if it is also an eigenvector of A^{-1} corresponding to $1/\lambda$.

6. **(a)** For $i = 1, \ldots n$, we have

$$
\begin{aligned}
B\mathbf{v}_i &= \left[A - \lambda_1\mathbf{v}_1\mathbf{v}_1^{\,t}\right]\mathbf{v}_i \\
&= A\mathbf{v}_i - \lambda_1\mathbf{v}_1\mathbf{v}_1^{\,t}\mathbf{v}_i \\
&= \lambda_i\mathbf{v}_i - \lambda_1\mathbf{v}_1\mathbf{v}_1^{\,t}\mathbf{v}_i \quad (\mathbf{v}_i \text{ is an eigenvector of } A)
\end{aligned}
$$

Observe that $\mathbf{v}_1^{\,t}\mathbf{v}_i = \begin{cases} 1 & \text{if } i = 1 \\ 0 & \text{if } i \neq 1 \end{cases}$ because the eigenvectors form an orthonormal set and the entry in the 1×1 matrix $\mathbf{v}_1^t\mathbf{v}_i$ is just $\mathbf{v}_1 \cdot \mathbf{v}_i$. Therefore,

$$
B\mathbf{v}_i = \begin{cases} \mathbf{0} & \text{if } i = 1 \\ \lambda_i\mathbf{v}_i & \text{if } i \neq 1 \end{cases}
$$

(b) We have

$$
\begin{aligned}
B^t &= \left[A - \lambda_1\mathbf{v}_1\mathbf{v}_1^{\,t}\right]^t \\
&= A^t - \left[\lambda_1\mathbf{v}_1\mathbf{v}_1^{\,t}\right]^t \\
&= A - \lambda_1\left[\mathbf{v}_1^{\,t}\right]^t\mathbf{v}_1^{\,t} \qquad (A \text{ is symmetric}) \\
&= A - \lambda_1\mathbf{v}_1\mathbf{v}_1^{\,t} \\
&= B
\end{aligned}
$$

(c) We have $B\mathbf{v} = \lambda\mathbf{v}$, or $\left[A - \lambda_1\mathbf{v}_1\mathbf{v}_1^{\,t}\right]\mathbf{v} = \lambda\mathbf{v}$. Therefore

$$
A\mathbf{v} = \lambda_1\mathbf{v}_1\mathbf{v}_1^{\,t}\mathbf{v} + \lambda\mathbf{v}
$$

But since B is symmetric, Theorem 6.3.2 guarantees that $\mathbf{v}_1$ and $\mathbf{v}$ are orthogonal; that is $\mathbf{v}_1^{\,t}\mathbf{v} = 0$. Hence $A\mathbf{v} = \lambda\mathbf{v}$, which completes the proof.

EXERCISE SET 10.1

3. **(b)** Since two complex numbers are equal if and only if both their real and imaginary parts are equal, we have

$$x + y = 3$$

and

$$x - y = 1$$

Thus $x = 2$ and $y = 1$.

4. **(a)**
$$z_1 + z_2 = 1 - 2i + 4 + 5i = 5 + 3i$$

(c)
$$4z_1 = 4(1 - 2i) = 4 - 8i$$

(e)
$$\begin{aligned} 3z_1 + 4z_2 &= 3(1 - 2i) + 4(4 + 5i) \\ &= 3 - 6i + 16 + 20i \\ &= 19 + 14i \end{aligned}$$

5. **(a)** Since complex numbers obey all the usual rules of algebra, we have

$$z = 3 + 2i - (1 - i) = 2 + 3i$$

(c) Since $(i - z) + (2z - 3i) = -2 + 7i$, we have

$$i + (-z + 2z) - 3i = -2 + 7i$$

or

$$z = -2 + 7i - i + 3i = -2 + 9i$$

6. **(a)**

$$z_1 + z_2 = 4 + 5i \qquad \text{and} \qquad z_1 - z_2 = 2 - 3i$$

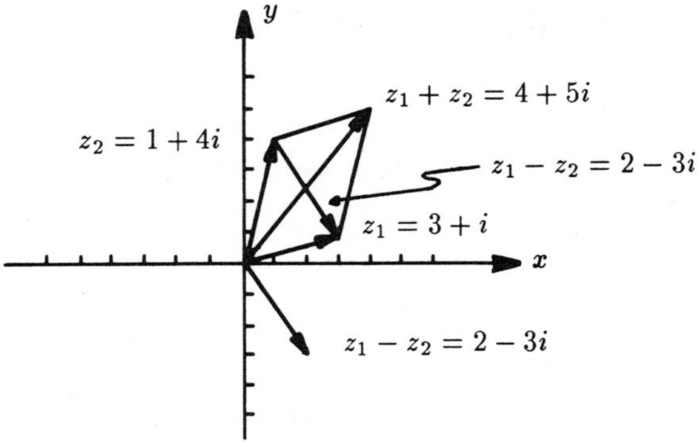

7. **(b)** $-2z = 6 + 8i$

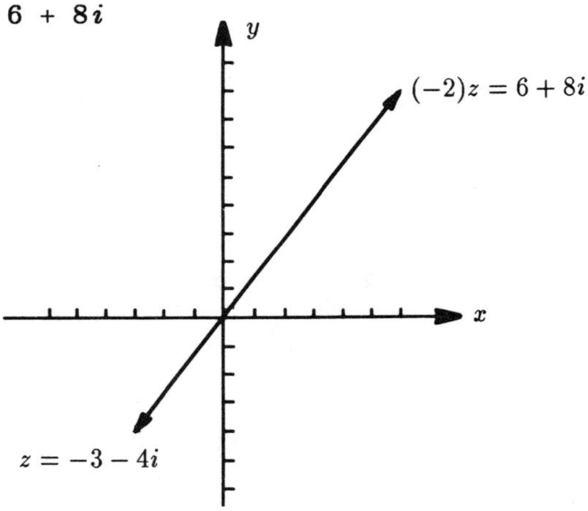

8. **(a)** We have $k_1 i + k_2(1 + i) = k_2 + \left[k_1 + k_2\right] i = 3 - 2i$. Hence, $k_2 = 3$ and $k_1 + k_2 = -2$ or $k_1 = -5$.

9. **(c)**

$$z_1 z_2 = \frac{1}{6}(2 + 4i)(1 - 5i) = \frac{2}{6}(1 + 2i)(1 - 5i)$$

$$= \frac{1}{3}(1 - 3i + 10) = \frac{11}{3} - i$$

$$z_1^2 = \frac{4}{9}(1 + 2i)^2 = \frac{4}{9}(-3 + 4i)$$

$$z_2^2 = \frac{1}{4}(1 - 5i)^2 = \frac{1}{4}(-24 - 10i) = -6 - \frac{5}{2}i$$

10. (a) First, $z_1 z_2 = (2 - 5i)(-1 - i) = -7 + 3i$. Thus,

$$z_1 - z_1 z_2 = (2 - 5i) - (-7 + 3i) = 9 - 8i$$

(c) Since $1 + z_2 = -i$, then $z_1 + \left[1 + z_2\right] = 2 - 6i$. Thus,

$$\left[z_1 + \left[1 + z_2\right]\right]^2 = (2 - 6i)^2 = 2^2(1 - 3i)^2$$
$$= 4(-8 - 6i) = -32 - 24i$$

11. Since $(4 - 6i)^2 = 2^2(2 - 3i)^2 = 4(-5 - 12i) = -4(5 + 12i)$, then

$$(1 + 2i)(4 - 6i)^2 = -4(1 + 2i)(5 + 12i)$$
$$= -4(-19 + 22i)$$
$$= 76 - 88i$$

13. Since $(1 - 3i)^2 = -8 - 6i = -2(4 + 3i)$, then

$$(1 - 3i)^3 = -2(1 - 3i)(4 + 3i) = -2(13 - 9i)$$

15. Since $(2 + i)\left[\frac{1}{2} + \frac{3}{4}i\right] = \frac{1}{4} + 2i$, then

$$\left[(2 + i)\left[\frac{1}{2} + \frac{3}{4}i\right]\right]^2 = \left[\frac{1}{4} + 2i\right]^2 = -\frac{63}{16} + i$$

17. Since $i^2 = -1$ and $i^3 = -i$, then $1 + i + i^2 + i^3 = 0$. Thus $\left[1 + i + i^2 + i^3\right]^{100} = 0$.

19. (a)

$$A + 3iB = \begin{bmatrix} 1 & i \\ -i & 3 \end{bmatrix} + \begin{bmatrix} 6i & -3 + 6i \\ 3 + 9i & 12i \end{bmatrix}$$

$$= \begin{bmatrix} 1 + 6i & -3 + 7i \\ 3 + 8i & 3 + 12i \end{bmatrix}$$

19. **(d)** $A^2 = \begin{bmatrix} 2 & 4i \\ -4i & 10 \end{bmatrix}$ and $B^2 = \begin{bmatrix} 11 + i & 12 + 6i \\ 18 - 6i & 23 + i \end{bmatrix}$

Hence

$$B^2 - A^2 = \begin{bmatrix} 9 + i & 12 + 2i \\ 18 - 2i & 13 + i \end{bmatrix}$$

20. **(a)**

$$A(BC) = \begin{bmatrix} 3 + 2i & 0 \\ -i & 2 \\ 1 + i & 1 - i \end{bmatrix} \begin{bmatrix} 5 + i & 4i & -11 \\ 3i & -2 & -5i \end{bmatrix}$$

$$= \begin{bmatrix} 13 + 13i & -8 + 12i & -33 - 22i \\ 1 + i & 0 & i \\ 7 + 9i & -6 + 6i & -16 - 16i \end{bmatrix}$$

(c) We have

$$CA = \begin{bmatrix} -6i & -1 - i \\ 6 + i & -5 + 9i \end{bmatrix} \quad \text{and} \quad B^2 = \begin{bmatrix} -1 & 0 \\ 0 & -1 \end{bmatrix}$$

Hence

$$(CA)B^2 = \begin{bmatrix} 6i & 1 + i \\ -6 - i & 5 - 9i \end{bmatrix}$$

21. **(a)** Let $z = x + iy$. Then

$$\begin{aligned} \text{Im}(iz) &= \text{Im}[i(x + iy)] = \text{Im}(-y + ix) = x \\ &= \text{Re}(x + iy) = \text{Re}(z) \end{aligned}$$

22. **(a)** We have

$$z^2 + 2z + 2 = 0 \iff z = \frac{-2 \pm \sqrt{4 - 8}}{2} = -1 \pm i$$

Let $z_1 = -1 + i$ and $z_2 = -1 - i$. Then

$$\begin{aligned} z_1^2 + 2z_1 + 2 &= (-1 + i)^2 + 2(-1 + i) + 2 \\ &= -2i - 2 + 2i + 2 \\ &= 0 \end{aligned}$$

Thus, z_1 is a root of $z^2 + 2z + 2$. The verification that z_2 is also a root is similar.

23. **(a)** We know that $i^1 = i$, $i^2 = -1$, $i^3 = -i$, and $i^4 = 1$. We also know that $i^{m+n} = i^m i^n$ and $i^{mn} = \left[i^m\right]^n$ where m and n are positive integers. The proof can be broken into four cases:

1. $n = 1, 5, 9, \cdots$ or $n = 4k + 1$

2. $n = 2, 6, 10, \cdots$ or $n = 4k + 2$

3. $n = 3, 7, 11, \cdots$ or $n = 4k + 3$

4. $n = 4, 8, 12, \cdots$ or $n = 4k + 4$

where $k = 0, 1, 2, \cdots$. In each case, $i^n = i^{4k+\ell}$ for some integer ℓ between 1 and 4. Thus

$$i^n = i^{4k} i^{\ell} = \left[i^4\right]^k i^{\ell} = 1^k i^{\ell} = i^{\ell}$$

This completes the proof.

(b) Since $2509 = 4 \cdot 627 + 1$, Case 1 of Part (a) applies, and hence $i^{2509} = i$.

24. Since $z_1 z_2 = z_2 z_1$ (see Exercise 27), we suppose, without loss of generality, that $z_1 \neq 0$. Observe that if $z = x + iy$ and $z \neq 0$, then

$$\left[\frac{x - iy}{x^2 + y^2}\right](x + iy) = 1$$

Thus, for every complex number $z \neq 0$, there is a number which we shall call $1/z$ such that

$$(1/z)z = 1$$

Hence if $z_1 \neq 0$ and $z_1 z_2 = 0$, then

$$\left[1/z_1\right]\left[z_1 z_2\right] = \left[1/z_1\right]0 = 0$$

But because

$$\left[1/z_1\right]\left[z_1 z_2\right] = \left[\left[1/z_1\right]z_1\right]z_2 = 1 \cdot z_2 = z_2$$

(see Exercise 27), this yields $z_2 = 0$. That is, if $z_1 \neq 0$, and $z_1 z_2 = 0$, then $z_2 = 0$. Similarly, as observed above, if $z_2 \neq 0$ then $z_1 = 0$, which completes the proof.

Alternate Solution: This result can also be obtained by letting $z_1 = x_1 + iy_1$ and $z_2 = x_2 + iy_2$ and observing that

$$z_1 z_2 = 0 \iff x_1 x_2 = y_1 y_2 \quad \text{and} \quad x_1 y_2 = -x_2 y_1$$

If we suppose that $x_1 \neq 0$, then we have

$$z_1 z_2 = 0 \iff x_2 = \frac{y_1 y_2}{x_1} \quad \text{and} \quad y_2 = \frac{-x_2 y_1}{x_1}$$

This implies that

$$y_2 = -\left[\frac{y_1}{x_1}\right]^2 y_2$$

But since $-\left[y_1/x_1\right]^2 \le 0$, the above equation can hold if and only if $y_2 = 0$. Recall that $x_2 = \left[y_1 y_2\right]/x_1$; then $y_2 = 0$ implies that $x_2 = 0$ and hence that $z_2 = 0$. Thus if $x_1 \neq 0$, then $z_2 = 0$. A similar argument shows that if $y_1 \neq 0$, then $z_2 = 0$. Therefore, if $z_1 \neq 0$, then $z_2 = 0$. Similarly, if $z_2 \neq 0$, then $z_1 = 0$.

25. Observe that $zz_1 = zz_2 \Leftrightarrow zz_1 - zz_2 = 0 \Leftrightarrow z\left[z_1 - z_2\right] = 0$. Since $z \neq 0$ by hypothesis, it follows from Exercise 24 that $z_1 - z_2 = 0$, i.e., that $z_1 = z_2$.

26. **(a)** Let $z_1 = x_1 + iy_1$ and $z_2 = x_2 + iy_2$. Then

$$\begin{aligned} z_1 + z_2 &= \left[x_1 + iy_1\right] + \left[x_2 + iy_2\right] \\ &= \left[x_1 + x_2\right] + i\left[y_1 + y_2\right] \\ &= \left[x_2 + x_1\right] + i\left[y_2 + y_1\right] \\ &= \left[x_2 + iy_2\right] + \left[x_1 + iy_2\right] \\ &= z_2 + z_1 \end{aligned}$$

27. **(a)** Let $z_1 = x_1 + iy_1$ and $z_2 = x_2 + iy_2$. Then

$$\begin{aligned} z_1 z_2 &= \left[x_1 + iy_1\right]\left[x_2 + iy_2\right] \\ &= \left[x_1 x_2 - y_1 y_2\right] + i\left[x_1 y_2 + x_2 y_1\right] \\ &= \left[x_2 x_1 - y_2 y_1\right] + i\left[y_2 x_1 + y_1 x_2\right] \\ &= \left[x_2 + iy_2\right]\left[x_1 + iy_1\right] \\ &= z_2 z_1 \end{aligned}$$

29. Let $z = x + iy$. Then $[\text{Re}(z)]^2 = x^2$ and $|z|^2 = x^2 + y^2$. Hence

$$[\text{Re}(z)]^2 \le |z|^2$$

Taking positive square roots gives $|\text{Re}(z)| \le |z|$. The inequality $|\text{Im}(z)| \le |z|$ may be proved in a similar way.

EXERCISE SET 10.2

2. **(a)** Since $i = 0 + i1$, then

$$|i| = \sqrt{0^2 + 1^2} = 1$$

(c) $|-3 - 4i| = \sqrt{(-3)^2 + (-4)^2} = \sqrt{25} = 5$

(e) $|-8| = |-8 + 0i| = \sqrt{(-8)^2 + 0^2} = 8$

3. **(a)** We have

$$z\bar{z} = (2 - 4i)(2 + 4i) = 20$$

On the other hand,

$$|z|^2 = 2^2 + (-4)^2 = 20$$

(b) We have

$$z\bar{z} = (-3 + 5i)(-3 - 5i) = 34$$

On the other hand,

$$|z|^2 = (-3)^2 + 5^2 = 34$$

4. **(a)** From Equation (5), we have

$$\frac{z_1}{z_2} = \frac{1 - 5i}{3 + 4i} = \frac{(1 - 5i)(3 - 4i)}{3^2 + 4^2} = \frac{-17 - 19i}{25}$$

4. **(c)** Again using Equation (5), we have

$$\frac{z_1}{z_2} = \frac{1 - 5i}{3 - 4i} = \frac{(1 - 5i)(3 + 4i)}{3^2 + (-4)^2} = \frac{23 - 11i}{25}$$

(e) Since $|z_2| = 5$, we have

$$\frac{z_1}{|z_2|} = \frac{1 - 5i}{5} = \frac{1}{5} - i$$

5. **(a)** Equation (5) with $z_1 = 1$ and $z_2 = i$ yields

$$\frac{1}{i} = \frac{1(-i)}{1} = -i$$

(c)

$$\frac{1}{z} = \frac{7}{-i} = \frac{7(i)}{1} = 7i$$

6. **(a)** Since

$$\frac{z_1}{z_2} = \frac{1 + i}{1 - 2i} = \frac{(1 + i)(1 + 2i)}{5} = -\frac{1}{5} + \frac{3}{5}i$$

then

$$z_1 - \frac{z_1}{z_2} = (1 + i) - \left[-\frac{1}{5} + \frac{3}{5}i\right] = \frac{6}{5} + \frac{2}{5}i$$

(c) Since

$$\frac{iz_1}{z_2} = \frac{i(1 + i)}{1 - 2i} = \frac{-1 + i}{1 - 2i} = \frac{(-1 + i)(1 + 2i)}{(1 - 2i)(1 + 2i)} = -\frac{3}{5} - \frac{1}{5}i$$

and

$$\bar{z}_1^{\,2} = 2i$$

then

$$\bar{z}_1^2 - \frac{iz_1}{z_2} = 2i - \left(-\frac{3}{5} - \frac{1}{5}i\right) = \frac{3}{5} + \frac{11}{5}i$$

7. Equation (5) with $z_1 = i$ and $z_2 = 1 + i$ gives

$$\frac{i}{1 + i} = \frac{i(1 - i)}{2} = \frac{1}{2} + \frac{1}{2}i$$

9. Since $(3 + 4i)^2 = -7 + 24i$, we have

$$\frac{1}{(3 + 4i)^2} = \frac{-7 - 24i}{(-7)^2 + (-24)^2} = \frac{-7 - 24i}{625}$$

11. Since

$$\frac{\sqrt{3} + i}{\sqrt{3} - i} = \frac{(\sqrt{3} + i)^2}{4} = \frac{2 + 2\sqrt{3}\,i}{4} = \frac{1}{2} + \frac{\sqrt{3}}{2}i$$

then

$$\frac{\sqrt{3} + i}{(1 - i)(\sqrt{3} - i)} = \frac{\frac{1}{2} + \frac{\sqrt{3}}{2}i}{1 - i} = \frac{\left[\frac{1}{2} + \frac{\sqrt{3}}{2}i\right](1 + i)}{2}$$

$$= \frac{1 - \sqrt{3}}{4} + \left[\frac{1 + \sqrt{3}}{4}\right]i$$

13. We have

$$\frac{i}{1 - i} = \frac{i(1 + i)}{2} = -\frac{1}{2} + \frac{1}{2}i$$

and

$$(1 - 2i)(1 + 2i) = 5$$

Thus

$$\frac{i}{(1 - i)(1 - 2i)(1 + 2i)} = \frac{-\frac{1}{2} + \frac{1}{2}i}{5} = -\frac{1}{10} + \frac{1}{10}i$$

15. **(a)** If $iz = 2 - i$, then

$$z = \frac{2 - i}{i} = \frac{(2 - i)(-i)}{1} = -1 - 2i$$

16. **(a)**

$$\overline{\bar{z} + 5i} = \bar{\bar{z}} + \overline{5i} = z - 5i$$

(c) The result follows easily from the fact that $\overline{i + \bar{z}} = \bar{i} + \bar{\bar{z}} = z - i$. This result holds whenever $z \neq i$.

17. **(a)** The set of points satisfying the equation $|z| = 2$ is the set of all points representing vectors of length 2. Thus, it is a circle of radius 2 and center at the origin.

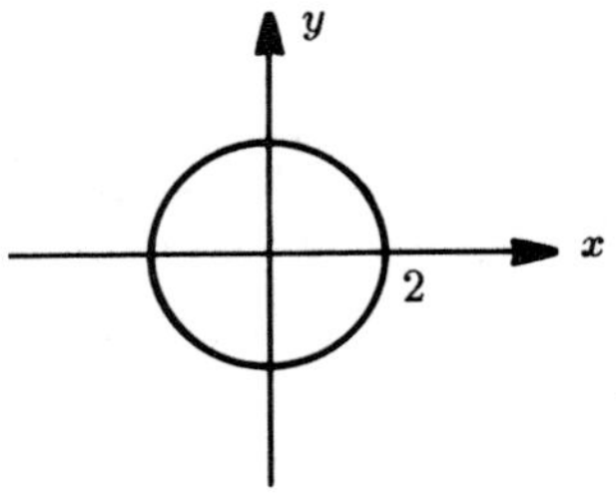

Analytically, if $z = x + iy$, then

$$|z| = 2 \Leftrightarrow \sqrt{x^2 + y^2} = 2$$
$$\Leftrightarrow x^2 + y^2 = 4$$

which is the equation of the above circle.

(c) The values of z which satisfy the equation $|z - i| = |z + i|$ are just those z whose distance from the point i is equal to their distance from the point $-i$. Geometrically, then, z can be any point on the real axis.

We now show this result analytically. Let $z = x + iy$. Then

$$\begin{aligned}|z - i| = |z + i| &\Leftrightarrow |z - i|^2 = |z + i|^2 \\ &\Leftrightarrow |x + i(y - 1)|^2 = |x + i(y + 1)|^2 \\ &\Leftrightarrow x^2 + (y - 1)^2 = x^2 + (y + 1)^2 \\ &\Leftrightarrow -2y = 2y \\ &\Leftrightarrow y = 0\end{aligned}$$

18. **(a)** This inequality represents the set of all points whose distance from the point $-i$ is at most 1.

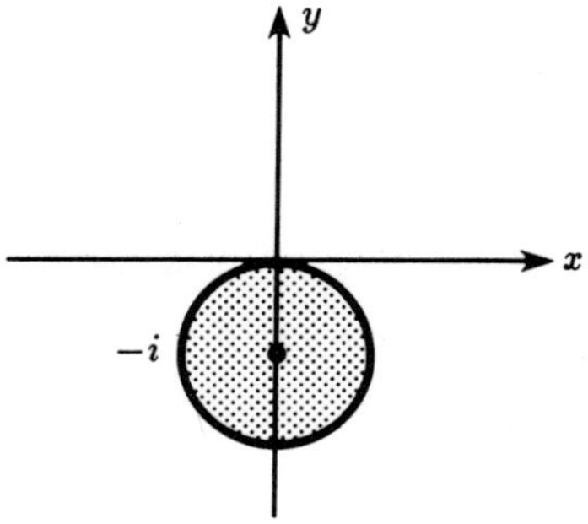

Analytically let $z = x + iy$. Then

$$\begin{aligned}|z + i| \le 1 &\Leftrightarrow |z + i|^2 \le 1 \\ &\Leftrightarrow x^2 + (y + 1)^2 \le 1\end{aligned}$$

Thus, the inequality represents the set of all points in or on the circle with center at $-i$ and radius 1.

(c) Since $|2z - 4i| < 1 \Leftrightarrow |z - 2i| < \frac{1}{2}$, this inequality represents the set of all points whose distance from the point $2i$ is less than 1/2.

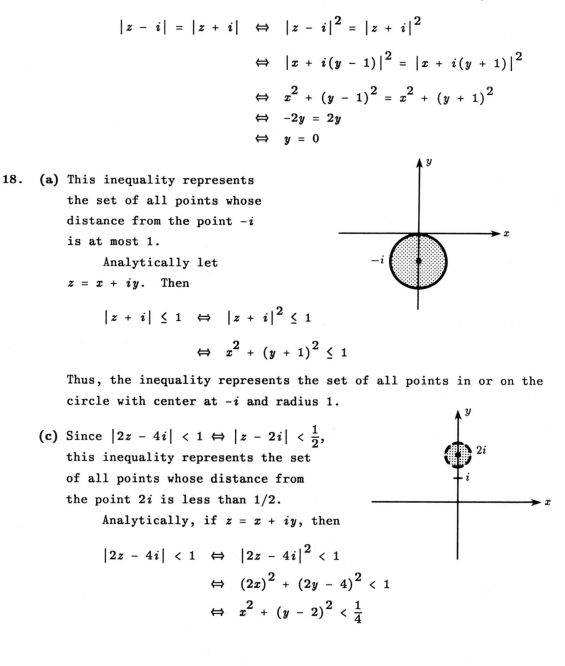

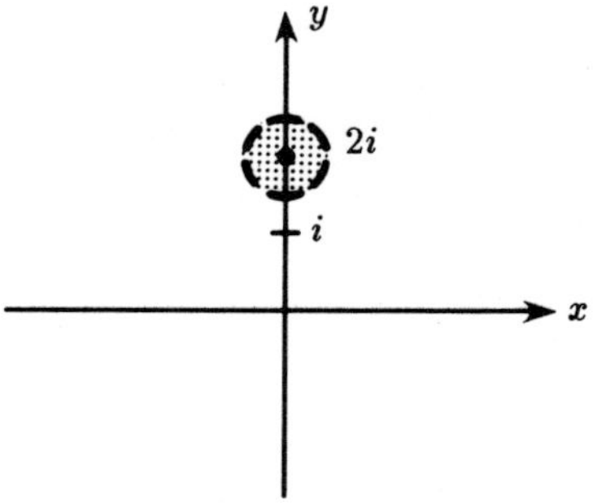

Analytically, if $z = x + iy$, then

$$\begin{aligned}|2z - 4i| < 1 &\Leftrightarrow |2z - 4i|^2 < 1 \\ &\Leftrightarrow (2x)^2 + (2y - 4)^2 < 1 \\ &\Leftrightarrow x^2 + (y - 2)^2 < \frac{1}{4}\end{aligned}$$

19. **(a)** $\mathrm{Re}(\overline{iz}) = \mathrm{Re}(\overline{i} \cdot \overline{z}) = \mathrm{Re}[(-i)(x - iy)] = \mathrm{Re}(-y - ix) = -y$

(c) $\mathrm{Re}(i\overline{z}) = \mathrm{Re}[i(x - iy)] = \mathrm{Re}(y + ix) = y$

20. **(a)** Since $1/i = -i$ and $(-i)^n = (-1)^n(i)^n$, this problem is just a variation on Exercise 23 of Section 10.1.

(b) Since $2509 = 4 \cdot 627 + 1$ and $(i)^4 = 1$, $(i)^{2509} = i$ and therefore $(1/i)^{2509} = (-1)^{2509} \cdot i = -i$.

21. **(a)** Let $z = x + iy$. Then

$$\frac{1}{2}(z + \overline{z}) = \frac{1}{2}[(x + iy) + (x - iy)]$$

$$= \frac{1}{2}(2x) = x = \mathrm{Re}(z)$$

22. Let $z = x + iy$. Then

$$z = \overline{z} \iff x + iy = x - iy \iff y = 0 \iff z \text{ is real.}$$

23. **(a)** Equation (5) gives

$$\frac{z_1}{z_2} = \frac{1}{|z_2|^2} z_1\overline{z}_2$$

$$= \frac{1}{x_2^2 + y_2^2}\left(x_1 + iy_1\right)\left(x_2 - iy_2\right)$$

$$= \frac{1}{x_2^2 + y_2^2}\left[\left(x_1x_2 + y_1y_2\right) + i\left(x_2y_1 - x_1y_2\right)\right]$$

Thus

$$\mathrm{Re}\left(\frac{z_1}{z_2}\right) = \frac{x_1x_2 + y_1y_2}{x_2^2 + y_2^2}$$

25. $|z| = \sqrt{x^2 + y^2} = \sqrt{x^2 + (-y)^2} = |\bar{z}|$

26. Let $z_1 = x_1 + iy_1$ and $z_2 = x_2 + iy_2$.

(a)
$$\begin{aligned}\overline{z_1 - z_2} &= \overline{(x_1 + iy_1) - (x_2 + iy_2)} \\ &= \overline{(x_1 - x_2) + i(y_1 - y_2)} \\ &= (x_1 - x_2) - i(y_1 - y_2) \\ &= (x_1 - iy_1) - (x_2 - iy_2) \\ &= \bar{z}_1 - \bar{z}_2\end{aligned}$$

(c)
$$\begin{aligned}\overline{(z_1/z_2)} &= \frac{1}{|z_2|^2}\,\overline{z_1\bar{z}_2} \\ &= \frac{1}{|z_2|^2}\,\bar{z}_1 \cdot \bar{\bar{z}}_2 && \text{[Part (b)]} \\ &= \bar{z}_1/\bar{z}_2 && \text{[Part (d)]}\end{aligned}$$

27. **(a)** $\overline{z^2} = \overline{zz} = \bar{z}\,\bar{z} = (\bar{z})^2$

(b) We use mathematical induction. In Part (a), we verified that the result holds when $n = 2$. Now, assume that $(\bar{z})^n = \overline{z^n}$. Then

$$\begin{aligned}(\bar{z})^{n+1} &= (\bar{z})^n\bar{z} \\ &= \overline{z^n}\;\bar{z} \\ &= \overline{z^{n+1}}\end{aligned}$$

and the result is proved.

28. Let A denote the matrix of the system, i.e.,

$$A = \begin{bmatrix} i & -i \\ 2 & 1 \end{bmatrix}$$

Thus $\det(A) = i - (-2i) = 3i$. Hence

$$x_1 = \frac{1}{3i}\begin{vmatrix} -2 & -i \\ i & 1 \end{vmatrix} = \frac{1}{3i}\left[-2 + i^2\right] = \frac{-1}{i} = i$$

and

$$x_2 = \frac{1}{3i}\begin{vmatrix} i & -2 \\ 2 & i \end{vmatrix} = \frac{1}{3i}\left[i^2 + 4\right] = \frac{1}{i} = -i$$

30. Let A denote the matrix of the system, i.e.,

$$A = \begin{bmatrix} 1 & 1 & 1 \\ 1 & 1 & -1 \\ 1 & -1 & 1 \end{bmatrix}$$

Then $\det(A) = -4$ and hence

$$x_1 = -\frac{1}{4}\begin{vmatrix} 3 & 1 & 1 \\ 2 + 2i & 1 & -1 \\ -1 & -1 & 1 \end{vmatrix} = \frac{1}{2} + i$$

$$x_2 = -\frac{1}{4}\begin{vmatrix} 1 & 3 & 1 \\ 1 & 2 + 2i & -1 \\ 1 & -1 & 1 \end{vmatrix} = 2$$

$$x_3 = -\frac{1}{4}\begin{vmatrix} 1 & 1 & 3 \\ 1 & 1 & 2 + 2i \\ 1 & -1 & -1 \end{vmatrix} = \frac{1}{2} - i$$

32. $$\begin{bmatrix} -1 & -1 - i \\ -1 + i & -2 \end{bmatrix} \to \begin{bmatrix} 1 & 1 + i \\ -1 + i & -2 \end{bmatrix} \to \begin{bmatrix} 1 & 1 + i \\ 0 & 0 \end{bmatrix}$$

Therefore $x_1 = -(1 + i)t$ and $x_2 = t$ where t is arbitrary.

34.

$$\begin{bmatrix} 1 & i & -i \\ -1 & 1-i & 2i \\ 2 & -1+2i & -3i \end{bmatrix} \longrightarrow \begin{bmatrix} 1 & i & -i \\ 0 & 1 & i \\ 0 & -1 & -i \end{bmatrix} \longrightarrow$$

$$\begin{bmatrix} 1 & i & -i \\ 0 & 1 & i \\ 0 & 0 & 0 \end{bmatrix} \longrightarrow \begin{bmatrix} 1 & 0 & 1-i \\ 0 & 1 & i \\ 0 & 0 & 0 \end{bmatrix}$$

If we let $x_3 = t$, where t is arbitrary, then $x_2 = -it$ and $x_1 = -(1 - i)t$.

35. **(a)**

$$A^{-1} = \frac{1}{i^2 + 2}\begin{bmatrix} i & 2 \\ -1 & i \end{bmatrix} = \begin{bmatrix} i & 2 \\ -1 & i \end{bmatrix}$$

It is easy to verify that $AA^{-1} = A^{-1}A = I$.

36. From Exercise 27.(b), we have that $(\bar{z})^n = \overline{z^n}$. Since a_k is real, it follows that $a_k(\overline{z^k}) = \overline{a_k z^k}$. Thus,

$$\begin{aligned} p(\bar{z}) &= a_0 + a_1\bar{z} + a_2(\bar{z})^2 + \cdots + a_n(\bar{z})^n \\ &= \bar{a}_0 + \overline{a_1 z} + \overline{a_2 z^2} + \cdots + \overline{a_n z^n} \\ &= \overline{a_0 + a_1 z + a_2 z^2 + \cdots + a_n z^n} \\ &= \overline{p(z)} \end{aligned}$$

Hence, $p(z) = 0 \implies p(\bar{z}) = \bar{0} = 0$.

38. **(a)** $$\left[\begin{array}{ccc|ccc} 1 & 1+i & 0 & 1 & 0 & 0 \\ 0 & 1 & i & 0 & 1 & 0 \\ -i & 1-2i & 2 & 0 & 0 & 1 \end{array}\right]$$

$$\left[\begin{array}{ccc|ccc} 1 & 1+i & 0 & 1 & 0 & 0 \\ 0 & 1 & i & 0 & 1 & 0 \\ 0 & -i & 2 & i & 0 & 1 \end{array}\right] \quad \boxed{R_3 \to R_3 + iR_1}$$

$$\left[\begin{array}{ccc|ccc} 1 & 1+i & 0 & 1 & 0 & 0 \\ 0 & 1 & i & 0 & 1 & 0 \\ 0 & 0 & 1 & i & i & 1 \end{array}\right] \quad \boxed{R_3 \to R_3 + iR_2}$$

$$\left[\begin{array}{ccc|ccc} 1 & 1+i & 0 & 1 & 0 & 0 \\ 0 & 1 & 0 & 1 & 2 & -i \\ 0 & 0 & 1 & i & i & 1 \end{array}\right] \quad \boxed{R_2 \to R_2 - iR_3}$$

$$\left[\begin{array}{ccc|ccc} 1 & 0 & 0 & -i & -2-2i & -1+i \\ 0 & 1 & 0 & 1 & 2 & -i \\ 0 & 0 & 1 & i & i & 1 \end{array}\right] \quad \boxed{R_2 \to R_1 - (1+i)R_2}$$

Thus

$$A^{-1} = \begin{bmatrix} -i & -2-2i & -1+i \\ 1 & 2 & -i \\ i & i & 1 \end{bmatrix}$$

EXERCISE SET 10.3

1. **(a)** If $z = 1$, then $\arg z = 2k\pi$ where $k = 0, \pm 1, \pm 2, \cdots$. Thus, $\text{Arg}(1) = 0$.

   (c) If $z = -i$, then $\arg z = \frac{3\pi}{2} + 2k\pi$ where $k = 0, \pm 1, \pm 2, \cdots$. Thus, $\text{Arg}(-i) = -\pi/2$.

   (e) If $z = -1 + \sqrt{3}\, i$, then $\arg z = \frac{2\pi}{3} + 2k\pi$ where $k = 0, \pm 1, \pm 2, \cdots$. Thus, $\text{Arg}(-1 + \sqrt{3}i) = \frac{2\pi}{3}$.

2. We have

$$\arg(1 - \sqrt{3}\, i) = \frac{5\pi}{3} + 2k\pi, \qquad k = 0, \pm 1, \pm 2, \cdots$$

   (a) Put $k = 0$ in the above equation.

   (b) Put $k = -1$ in the above equation.

3. **(a)** Since $|2i| = 2$ and $\text{Arg}(2i) = \pi/2$, we have

$$2i = 2\left[\cos\left[\frac{\pi}{2}\right] + i \sin\left[\frac{\pi}{2}\right]\right]$$

   (c) Since $|5 + 5i| = \sqrt{50} = 5\sqrt{2}$ and $\text{Arg}(5 + 5i) = \pi/4$, we have

$$5 + 5i = 5\sqrt{2}\left[\cos\left[\frac{\pi}{4}\right] + i \sin\left[\frac{\pi}{4}\right]\right]$$

3. **(e)** since $|-3 - 3i| = \sqrt{18} = 3\sqrt{2}$ and $\text{Arg}(-3 - 3i) = -\frac{3\pi}{4}$, we have

$$-3 - 3i = 3\sqrt{2}\left[\cos\left(-\frac{3\pi}{4}\right) + i\sin\left(-\frac{3\pi}{4}\right)\right]$$

4. We have $|z_1| = 2$, $\text{Arg}(z_1) = \frac{\pi}{4}$, $|z_2| = 3$, and $\text{Arg}(z_2) = \frac{\pi}{6}$.

(a) Here $|z_1 z_2| = |z_1|\,|z_2| = 6$ and $\text{Arg}(z) = \text{Arg}(z_1) + \text{Arg}(z_2) = \frac{5\pi}{12}$.

Hence

$$z_1 z_2 = 6\left[\cos\left(\frac{5\pi}{12}\right) + i\sin\left(\frac{5\pi}{12}\right)\right]$$

(c) Here $|z_2/z_1| = |z_2|/|z_1| = 3/2$ and

$\text{Arg}(z_2/z_1) = \text{Arg}(z_2) - \text{Arg}(z_1) = -\frac{\pi}{12}$. Hence

$$\frac{z_2}{z_1} = \frac{3}{2}\left[\cos\left(-\frac{\pi}{12}\right) + i\sin\left(-\frac{\pi}{12}\right)\right]$$

5. We have $|z_1| = 1$, $\text{Arg}(z_1) = \frac{\pi}{2}$, $|z_2| = 2$, $\text{Arg}(z_2) = -\frac{\pi}{3}$, $|z_3| = 2$, and $\text{Arg}(z_3) = \frac{\pi}{6}$. So

$$\left|\frac{z_1 z_2}{z_3}\right| = \frac{|z_1|\,|z_2|}{|z_3|} = 1$$

and

$$\text{Arg}\left(\frac{z_1 z_2}{z_3}\right) = \text{Arg}(z_1) + \text{Arg}(z_2) - \text{Arg}(z_3) = 0$$

Therefore

$$\frac{z_1 z_2}{z_3} = \cos(0) + i\sin(0) = 1$$

6. **(a)** We have $r = \sqrt{2}$, $\theta = \frac{\pi}{4}$, and $n = 12$. Thus

$$(1 + i)^{12} = 2^6[\cos(3\pi) + i\sin(3\pi)] = -64$$

(c) We have $r = 2$, $\theta = \frac{\pi}{6}$, and $n = 7$. Thus

$$(\sqrt{3} + i)^7 = 2^7\left[\cos\left[\frac{7\pi}{6}\right] + i\sin\left[\frac{7\pi}{6}\right]\right]$$

$$= 2^7\left[-\frac{\sqrt{3}}{2} - i\,\frac{1}{2}\right] = -64\sqrt{3} - 64i$$

7. We use Formula (10).

(a) We have $r = 1$, $\theta = -\frac{\pi}{2}$, and $n = 2$. Thus

$$(-i)^{1/2} = \cos\left[-\frac{\pi}{4} + k\pi\right] + i\sin\left[-\frac{\pi}{4} + k\pi\right] \qquad k = 0,\ 1$$

Thus, the two square roots of $-i$ are:

$$\cos\left[-\frac{\pi}{4}\right] + i\sin\left[-\frac{\pi}{4}\right] = \frac{1}{\sqrt{2}} - \frac{1}{\sqrt{2}}\,i$$

$$\cos\left[\frac{3\pi}{4}\right] + i\sin\left[\frac{3\pi}{4}\right] = -\frac{1}{\sqrt{2}} + \frac{1}{\sqrt{2}}\,i$$

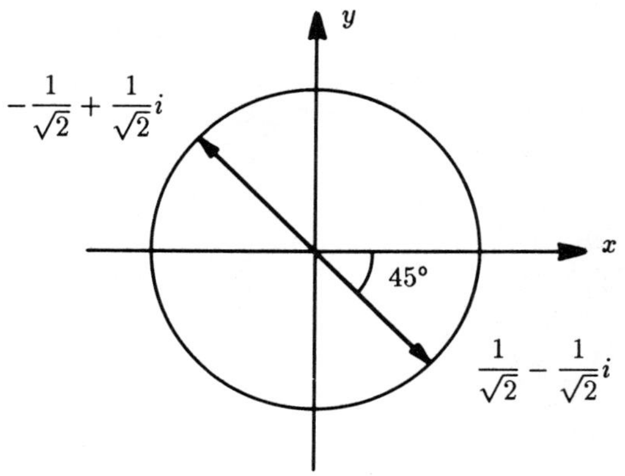

7. **(c)** We have $r = 27$, $\theta = \pi$, and $n = 3$. Thus

$$(-27)^{1/3} = 3\left[\cos\left[\frac{\pi}{3} + \frac{2k\pi}{3}\right] + i\sin\left[\frac{\pi}{3} + \frac{2k\pi}{3}\right]\right] \qquad k = 0,1,2$$

Therefore, the three cube roots of -27 are:

$$3\left[\cos\left[\frac{\pi}{3}\right] + i\sin\left[\frac{\pi}{3}\right]\right] = \frac{3}{2} + \frac{3\sqrt{3}}{2}\,i$$

$$3[\cos(\pi) + i\sin(\pi)] = -3$$

$$3\left[\cos\left[\frac{5\pi}{3}\right] + i\sin\left[\frac{5\pi}{3}\right]\right] = \frac{3}{2} - \frac{3\sqrt{3}}{2}\,i$$

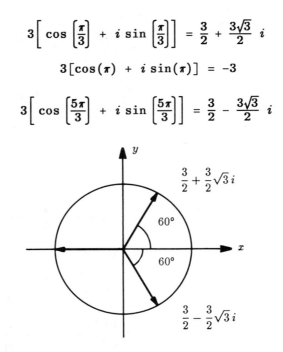

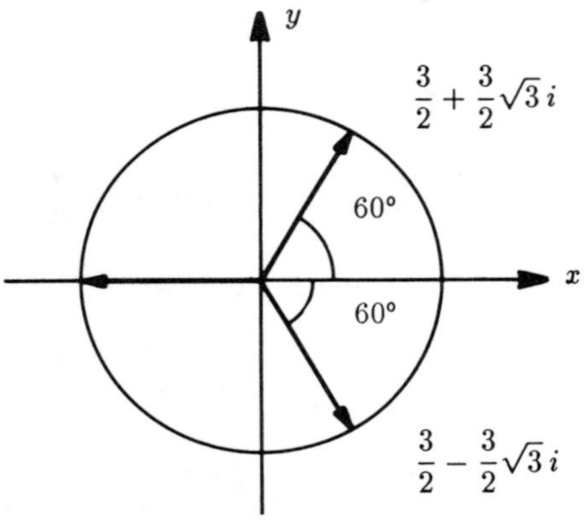

(e) Here $r = 1$, $\theta = \pi$, and $n = 4$. Thus

$$(-1)^{1/4} = \cos\left[\frac{\pi}{4} + \frac{k\pi}{2}\right] + i\sin\left[\frac{\pi}{4} + \frac{k\pi}{2}\right] \qquad k = 0,1,2,3$$

Therefore the four fourth roots of -1 are:

$$\cos\frac{\pi}{4} + i\sin\frac{\pi}{4} = \frac{1}{\sqrt{2}} + \frac{1}{\sqrt{2}}\,i$$

$$\cos\frac{3\pi}{4} + i\sin\frac{3\pi}{4} = -\frac{1}{\sqrt{2}} + \frac{1}{\sqrt{2}}\,i$$

$$\cos \frac{5\pi}{4} + i \sin \frac{5\pi}{4} = -\frac{1}{\sqrt{2}} - \frac{1}{\sqrt{2}} i$$

$$\cos \frac{7\pi}{4} + i \sin \frac{7\pi}{4} = \frac{1}{\sqrt{2}} - \frac{1}{\sqrt{2}} i$$

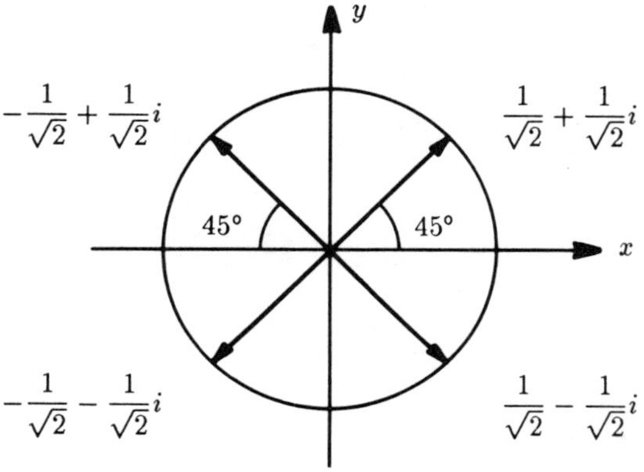

9. We observe that $w = 1$ is one sixth root of 1. Since the remaining 5 must be equally spaced around the unit circle, any two roots must be separated from one another by an angle of $\frac{2\pi}{6} = \frac{\pi}{3} = 60^{\circ}$. We show all six sixth roots in the diagram.

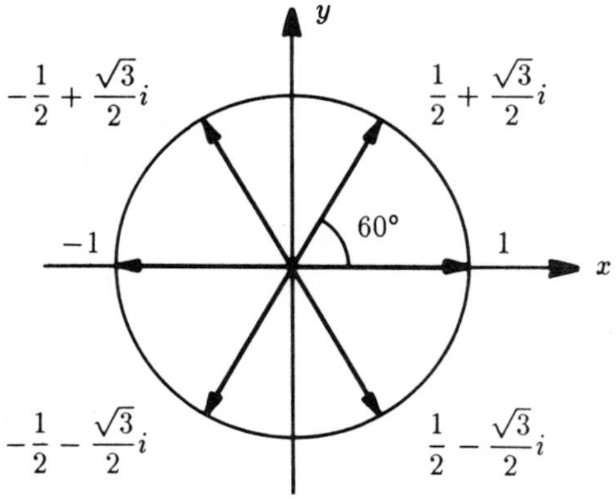

11. **(a)** We have $z^4 = 16 \Leftrightarrow z = 16^{1/4}$. The fourth roots of 16 are 2, $2i$, -2, and $-2i$.

12. The fourth roots of −8 may be found by using Formula (10) with $r = 8$, $\theta = \pi$, and $n = 4$. This yields

$$z_1 = 8^{1/4}\left[\cos\left[\frac{\pi}{4}\right] + i\sin\left[\frac{\pi}{4}\right]\right] = 2^{1/4} + 2^{1/4}\, i$$

$$z_2 = 8^{1/4}\left[\cos\left[\frac{3\pi}{4}\right] + i\sin\left[\frac{3\pi}{4}\right]\right] = -2^{1/4} + 2^{1/4}\, i$$

$$z_3 = 8^{1/4}\left[\cos\left[\frac{5\pi}{4}\right] + i\sin\left[\frac{5\pi}{4}\right]\right] = -2^{1/4} - 2^{1/4}\, i$$

$$z_4 = 8^{1/4}\left[\cos\left[\frac{7\pi}{4}\right] + i\sin\left[\frac{7\pi}{4}\right]\right] = 2^{1/4} - 2^{1/4}\, i$$

Since $z^4 + 8$ has exactly four zeros, it follows that

$$z^4 + 8 = (z - z_1)(z - z_2)(z - z_3)(z - z_4)$$

But

$$(z - z_1)(z - z_4) = z^2 + 2^{5/4}z + 2^{3/2}$$

and

$$(z - z_2)(z - z_3) = z^2 + 2^{5/4}z + 2^3/2$$

Thus,

$$z^4 + 8 = \left[z^2 - 2^{5/4}z + 2^{3/2}\right]\left[z^2 + 2^{5/4}z + 2^{3/2}\right]$$

14. **(a)** We have $r = \sqrt{2}$, $\theta = \frac{\pi}{4}$, and $n = 8$. Thus

$$(1 + i)^8 = 2^4[\cos(2\pi) + i\sin(2\pi)] = 16$$

15. **(a)** Since

$$z = 3e^{i\pi} = 3[\cos(\pi) + i\sin(\pi)] = -3$$

then $\text{Re}(z) = -3$ and $\text{Im}(z) = 0$.

(c) Since

$$\bar{z} = \sqrt{2}\, e^{i\pi/2} = \sqrt{2}\left[\cos\left[\frac{\pi}{2}\right] + i \sin\left[\frac{\pi}{2}\right]\right] = \sqrt{2}\, i$$

then $z = -\sqrt{2}\, i$ and hence $\text{Re}(z) = 0$ and $\text{Im}(z) = -\sqrt{2}$.

16. The values of $z^{1/n}$ given in Formula (10) are all of the form

$$z_k = r^{1/n}\left[\cos \theta_k + i \sin \theta_k\right]$$

where

$$\theta_k = \frac{\theta}{n} + \frac{2k\pi}{n}$$

Thus, two roots, z_ℓ and z_m, can be equal if and only if $\cos \theta_\ell + i \sin \theta_\ell = \cos \theta_m + i \sin \theta_m$. This can occur if and only if $\cos \theta_\ell = \cos \theta_m$ and $\sin \theta_\ell = \sin \theta_m$, which, in turn, can occur if and only if θ_ℓ and θ_m differ by an integral multiple of 2π. But

$$\theta_\ell - \theta_m = \frac{2\ell m}{n} - \frac{2m\pi}{n} = \frac{(\ell - m)}{n}(2\pi)$$

Thus we have:

(a) If $\ell \neq m$ and ℓ and m are integers between 0 and $n - 1$, then $0 < \left|\frac{\ell - m}{n}\right| < 1$ so that $\frac{(\ell - m)}{n}(2\pi)$ cannot be an integral multiple of 2π and hence the two resulting values of $z^{1/n}$ are different.

(b) If, say, ℓ is not in the range 0 to $n - 1$, then we can write ℓ in the form

$$\ell = nk + m$$

for some integer m between 0 and $n - 1$ and for some integer $k \neq 0$.

Thus

$$\theta_\ell - \theta_m = \frac{(\ell - m)}{n}(2\pi) = k(2\pi)$$

which guarantees that $z_\ell = z_m$.

17. Case 1. Suppose that $n = 0$. Then

$$(\cos\theta + i\sin\theta)^n = 1 = \cos(0) + i\sin(0)$$

So Formula (7) is valid if $n = 0$.

Case 2. In order to verify that Formula (7) holds if n is a negative integer, we first let $n = -1$. Then

$$\begin{aligned}(\cos\theta + i\sin\theta)^{-1} &= \frac{1}{\cos\theta + i\sin\theta}\\ &= \cos\theta - i\sin\theta\\ &= \cos(-\theta) + i\sin(-\theta)\end{aligned}$$

Thus, Formula (7) is valid if $n = -1$.

Now suppose that n is a positive integer (and hence that $-n$ is a negative integer). Then

$$\begin{aligned}(\cos\theta + i\sin\theta)^{-n} &= \left[(\cos\theta + i\sin\theta)^{-1}\right]^n\\ &= [\cos(-\theta) + i\sin(-\theta)]^n\\ &= \cos(-n\theta) + i\sin(-n\theta) \quad \text{[By (7)]}\end{aligned}$$

This completes the proof.

19. We have $z_1 = r_1 e^{i\theta_1}$ and $z_2 = r_2 e^{i\theta_2}$. But (see Exercise 17)

$$\frac{1}{z_2} = \frac{1}{r_2 e^{i\theta_2}} = \frac{1}{r_2}e^{-i\theta_2}$$

If we replace z_2 by $1/z_2$ in Formula (3), we obtain

$$\begin{aligned}\frac{z_1}{z_2} &= z_1\left[\frac{1}{z_2}\right]\\ &= \frac{r_1}{r_2}\left[\cos\left[\theta_1 + (-\theta_2)\right] + i\sin\left[\theta_1 + (-\theta_2)\right]\right]\\ &= \frac{r_1}{r_2}\left[\cos\left[\theta_1 - \theta_2\right] + i\sin\left[\theta_1 - \theta_2\right]\right]\end{aligned}$$

which is Formula (5).

EXERCISE SET 10.4

1. (a) $\mathbf{u} - \mathbf{v} = (2i - (-i), 0 - i, -1 - (1 + i), 3 - (-1))$
$= (3i, -i, -2 - i, 4)$

(c) $-\mathbf{w} + \mathbf{v} = (-(1 + i) - i, i + i, -(-1 + 2i) + (1 + i), 0 + (-1))$
$= (-1 - 2i, 2i, 2 - i, -1)$

(e) $-i\mathbf{v} = (-1, 1, 1 - i, i)$ and $2i\mathbf{w} = (-2 + 2i, 2, -4 - 2i, 0)$. Thus

$$-i\mathbf{v} + 2i\mathbf{w} = (-3 + 2i, 3, -3 - 3i, i)$$

2. Observe that $i\mathbf{x} = \mathbf{u} - \mathbf{v} - \mathbf{w}$. Thus

$$\mathbf{x} = -i(\mathbf{u} - \mathbf{v} - \mathbf{w})$$

3. Consider the equation $c_1\mathbf{u}_1 + c_2\mathbf{u}_2 + c_3\mathbf{u}_3 = (-3 + i, 3 + 2i, 3 - 4i)$. The augmented matrix for this system of equations is

$$\begin{bmatrix} 1-i & 2i & 0 & -3+i \\ i & 1+i & 2i & 3+2i \\ 0 & 1 & 2-i & 3-4i \end{bmatrix}$$

The row-echelon form for the above matrix is

$$\begin{bmatrix} 1 & -1+i & 0 & -2-i \\ 0 & 1 & \frac{1}{2} + \frac{1}{2}i & \frac{3}{2} + \frac{1}{2}i \\ 0 & 0 & 1 & 2-i \end{bmatrix}$$

Hence, $c_3 = 2 - i$, $c_2 = \frac{3}{2} + \frac{1}{2}i - \left(\frac{1}{2} + \frac{1}{2}i\right)c_3 = 0$, and $c_1 = -2 - i$.

5. **(a)** $\|\mathbf{v}\| = \sqrt{|1|^2 + |i|^2} = \sqrt{2}$

(c) $\|\mathbf{v}\| = \sqrt{|2i|^2 + |0|^2 + |2i + 1|^2 + |(-1)|^2}$

$= \sqrt{4 + 0 + 5 + 1} = \sqrt{10}$

6. **(a)** Since $\mathbf{u} + \mathbf{v} = (3i, 3 + 4i, -3i)$, then

$$\|\mathbf{u} + \mathbf{v}\| = \sqrt{|3i|^2 + |3 + 4i|^2 + |-3i|^2}$$
$$= \sqrt{9 + 25 + 9} = \sqrt{43}$$

(c) We have $\|\mathbf{u}\| = \sqrt{|3i|^2 + |-i|^2} = \sqrt{9 + 1} = \sqrt{10}$, $-i\mathbf{u} = (3, 0, -1)$ and $\|-i\mathbf{u}\| = \sqrt{10}$. It follows that

$$\|-i\mathbf{u}\| + i\|\mathbf{u}\| = \sqrt{10} + i\sqrt{10}$$

(e) Since $\|\mathbf{w}\| = \sqrt{|1 + i|^2 + |2i|^2} = \sqrt{2 + 4} = \sqrt{6}$, then

$$\frac{1}{\|\mathbf{w}\|}\mathbf{w} = \left(\frac{1 + i}{\sqrt{6}}, \frac{2i}{\sqrt{6}}, 0\right)$$

8. Since $k\mathbf{v} = (3ki, 4ki)$, then

$$\|k\mathbf{v}\| = \sqrt{|3ki|^2 + |4ki|^2} = 5|k|.$$

Thus $\|k\mathbf{v}\| = 1 \Leftrightarrow k = \pm 1/5$.

9. **(a)** $\mathbf{u} \cdot \mathbf{v} = (-i)(-3i) + (3i)(-2i) = -3 + 6 = 3.$

(c) $\mathbf{u} \cdot \mathbf{v} = (1 - i)(4 - 6i) + (1 + i)(5i) + (2i)(-1 - i) + (3)(-i)$

$= (-2 - 10i) + (-5 + 5i) + (2 - 2i) + (-3i)$

$= -5 - 10i$

11. Let V denote the set and let

$$\mathbf{u} = \begin{bmatrix} u & 0 \\ 0 & \bar{u} \end{bmatrix} \quad \text{and} \quad \mathbf{v} = \begin{bmatrix} v & 0 \\ 0 & \bar{v} \end{bmatrix}$$

We check the axioms listed in the definition of a vector space (see Section 4.2).

(1) $$\mathbf{u} + \mathbf{v} = \begin{bmatrix} u+v & 0 \\ 0 & \bar{u}+\bar{v} \end{bmatrix} = \begin{bmatrix} u+v & 0 \\ 0 & \overline{u+v} \end{bmatrix}$$

So $\mathbf{u} + \mathbf{v}$ belongs to V.

(2) Since $u + v = v + u$ and $\bar{u} + \bar{v} = \bar{v} + \bar{u}$, it follows that $\mathbf{u} + \mathbf{v} = \mathbf{v} + \mathbf{u}$.

(3) Axiom (3) follows by a routine, if tedious, check.

(4) The matrix $\begin{bmatrix} 0 & 0 \\ 0 & 0 \end{bmatrix}$ serves as the zero vector.

(5) Let $-\mathbf{u} = \begin{bmatrix} -u & 0 \\ 0 & -\bar{u} \end{bmatrix} = \begin{bmatrix} -u & 0 \\ 0 & \overline{-u} \end{bmatrix}$. Then $\mathbf{u} + (-\mathbf{u}) = \mathbf{0}$.

(6) Since $k\mathbf{u} = \begin{bmatrix} ku & 0 \\ 0 & k\bar{u} \end{bmatrix}$, $k\mathbf{u}$ will be in V if and only if $k\bar{u} = \overline{ku}$, which is true if and only if k is real or $u = 0$. Thus Axiom (6) fails.

(7) - (9) These axioms all hold by virtue of the properties of matrix addition and scalar multiplication. However, as seen above, the closure property of scalar multiplication may fail, so the vectors need not be in V.

(10) Clearly $1\mathbf{u} = \mathbf{u}$.

Thus, this set is not a vector space because Axiom (6) fails.

12. Let $\mathbf{x} = (1, 1, \ldots, 1)$ and let $k = i$. Then

$$k\mathbf{x} = i\mathbf{x} = (i, i, \ldots, i)$$

is not in R^n although $\mathbf{x}$ clearly does belong to R^n. Hence R^n is not closed under scalar multiplication and is therefore not a subspace of C^n.

13. **(a)** Since

$$(z_1, 0, 0) + (z_2, 0, 0) = (z_1 + z_2, 0, 0)$$

and

$$k(z, 0, 0) = (kz, 0, 0)$$

this set is closed under both addition and scalar multiplication. Hence it is a subspace.

(b) Since

$$(z_1, i, i) + (z_2, i, i) = (z_1 + z_2, 2i, 2i)$$

this is not a subspace. Part (b) of Theorem 4.3.1 also is not satisfied.

14. **(a)** While this set is closed under vector addition, it is not closed under scalar multiplication since the real entries, when multiplied by a non-real scalar, need not remain real. Hence it does not form a subspace.

(b) This set does form a subspace since

$$\begin{bmatrix} u_1 & u_2 \\ u_3 & -u_1 \end{bmatrix} + \begin{bmatrix} v_1 & v_2 \\ v_3 & -v_1 \end{bmatrix} = \begin{bmatrix} u_1+v_1 & u_2+v_2 \\ u_3+v_3 & -(u_1+v_1) \end{bmatrix}$$

and

$$k\begin{bmatrix} z_1 & z_2 \\ z_3 & -z_1 \end{bmatrix} = \begin{bmatrix} kz_1 & kz_2 \\ kz_3 & -(kz_1) \end{bmatrix}$$

That is, the set is closed under both operations.

15. **(a)** Since

$$(f + g)(1) = f(1) + g(1) = 0 + 0 = 0$$

and

$$kf(1) = k(0) = 0$$

for all functions f and g in the set and for all scalars k, this set forms a subspace.

(c) Since

$$\begin{aligned}(f + g)(-x) &= f(-x) + g(-x) = \overline{f(x)} + \overline{g(x)} \\ &= \overline{f(x) + g(x)} = \overline{(f + g)(x)}\end{aligned}$$

the set is closed under vector addition. It is closed under scalar multiplication by a real scalar, but not by a complex scalar. For instance, if $f(x) = xi$, then $f(x)$ is in the set but $if(x)$ is not.

16. **(a)** Consider the equation $k_1\mathbf{u} + k_2\mathbf{v} = (3i, 3i, 3i)$. Equating components yields

$$\begin{aligned} k_1 i + k_2 2i &= 3i \\ -k_1 i + k_2 4i &= 3i \\ k_1 3i \qquad &= 3i \end{aligned}$$

The above system has the solution $k_1 = k_2 = 1$. Thus, $(3i, 3i, 3i)$ is a linear combination of $\mathbf{u}$ and $\mathbf{v}$.

16. **(c)** Consider the equation $k_1\mathbf{u} + k_2\mathbf{v} = (i, 5i, 6i)$. Equating components yields

$$k_1 i + k_2 2i = i$$
$$-k_1 i + k_2 4i = 5i$$
$$k_1 3i \qquad = 6i$$

The above system is inconsistent. Hence, $(i, 5i, 6i)$ is not a linear combination of **u** and **v**.

17. **(a)** Consider the equation $k_1\mathbf{u} + k_2\mathbf{v} + k_3\mathbf{w} = (1, 1, 1)$. Equating components yields

$$k_1 + (1 + i)k_2 \qquad = 1$$
$$k_2 + ik_3 = 1$$
$$-ik_1 + (1 - 2i)k_2 + 2k_3 = 1$$

Solving the system yields $k_1 = -3 - 2i$, $k_2 = 3 - i$, and $k_3 = 1 + 2i$.

(c) Let A be the matrix whose first, second and third columns are the components of **u**, **v**, and **w**, respectively. By Part (a), we know that $\det(A) \neq 0$. Hence, $k_1 = k_2 = k_3 = 0$.

18. We let A denote the matrix whose first, second and third columns are the components of $\mathbf{v}_1$, $\mathbf{v}_2$, and $\mathbf{v}_3$, respectively.

(a) Since $\det(A) = 6i \neq 0$, it follows that $\mathbf{v}_1$, $\mathbf{v}_2$, and $\mathbf{v}_3$ span C^3.

(d) Since $\det(A) = 0$, it follows that $\mathbf{v}_1$, $\mathbf{v}_2$, and $\mathbf{v}_3$ do not span C^3.

19. **(a)** Recall that $e^{ix} = \cos x + i \sin x$ and that $e^{-ix} = \cos(-x) + i \sin(-x) = \cos x - i \sin x$. Therefore,

$$\cos x = \frac{e^{ix} + e^{-ix}}{2} = \frac{1}{2}\mathbf{f} + \frac{1}{2}\mathbf{g}$$

and so $\cos x$ lies in the space spanned by **f** and **g**.

(b) If $a\mathbf{f} + b\mathbf{g} = \sin x$, then (see Part (a))

$$(a + b)\cos x + (a - b)i \sin x = \sin x$$

Thus, since the sine and cosine functions are linearly independent, we have

$$a + b = 0$$

and

$$a - b = -i$$

This yields $a = -i/2$, $b = i/2$, so again the vector lies in the space spanned by **f** and **g**.

(c) If $a\mathbf{f} + b\mathbf{g} = \cos x + 3i \sin x$, then (see Part (a))

$$a + b = 1$$

and

$$a - b = 3$$

Hence $a = 2$ and $b = -1$ and thus the given vector does lie in the space spanned by **f** and **g**.

20. **(a)** Note that $i\mathbf{u}_1 = \mathbf{u}_2$.

21. Let A denote the matrix whose first, second, and third columns are the components of $\mathbf{u}_1$, $\mathbf{u}_2$, and $\mathbf{u}_3$, respectively.

(a) Since the last row of A consists entirely of zeros, it follows that $\det(A) = 0$ and hence $\mathbf{u}_1$, $\mathbf{u}_2$, and $\mathbf{u}_3$, are linearly dependent.

(c) Since $\det(A) = i \neq 0$, then $\mathbf{u}_1$, $\mathbf{u}_2$, and $\mathbf{u}_3$ are linearly independent.

22. Observe that $\mathbf{f} - 3\mathbf{g} - 3\mathbf{h} = \mathbf{0}$.

23. **(a)** Since the dimension of C^2 is two, any basis for C^2 must contain precisely two vectors.

24. **(a)** Since $\begin{vmatrix} 2i & 4i \\ -i & 0 \end{vmatrix} = -4 \neq 0$, the vectors are linearly independent and hence form a basis for C^2.

(d) Since $\begin{vmatrix} 2-3i & 3+2i \\ i & -1 \end{vmatrix} = 0$, the vectors are linearly dependent and hence are not a basis for C^2.

25. Since the number of vectors is the same as the dimension of C^3, the vectors will form a basis if and only if they are linearly independent.

(a) Since $\begin{vmatrix} i & i & i \\ 0 & i & i \\ 0 & 0 & i \end{vmatrix} = -i \neq 0$, the vectors are linearly independent.

Hence, they form a basis.

(c) From Problem 21(c), we know that the vectors are linearly independent. Hence, they form a basis.

26. The row-echelon form of the matrix of the system is

$$\begin{bmatrix} 1 & 1+i \\ 0 & 0 \end{bmatrix}$$

So x_2 is arbitrary and $x_1 = -(1 + i)x_2$. Hence, the dimension of the solution space is one and $\begin{bmatrix} -(1 + i) \\ 1 \end{bmatrix}$ is a basis for that space.

28. The reduced row-echelon form of the matrix of the system is

$$\begin{bmatrix} 1 & 0 & -3-6i \\ 0 & 1 & 3i \\ 0 & 0 & 0 \end{bmatrix}$$

So x_3 is arbitrary, $x_2 = (-3i)x_3$, and $x_1 = (3 + 6i)x_3$. Hence, the dimension of the solution space is one and $\begin{bmatrix} 3+6i \\ -3i \\ 1 \end{bmatrix}$ is a basis for that space.

30. Let $\mathbf{u} = (u_1, u_2, \ldots, u_n)$ and $\mathbf{v} = (v_1, v_2, \ldots, v_n)$. From the definition of the Euclidean inner product in C^n, we have

$$\begin{aligned} \mathbf{u} \cdot (k\mathbf{v}) &= u_1(\overline{kv_1}) + u_2(\overline{kv_2}) + \cdots + u_n(\overline{kv_n}) \\ &= u_1(\bar{k}\bar{v}_1) + u_2(\bar{k}\bar{v}_2) + \cdots + u_n(\bar{k}\bar{v}_n) \\ &= \bar{k}(u_1\bar{v}_1) + \bar{k}(u_2\bar{v}_2) + \cdots + \bar{k}(u_n\bar{v}_n) \\ &= \bar{k}[u_1\bar{v}_1 + u_2\bar{v}_2 + \cdots + u_n\bar{v}_n] \\ &= \bar{k}(\mathbf{u} \cdot \mathbf{v}) \end{aligned}$$

31. (a) Let $\mathbf{u} = (u_1, u_2, \ldots, u_n)$, $\mathbf{v} = (v_1, v_2, \ldots, v_n)$ and $\mathbf{w} = (w_1, w_2, \ldots, w_n)$. Then, since we have $\mathbf{u} + \mathbf{v} = (u_1 + v_1, u_2 + v_2, \ldots, u_n + v_n)$,

$$\begin{aligned} (\mathbf{u} + \mathbf{v}) \cdot \mathbf{w} &= (u_1+v_1)\bar{w}_1 + (u_2+v_2)\bar{w}_2 + \cdots + (u_n+v_n)\bar{w}_n \\ &= [u_1\bar{w}_1+u_2\bar{w}_2+\cdots+u_n\bar{w}_n] + [v_1\bar{w}_1+v_2\bar{w}_2+\cdots+v_n\bar{w}_n] \\ &= \mathbf{u} \cdot \mathbf{w} + \mathbf{v} \cdot \mathbf{w} \end{aligned}$$

32. Let $f(x) = f_1(x) + if_2(x)$ and $g(x) = g_1(x) + ig_2(x)$. Then

$$\begin{aligned}(f + g)(x) &= f(x) + g(x) \\ &= [f_1(x) + g_1(x)] + i[f_2(x) + g_2(x)]\end{aligned}$$

and

$$\begin{aligned}(kf)(x) &= kf(x) \\ &= kf_1(x) + ikf_2(x)\end{aligned}$$

Thus, if f and g are continuous, then $f + g$ is continuous; similarly, if f is continuous, then kf is continuous. Hence, the result follows from Theorem 4.3.1.

33. Hint: Show that

$$\|\mathbf{u} + k\mathbf{v}\|^2 = \|\mathbf{u}\|^2 + k(\mathbf{v} \cdot \mathbf{u}) + \bar{k}(\mathbf{u} \cdot \mathbf{v}) + k\bar{k}\|\mathbf{v}\|^2$$

and apply this result to each term on the right-hand side of the identity.

EXERCISE SET 10.5

1. Let $\mathbf{u} = (u_1, u_2)$, $\mathbf{v} = (v_1, v_2)$, and $\mathbf{w} = (w_1, w_2)$. We proceed to check the four axioms

(1)
$$\overline{\langle \mathbf{v},\mathbf{u} \rangle} = \overline{3v_1\overline{u}_1 + 2v_2\overline{u}_2}$$
$$= 3u_1\overline{v}_1 + 2u_2\overline{v}_2 = \langle \mathbf{u},\mathbf{v} \rangle$$

(2)
$$\langle \mathbf{u} + \mathbf{v}, \mathbf{w} \rangle = 3(u_1 + v_1)\overline{w}_1 + 2(u_2 + v_2)\overline{w}_2$$
$$= [3u_1\overline{w}_1 + 2u_2\overline{w}_2] + [3v_1\overline{w}_1 + 2v_2\overline{w}_2]$$
$$= \langle \mathbf{u},\mathbf{w} \rangle + \langle \mathbf{v},\mathbf{w} \rangle$$

(3)
$$\langle k\mathbf{u},\mathbf{v} \rangle = 3(ku_1)\overline{v}_1 + 2(ku_2)\overline{v}_2$$
$$= k[3u_1\overline{v}_1 + 2u_2\overline{v}_2] = k\langle \mathbf{u},\mathbf{v} \rangle$$

(4)
$$\langle \mathbf{u},\mathbf{u} \rangle = 3u_1\overline{u}_1 + 2u_2\overline{u}_2$$
$$= 3|u_1|^2 + 2|u_2|^2 \qquad \text{(Theorem 10.2.1)}$$
$$\geq 0$$

Indeed, $\langle \mathbf{u},\mathbf{u} \rangle = 0 \Leftrightarrow u_1 = u_2 = 0 \Leftrightarrow \mathbf{u} = \mathbf{0}$.

Hence, this is an inner product on C^2.

2. **(a)** $\langle \mathbf{u},\mathbf{v}\rangle = 3(2i)(\overline{-i}) + 2(-i)(\overline{3i}) = -6 - 6 = -12$

(c) $\langle \mathbf{u},\mathbf{v}\rangle = 3(1 + i)(\overline{1 - i}) + 2(1 - i)(\overline{1 + i})$

$= 3(2i) + 2(-2i) = 2i$

3. Let $\mathbf{u} = (u_1, u_2)$ and $\mathbf{v} = (v_1, v_2)$. We check Axioms 1 and 4, leaving 2 and 3 to you.

(1)

$$\begin{aligned}\overline{\langle \mathbf{v},\mathbf{u}\rangle} &= \overline{v_1\bar{u}_1} + \overline{(1 + i)v_1\bar{u}_2} + \overline{(1 - i)v_2\bar{u}_1} + \overline{3v_2\bar{u}_2} \\ &= u_1\bar{v}_1 + (1 - i)u_2\bar{v}_1 + (1 + i)u_1\bar{v}_2 + 3u_2\bar{v}_2 \\ &= \langle \mathbf{u},\mathbf{v}\rangle\end{aligned}$$

(4) Recall that $|\mathrm{Re}(z)| \le |z|$ by Problem 29 of Section 10.1. Now

$$\begin{aligned}\langle \mathbf{u},\mathbf{u}\rangle &= u_1\bar{u}_1 + (1 + i)u_1\bar{u}_2 + (1 - i)u_2\bar{u}_1 + 3u_2\bar{u}_2 \\ &= |u_1|^2 + (1 + i)u_1\bar{u}_2 + \overline{(1 + i)u_1\bar{u}_2} + 3|u_2|^2 \\ &= |u_1|^2 + 2\mathrm{Re}((1 + i)u_1\bar{u}_2) + 3|u_2|^2 \\ &\ge |u_1|^2 - 2|(1 + i)u_1\bar{u}_2| + 3|u_2|^2 \\ &= |u_1|^2 - 2\sqrt{2}\,|u_1||u_2| + 3|u_2|^2 \\ &= \left[|u_1| - \sqrt{2}\,|u_2|\right]^2 + |u_2|^2 \\ &\ge 0\end{aligned}$$

Moreover, $\langle \mathbf{u},\mathbf{u}\rangle = 0$ if and only if both $|u_2|$ and $|u_1| - \sqrt{2}\,|u_2| = 0$, or $\mathbf{u} = \mathbf{0}$.

4. (a) $\langle \mathbf{u},\mathbf{v}\rangle = (2i)i + (1+i)(2i)(-3i) + (1-i)(-i)(i) + 3(-i)(-3i)$
$= -4 + 5i$

(c) $\langle \mathbf{u},\mathbf{v}\rangle = (1 + i)(1 + i) + (1 + i)(1 + i)(1 - i)$
$+ (1 - i)(1 - i)(1 + i) + 3(1 - i)(1 - i)$
$= 4 - 4i$

5. (a) This is not an inner product on C^2. Axioms 1 – 3 are easily checked. Moreover,

$$\langle \mathbf{u},\mathbf{u}\rangle = u_1\bar{u}_1 = |u_1| \geq 0$$

However, $\langle \mathbf{u},\mathbf{u}\rangle = 0 \Leftrightarrow u_1 = 0 \Leftrightarrow \mathbf{u} = \mathbf{0}$. For example, $\langle i,i\rangle = 0$ although $i \neq 0$. Hence, Axiom 4 fails.

(c) This is not an inner product on C^2. Axioms 1 and 4 are easily checked. However, for $\mathbf{w} = (w_1, w_2)$, we have

$$\begin{aligned}\langle \mathbf{u} + \mathbf{v}, \mathbf{w}\rangle &= |u_1 + v_1|^2|w_1|^2 + |u_2 + v_2|^2|w_2|^2 \\ &\neq \left[|u_1|^2+|v_1|^2\right]|w_1|^2 + \left[|u_2|^2+|v_2|^2\right]|w_2|^2 \\ &= \langle \mathbf{u},\mathbf{w}\rangle + \langle \mathbf{v},\mathbf{w}\rangle\end{aligned}$$

For instance, $\langle 1 + 1, 1\rangle = 4$, but $\langle 1,1\rangle + \langle 1,1\rangle = 2$. Moreover, $\langle k\mathbf{u},\mathbf{v}\rangle = |k|^2 \langle \mathbf{u},\mathbf{v}\rangle$, so that $\langle k\mathbf{u},\mathbf{v}\rangle \neq k\langle \mathbf{u},\mathbf{v}\rangle$ for most values of k, $\mathbf{u}$, and $\mathbf{v}$. Thus both Axioms 2 and 3 fail.

(e) Axiom 1 holds since

$$\begin{aligned}\overline{\langle \mathbf{v},\mathbf{u}\rangle} &= \overline{2v_1\bar{u}_1 + iv_1\bar{u}_2 - iv_2\bar{u}_1 + 2v_2\bar{u}_2} \\ &= 2u_1\bar{v}_1 - iu_2\bar{v}_1 + iu_1\bar{v}_2 + 2u_2\bar{v}_2 \\ &= \langle \mathbf{u},\mathbf{v}\rangle\end{aligned}$$

A similar argument serves to verify Axiom 2 and Axiom 3 holds by inspection. Finally, using the result of Problem 29 of Section 10.1, we have

$$\begin{aligned}\langle \mathbf{u},\mathbf{u}\rangle &= 2u_1\bar{u}_1 + iu_1\bar{u}_2 - iu_2\bar{u}_1 + 2u_2\bar{u}_2 \\ &= 2|u_1|^2 + 2\mathrm{Re}(iu_1\bar{u}_2) + 2|u_2|^2 \\ &\geq 2|u_1|^2 - 2|iu_1\bar{u}_2| + 2|u_2|^2 \\ &= \Big[|u_1| - |u_2|\Big]^2 + |u_1|^2 + |u_2|^2 \\ &\geq 0\end{aligned}$$

Moreover, $\langle \mathbf{u},\mathbf{u}\rangle = 0 \Leftrightarrow u_1 = u_2 = 0$, or $\mathbf{u} = \mathbf{0}$. Thus all four axioms hold.

6. $\langle \mathbf{u},\mathbf{v}\rangle = (-i)(3) + (1 + i)(-2 + 3i) + (1 - i)(-4i) + i(1)$

$= -9 - 5i$

8. First,

$$\begin{aligned}\langle \mathbf{g},\mathbf{f}\rangle &= \overline{(g_1(0) + ig_2(0))\overline{(f_1(0) + if_2(0))}} \\ &= (f_1(0) + if_2(0))\overline{(g_1(0) + ig_2(0))} \\ &= \langle \mathbf{f},\mathbf{g}\rangle\end{aligned}$$

Second, if $\mathbf{h} = h_1(x) + ih_2(x)$, then

$$\begin{aligned}\langle \mathbf{f} + \mathbf{g}, \mathbf{h}\rangle &= (f_1(0)+g_1(0)+i[f_2(0)+g_2(0)])\cdot\overline{(h_1(0)+ih_2(0)} \\ &= \langle \mathbf{f},\mathbf{h}\rangle + \langle \mathbf{g},\mathbf{h}\rangle\end{aligned}$$

Third, it is easily checked that $<k\mathbf{f},\mathbf{g}> = k<\mathbf{f},\mathbf{g}>$. And finally,

$$\begin{aligned} <\mathbf{f},\mathbf{f}> &= (f_1(0) + if_2(0))\overline{(f_1(0) + if_2(0))} \\ &= |f_1(0)|^2 + |f_2(0)|^2 \\ &\geq 0 \end{aligned}$$

However, $<\mathbf{f},\mathbf{f}> = 0 \Leftrightarrow f_1(0) = f_2(0) = 0 \nLeftrightarrow \mathbf{f} = \mathbf{0}$. For instance, if $\mathbf{f} = ix$, then $<\mathbf{f},\mathbf{f}> = 0$, but $\mathbf{f} \neq \mathbf{0}$. Therefore, this is not an inner product.

9. (a) $\|\mathbf{w}\| = [3(-i)(i) + 2(3i)(-3i)]^{1/2} = \sqrt{21}$

(c) $\|\mathbf{w}\| = [3(0)(0) + 2(2 - i)(2 + i)]^{1/2} = \sqrt{10}$

10. (a) $\|\mathbf{w}\| = [|-i|^2 + |3i|^2]^{1/2} = \sqrt{10}$

(c) $\|\mathbf{w}\| = [|0|^2 + |2 - i|^2]^{1/2} = \sqrt{5}$

11. (a) $\|\mathbf{w}\| = [(1)(1) + (1 + i)(1)(i) + (1 - i)(-i)(1) + 3(-i)(i)]^{1/2}$
$= \sqrt{2}$

(c) $\|\mathbf{w}\| = [(3 - 4i)(3 + 4i)]^{1/2} = 5$

12. (a) $\|A\| = [(-i)(i) + (7i)(-7i) + (6i)(-6i) + (2i)(-2i)]^{1/2}$
$= 3\sqrt{10}$

13. (a) Since $\mathbf{u} - \mathbf{v} = (1 - i,\ 1 + i)$, then

$$\begin{aligned} d(\mathbf{u},\mathbf{v}) &= [3(1 - i)(1 + i) + 2(1 + i)(1 - i)]^{1/2} \\ &= \sqrt{10} \end{aligned}$$

14. **(a)** Since $\mathbf{u} - \mathbf{v} = (1 - i,\ 1 + i)$,

$$d(\mathbf{u},\mathbf{v}) = [(1 - i)(1 + i) + (1 + i)(1 - i)]^{1/2} = 2$$

15. **(a)** Since $\mathbf{u} - \mathbf{v} = (1 - i,\ 1 + i)$,

$$\begin{aligned} d(\mathbf{u},\mathbf{v}) &= [(1 - i)(1 + i) + (1 + i)(1 - i)(1 - i) \\ &\quad + (1 - i)(1 + i)(1 + i) + 3(1 + i)(1 - i)]^{1/2} \\ &= 2\sqrt{3} \end{aligned}$$

16. **(a)** Since $A - B = \begin{bmatrix} 6i & 5i \\ i & 6i \end{bmatrix}$,

$$\begin{aligned} d(A,B) &= [(6i)(-6i) + (5i)(-5i) + (i)(-i) + (6i)(-6i)]^{1/2} \\ &= \sqrt{98} = 7\sqrt{2} \end{aligned}$$

17. **(a)** Since $\mathbf{u} \cdot \mathbf{v} = (2i)(-i) + (i)(-6i) + (3i)(\overline{k})$, then $\mathbf{u} \cdot \mathbf{v} = 0 \Leftrightarrow 8 + 3i\overline{k} = 0$ or $k = -8i/3$.

18. **(a)** Let B denote the given matrix. Then

$$\begin{aligned} \langle A,B\rangle &= (2i)(-3) + (i)(1 + i) + (-i)(1 + i) + (3i)(2) \\ &= 0 \end{aligned}$$

Thus, A and B are orthogonal.

(d) Let B denote the given matrix. Then

$$\langle A,B\rangle = i + (-i)(3 + i) = 1 - 2i$$

Thus, A and B are not orthogonal.

19. Since $\mathbf{x} = \frac{1}{\sqrt{3}} e^{i\theta}(i,1,1)$, we have

$$\|\mathbf{x}\| = \frac{1}{\sqrt{3}} \|e^{i\theta}\| \, \|(i,1,1)\| = \frac{1}{\sqrt{3}} (1)(1 + 1 + 1)^{1/2} = 1$$

Also

$$\begin{aligned} \langle \mathbf{x}, (1,i,0)\rangle &= \frac{1}{\sqrt{3}} e^{i\theta}(i,1,1) \cdot (1,i,0) \\ &= \frac{1}{\sqrt{3}} e^{i\theta}(i - i + 0) \\ &= 0 \end{aligned}$$

and

$$\begin{aligned} \langle \mathbf{x}, (0,i,-i)\rangle &= \frac{1}{\sqrt{3}} e^{i\theta}(i,1,1) \cdot (0,i,-i) \\ &= \frac{1}{\sqrt{3}} e^{i\theta}(0 - i + i) \\ &= 0 \end{aligned}$$

20. (a) Since $\|(0, 1 - i)\| = \sqrt{2}$, the set of vectors is not orthonormal. (However, the set is orthogonal.)

(b) Since $\left(\frac{i}{\sqrt{2}}, -\frac{i}{\sqrt{2}}\right) \cdot \left(\frac{i}{\sqrt{2}}, \frac{i}{\sqrt{2}}\right) = -\frac{i^2}{2} + \frac{i^2}{2} = 0$ and the norm of each vector is 1, the set is orthonormal.

21. (a) Call the vectors $\mathbf{u}_1$, $\mathbf{u}_2$, and $\mathbf{u}_3$, respectively. Then

$\|\mathbf{u}_1\| = \|\mathbf{u}_2\| = \|\mathbf{u}_3\| = 1$ and $\mathbf{u}_1 \cdot \mathbf{u}_2 = \mathbf{u}_1 \cdot \mathbf{u}_3 = 0$. However, $\mathbf{u}_2 \cdot \mathbf{u}_3 = \frac{i^2}{\sqrt{6}} + \frac{(-i)^2}{\sqrt{6}} = -\frac{2}{\sqrt{6}} \neq 0$. Hence the set is not orthonormal.

22. If we use the Euclidean inner product, then $\|\mathbf{x}\| = \sqrt{2/5} \neq 1$. Hence, the vectors cannot be orthonormal with respect to the Euclidean inner product. In fact, neither vector is normal and the vectors are not orthogonal.

We now let $\langle \mathbf{u},\mathbf{v}\rangle = 3u_1\bar{v}_1 + 2u_2\bar{v}_2$. Then

$$\|\mathbf{x}\|^2 = 3\left[\frac{i}{\sqrt{5}}\right]\left[-\frac{i}{\sqrt{5}}\right] + 2\left[-\frac{i}{\sqrt{5}}\right]\left[\frac{i}{\sqrt{5}}\right] = 1$$

$$\|\mathbf{y}\|^2 = 3\left[\frac{2i}{\sqrt{30}}\right]\left[-\frac{2i}{\sqrt{30}}\right] + 2\left[\frac{3i}{\sqrt{30}}\right]\left[-\frac{3i}{\sqrt{30}}\right] = 1$$

and

$$\langle \mathbf{x},\mathbf{y}\rangle = 3\left[\frac{i}{\sqrt{5}}\right]\left[-\frac{2i}{\sqrt{30}}\right] + 2\left[-\frac{i}{\sqrt{5}}\right]\left[-\frac{3i}{\sqrt{30}}\right] = 0$$

Hence $\mathbf{x}$ and $\mathbf{y}$ are orthonormal with respect to this inner product.

24. **(a)** We have

$$\mathbf{v}_1 = \frac{\mathbf{u}_1}{\|\mathbf{u}_1\|} = \frac{(i,\ -3i)}{\sqrt{10}} = \left(\frac{i}{\sqrt{10}},\ -\frac{3i}{\sqrt{10}}\right)$$

and, since $\mathbf{u}_2 \cdot \mathbf{v}_1 = -4/\sqrt{10}$, it follows that

$$\mathbf{u}_2 - (\mathbf{u}_2 \cdot \mathbf{v}_1)\mathbf{v}_1 = (2i,\ 2i) + \frac{4}{\sqrt{10}}\left(\frac{i}{\sqrt{10}},\ \frac{-3i}{\sqrt{10}}\right)$$

$$= \left(\frac{12i}{5},\ \frac{4i}{5}\right)$$

Because $\left\|\left(\frac{12i}{5}, \frac{4i}{5}\right)\right\| = \frac{4\sqrt{10}}{5}$, then

$$\mathbf{v}_2 = \left(\frac{3i}{\sqrt{10}}, \frac{i}{\sqrt{10}}\right)$$

25. **(a)** We have

$$\mathbf{v}_1 = \left(\frac{i}{\sqrt{3}}, \frac{i}{\sqrt{3}}, \frac{i}{\sqrt{3}}\right)$$

and since $\mathbf{u}_2 \cdot \mathbf{v}_1 = 0$, then $\mathbf{u}_2 - (\mathbf{u}_2 \cdot \mathbf{v}_1)\mathbf{v}_1 = \mathbf{u}_2$. Thus,

$$\mathbf{v}_2 = \left(-\frac{i}{\sqrt{2}}, \frac{i}{\sqrt{2}}, 0\right)$$

Also, $\mathbf{u}_3 \cdot \mathbf{v}_1 = 4/\sqrt{3}$ and $\mathbf{u}_3 \cdot \mathbf{v}_2 = 1/\sqrt{2}$ and hence

$$\mathbf{u}_3 - (\mathbf{u}_3 \cdot \mathbf{v}_1)\mathbf{v}_1 - (\mathbf{u}_3 \cdot \mathbf{v}_2)\mathbf{v}_2 = \left(\frac{i}{6}, \frac{i}{6}, \frac{-i}{3}\right)$$

Since the norm of the above vector is $1/\sqrt{6}$, we have

$$\mathbf{v}_3 = \left(\frac{i}{\sqrt{6}}, \frac{i}{\sqrt{6}}, \frac{-2i}{\sqrt{6}}\right)$$

27. Let $\mathbf{u}_1 = (0, i, 1 - i)$ and $\mathbf{u}_2 = (-i, 0, 1 + i)$. We shall apply the Gram-Schmidt process to $\{\mathbf{u}_1, \mathbf{u}_2\}$. Since $\|\mathbf{u}_1\| = \sqrt{3}$, it follows that

$$\mathbf{v}_1 = \left(0, \frac{i}{\sqrt{3}}, \frac{1 - i}{\sqrt{3}}\right)$$

Because $\mathbf{u}_2 \cdot \mathbf{v}_1 = 2i/\sqrt{3}$, then

$$\mathbf{u}_2 - (\mathbf{u}_2 \cdot \mathbf{v}_1)\mathbf{v}_1 = (-i,\ 0,\ 1 + i) - \left(0,\ -\frac{2}{3},\ \frac{2}{3} + \frac{2}{3}\,i\right)$$
$$= \left(-i,\ \frac{2}{3},\ \frac{1}{3} + \frac{1}{3}\,i\right)$$

and because the norm of the above vector is $\sqrt{15}/3$, we have

$$\mathbf{v}_2 = \left(\frac{-3i}{\sqrt{15}},\ \frac{2}{\sqrt{15}},\ \frac{1 + i}{\sqrt{15}}\right)$$

28. Following Theorem 5.3.6, we orthonormalize the set $\{\mathbf{u}_1, \mathbf{u}_2\}$. This gives us

$$\mathbf{v}_1 = \frac{\mathbf{u}_1}{\|\mathbf{u}_1\|} = \left(-\frac{i}{\sqrt{6}},\ 0,\ \frac{i}{\sqrt{6}},\ \frac{2i}{\sqrt{6}}\right)$$

and, since $\mathbf{u}_2 \cdot \mathbf{v}_1 = 2/\sqrt{6}$,

$$\mathbf{u}_2 - (\mathbf{u}_2 \cdot \mathbf{v}_1)\mathbf{v}_1 = (0,\ i,\ 0,\ i) - \frac{1}{3}\,(-i,\ 0,\ i,\ 2i)$$
$$= \left(\frac{i}{3},\ i,\ -\frac{i}{3},\ \frac{i}{3}\right)$$

so that

$$\mathbf{v}_2 = \left(\frac{i}{2\sqrt{3}},\ \frac{3i}{2\sqrt{3}},\ -\frac{i}{2\sqrt{3}},\ \frac{i}{2\sqrt{3}}\right)$$

Therefore,

$$\mathbf{w}_1 = (\mathbf{w} \cdot \mathbf{v}_1)\mathbf{v}_1 + (\mathbf{w} \cdot \mathbf{v}_2)\mathbf{v}_2$$
$$= \frac{7}{\sqrt{6}}\,\mathbf{v}_1 + \frac{-1}{2\sqrt{3}}\,\mathbf{v}_2$$
$$= \left(-\frac{5}{4}\,i,\ -\frac{1}{4}\,i,\ \frac{5}{4}\,i,\ \frac{9}{4}\,i\right)$$

and

$$\mathbf{w}_2 = \mathbf{w} - \mathbf{w}_1$$
$$= \left(\frac{1}{4}\,i,\ \frac{9}{4}\,i,\ \frac{19}{4}\,i,\ -\frac{9}{4}\,i\right)$$

29. **(a)** By Axioms (2) and (3) for inner products,

$$\langle \mathbf{u} - k\mathbf{v}, \mathbf{u} - k\mathbf{v}\rangle = \langle \mathbf{u}, \mathbf{u} - k\mathbf{v}\rangle + \langle -k\mathbf{v}, \mathbf{u} - k\mathbf{v}\rangle$$
$$= \langle \mathbf{u}, \mathbf{u} - k\mathbf{v}\rangle - k\langle \mathbf{v}, \mathbf{u} - k\mathbf{v}\rangle$$

If we use Properties (ii) and (iii) of inner products, then we obtain

$$\langle \mathbf{u} - k\mathbf{v}, \mathbf{u} - k\mathbf{v}\rangle = \langle \mathbf{u},\mathbf{u}\rangle - \bar{k}\langle \mathbf{u},\mathbf{v}\rangle - k\langle \mathbf{v},\mathbf{u}\rangle + k\bar{k}\langle \mathbf{v},\mathbf{v}\rangle$$

Finally, Axiom (1) yields

$$\langle \mathbf{u} - k\mathbf{v}, \mathbf{u} - k\mathbf{v}\rangle = \langle \mathbf{u},\mathbf{u}\rangle - \bar{k}\langle \mathbf{u},\mathbf{v}\rangle - k\overline{\langle \mathbf{u},\mathbf{v}\rangle} + k\bar{k}\langle \mathbf{v},\mathbf{v}\rangle$$

and the result is proved.

(b) This follows from Part (a) and Axiom (4) for inner products.

30. Suppose that $\mathbf{v} \neq \mathbf{0}$ and hence that $\langle \mathbf{v},\mathbf{v}\rangle \neq 0$. Using the hint, we obtain

$$0 \le \langle \mathbf{u},\mathbf{u}\rangle - \left[\frac{\overline{\langle \mathbf{u},\mathbf{v}\rangle}}{\langle \mathbf{v},\mathbf{v}\rangle}\right]\langle \mathbf{u},\mathbf{v}\rangle - \left[\frac{\langle \mathbf{u},\mathbf{v}\rangle}{\langle \mathbf{v},\mathbf{v}\rangle}\right]\overline{\langle \mathbf{u},\mathbf{v}\rangle} + \left|\frac{\langle \mathbf{u},\mathbf{v}\rangle}{\langle \mathbf{v},\mathbf{v}\rangle}\right|^2\langle \mathbf{v},\mathbf{v}\rangle$$

Since $\langle \mathbf{v},\mathbf{v}\rangle$ is real, this yields the inequality

$$0 \le \langle \mathbf{u},\mathbf{u}\rangle - \frac{|\langle \mathbf{u},\mathbf{v}\rangle|^2}{\langle \mathbf{v},\mathbf{v}\rangle} - \frac{|\langle \mathbf{u},\mathbf{v}\rangle|^2}{\langle \mathbf{v},\mathbf{v}\rangle} + \frac{|\langle \mathbf{u},\mathbf{v}\rangle|^2}{\langle \mathbf{v},\mathbf{v}\rangle}$$

or

$$\frac{|\langle \mathbf{u},\mathbf{v}\rangle|^2}{\langle \mathbf{v},\mathbf{v}\rangle} \le \langle \mathbf{u},\mathbf{u}\rangle$$

or

$$|\langle \mathbf{u},\mathbf{v}\rangle|^2 \leq \langle \mathbf{u},\mathbf{u}\rangle \; \langle \mathbf{v},\mathbf{v}\rangle$$

as desired.

In case $\mathbf{v} = \mathbf{0}$, Property (i) for inner products guarantees that both sides of the Cauchy-Schwarz inequality are zero, and thus equality holds.

32. If $\mathbf{u}$ and $\mathbf{v}$ are linearly dependent, then either $\mathbf{v} = \mathbf{0}$ or $\mathbf{u}$ is a scalar multiple of $\mathbf{v}$. In either case, a routine computation using properties of the inner product shows that equality holds.

Conversely, if equality holds, then we shall show that $\mathbf{u}$ and $\mathbf{v}$ must be linearly dependent. For suppose that they are not. Then $\mathbf{u} - k\mathbf{v} \neq \mathbf{0}$ for every scalar k, so that $\langle \mathbf{u} - k\mathbf{v}, \mathbf{u} - k\mathbf{v}\rangle$ is greater than zero for all k. Thus, strict inequality must hold in Problem 29(b). Now if we choose k as in Problem 30 and rederive the Cauchy-Schwarz inequality, we find that strict inequality must hold there also. That is, if $\mathbf{u}$ and $\mathbf{v}$ are linearly independent, then the inequality must hold. Hence, if the equality holds, then we are forced to conclude that $\mathbf{u}$ and $\mathbf{v}$ are linearly dependent.

33. Hint: Let $\mathbf{v}$ be any nonzero vector, and consider the quantity $\langle \mathbf{v},\mathbf{v}\rangle + \langle \mathbf{0},\mathbf{v}\rangle$.

34. We have

$$\begin{aligned}
\langle \mathbf{u}, \mathbf{v} + \mathbf{w}\rangle &= \overline{\overline{\langle \mathbf{u}, \mathbf{v} + \mathbf{w}\rangle}} \\
&= \overline{\langle \mathbf{v} + \mathbf{w}, \mathbf{u}\rangle} && \text{(Axiom 1)} \\
&= \overline{\langle \mathbf{v},\mathbf{u}\rangle + \langle \mathbf{w},\mathbf{u}\rangle} && \text{(Axiom 2)} \\
&= \overline{\langle \mathbf{v},\mathbf{u}\rangle} + \overline{\langle \mathbf{w},\mathbf{u}\rangle} \\
&= \langle \mathbf{u},\mathbf{v}\rangle + \langle \mathbf{u},\mathbf{w}\rangle && \text{(Axiom 1)}
\end{aligned}$$

35. **(d)** Observe that $\|\mathbf{u} + \mathbf{v}\|^2 = \langle\mathbf{u} + \mathbf{v}, \mathbf{u} + \mathbf{v}\rangle$. As in Exercise 29,

$$\langle\mathbf{u} + \mathbf{v}, \mathbf{u} + \mathbf{v}\rangle = \langle\mathbf{u},\mathbf{u}\rangle + 2\ \mathrm{Re}(\langle\mathbf{u},\mathbf{v}\rangle) + \langle\mathbf{v},\mathbf{v}\rangle$$

Since (see Exercise 29 of Section 10.1)

$$|\mathrm{Re}(\langle\mathbf{u},\mathbf{v}\rangle)| \le |\langle\mathbf{u},\mathbf{v}\rangle|$$

this yields

$$\langle\mathbf{u} + \mathbf{v}, \mathbf{u} + \mathbf{v}\rangle \le \langle\mathbf{u},\mathbf{u}\rangle + 2|\langle\mathbf{u},\mathbf{v}\rangle| + \langle\mathbf{v},\mathbf{v}\rangle$$

By the Cauchy-Schwarz inequality and the definition of norm, this becomes

$$\|\mathbf{u} + \mathbf{v}\|^2 \le \|\mathbf{u}\|^2 + 2\|\mathbf{u}\|\ \|\mathbf{v}\| + \|\mathbf{v}\|^2 = (\|\mathbf{u}\| + \|\mathbf{v}\|)^2$$

which yields the desired result.

(h) Replace **u** by **u** − **w** and **v** by **w** − **v** in *L*4.

37. Observe that for any complex number k,

$$\begin{aligned}\|\mathbf{u} + k\mathbf{v}\|^2 &= \langle\mathbf{u} + k\mathbf{v}, \mathbf{u} + k\mathbf{v}\rangle \\ &= \langle\mathbf{u},\mathbf{u}\rangle + k\langle\mathbf{v},\mathbf{u}\rangle + \bar{k}\langle\mathbf{u},\mathbf{v}\rangle + k\bar{k}\langle\mathbf{v},\mathbf{v}\rangle \\ &= \langle\mathbf{u},\mathbf{u}\rangle + 2\ \mathrm{Re}(k\langle\mathbf{v},\mathbf{u}\rangle) + |k|^2\langle\mathbf{v},\mathbf{v}\rangle\end{aligned}$$

Therefore,

$$\begin{aligned}&\|\mathbf{u} + \mathbf{v}\|^2 - \|\mathbf{u} - \mathbf{v}\|^2 + i\|\mathbf{u} + i\mathbf{v}\|^2 - i\|\mathbf{u} - i\mathbf{v}\|^2 \\ &\quad= (1 - 1 + i - i)\langle\mathbf{u},\mathbf{u}\rangle + 2\ \mathrm{Re}(\langle\mathbf{v},\mathbf{u}\rangle) \\ &\qquad - 2\ \mathrm{Re}(-\langle\mathbf{v},\mathbf{u}\rangle) + 2i\ \mathrm{Re}(i\langle\mathbf{v},\mathbf{u}\rangle) \\ &\qquad\quad - 2i\ \mathrm{Re}(-i\langle\mathbf{v},\mathbf{u}\rangle) + (1 - 1 + i - i)\langle\mathbf{v},\mathbf{v}\rangle \\ &\quad= 4\ \mathrm{Re}(\langle\mathbf{v},\mathbf{u}\rangle) + 4i\ \mathrm{Im}(\langle\mathbf{v},\mathbf{u}\rangle) \\ &\quad= 4\langle\mathbf{v},\mathbf{u}\rangle\end{aligned}$$

38. By the hint, we have $\mathbf{u} = \sum_{j=1}^{n} \langle \mathbf{u},\mathbf{v}_j\rangle \mathbf{v}_j$ and $\mathbf{w} = \sum_{k=1}^{n} \langle \mathbf{w},\mathbf{v}_k\rangle \mathbf{v}_k$. Therefore,

$$\langle \mathbf{u},\mathbf{w}\rangle = \sum_{j,k=1}^{n} \langle \mathbf{u},\mathbf{v}_j\rangle \overline{\langle \mathbf{w},\mathbf{v}_k\rangle} \langle \mathbf{v}_j,\mathbf{v}_k\rangle$$

$$= \sum_{j=1}^{n} \langle \mathbf{u},\mathbf{v}_j\rangle \overline{\langle \mathbf{w},\mathbf{v}_j\rangle}$$

The last step uses the fact that the basis vectors $\mathbf{v}_j$ form an orthonormal set.

39. We check Axioms 2 and 4. For Axiom 2, we have

$$\langle \mathbf{f} + \mathbf{g},\ \mathbf{h}\rangle = \int_a^b (\mathbf{f} + \mathbf{g})\overline{\mathbf{h}}\, dx$$

$$= \int_a^b \mathbf{f}\overline{\mathbf{h}}\, dx + \int_a^b \mathbf{g}\overline{\mathbf{h}}\, dx$$

$$= \langle \mathbf{f},\mathbf{h}\rangle + \langle \mathbf{g},\mathbf{h}\rangle$$

For Axiom 4, we have

$$\langle \mathbf{f},\mathbf{f}\rangle = \int_a^b \mathbf{f}\overline{\mathbf{f}}\, dx = \int_a^b |\mathbf{f}|^2\, dx$$

$$= \int_a^b \left[\left[f_1(x)\right]^2 + \left[f_2(x)\right]^2\right] dx$$

Since $|\mathbf{f}|^2 \geq 0$ and $a < b$, then $\langle\mathbf{f},\mathbf{f}\rangle \geq 0$. Moreover, since f is continuous, $\int_a^b |\mathbf{f}|^2 dx > 0$ unless $\mathbf{f} = \mathbf{0}$ on $[a,b]$. [That is, the integral of a nonnegative, real-valued, continuous function (which represents the area under that curve and above the x-axis from a to b) is positive unless the function is identically zero.]

40. **(a)** Let $\mathbf{v}_m = e^{2\pi imx} = \cos(2\pi mx) + i\sin(2\pi mx)$. Then if $m \neq n$, we have

$$\begin{aligned}\langle\mathbf{v}_m,\mathbf{v}_n\rangle &= \int_0^1 [\cos(2\pi mx) + i\sin(2\pi mx)] \cdot \\ &\qquad [\cos(2\pi nx) - i\sin(2\pi nx)]\,dx \\ &= \int_0^1 [\cos(2\pi mx)\cos(2\pi nx) + \sin(2\pi mx)\sin(2\pi nx)]\,dx \\ &\qquad + i\int_0^1 [\sin(2\pi mx)\cos(2\pi nx) - \cos(2\pi mx)\sin(2\pi nx)]\,dx \\ &= \int_0^1 \cos[2\pi(m-n)x]\,dx + i\int_0^1 \sin[2\pi(m-n)x]\,dx \\ &= \frac{1}{2\pi(m-n)}\sin[2\pi(m-n)x]\Big]_0^1 \\ &\qquad - \frac{i}{2\pi(m-n)}\cos[2\pi(m-n)x]\Big]_0^1 \\ &= -\frac{i}{2\pi(m-n)} + \frac{i}{2\pi(m-n)} \\ &= 0\end{aligned}$$

Thus the vectors are orthogonal.

40. **(b)** If we let $m = n$ in Part (a), then we have

$$\langle \mathbf{v}_m, \mathbf{v}_m \rangle = \int_0^1 1\,dx + i\int_0^1 0\,dx = 1$$

Thus the set is already normalized.

EXERCISE SET 10.6

4. **(b)** The row vectors of the matrix are

$$\mathbf{r}_1 = \left(\frac{i}{\sqrt{2}}, \frac{1}{\sqrt{2}}\right) \quad \text{and} \quad \mathbf{r}_2 = \left(-\frac{i}{\sqrt{2}}, \frac{1}{\sqrt{2}}\right)$$

Since $\|\mathbf{r}_1\| = \|\mathbf{r}_2\| = 1$ and $\mathbf{r}_1 \cdot \mathbf{r}_2 = 0$, it follows that the matrix is unitary.

(c) The row vectors of the matrix are

$$\mathbf{r}_1 = (1 + i, 1 + i) \quad \text{and} \quad \mathbf{r}_2 = (1 - i, -1 + i)$$

Although $\mathbf{r}_1 \cdot \mathbf{r}_2 = (1 + i)(1 + i) + (1 + i)(-1 - i) = 0$ (from which it follows that $\mathbf{r}_1$ and $\mathbf{r}_2$ are orthogonal), they are not orthonormal because $\|\mathbf{r}_1\| = \|\mathbf{r}_2\| = 2$. Hence, the matrix is not unitary.

5. **(b)** The row vectors of the matrix are

$$\mathbf{r}_1 = \left(\frac{1}{\sqrt{2}}, \frac{1}{\sqrt{2}}\right) \quad \text{and} \quad \mathbf{r}_2 = \left(-\frac{1 + i}{2}, \frac{1 + i}{2}\right)$$

Since $\|\mathbf{r}_1\| = \|\mathbf{r}_2\| = 1$ and

$$\mathbf{r}_1 \cdot \mathbf{r}_2 = \frac{1}{\sqrt{2}}\left[\frac{-1 + i}{2}\right] + \frac{1}{\sqrt{2}}\left[\frac{1 - i}{2}\right] = 0$$

the matrix is unitary by Theorem 10.6.2. Hence,

$$A^{-1} = A^{*} = \bar{A}^{t} = \begin{bmatrix} \frac{1}{\sqrt{2}} & \frac{-1 + i}{2} \\ \frac{1}{\sqrt{2}} & \frac{1 - i}{2} \end{bmatrix}$$

5. **(d)** The row vectors of the matrix are

$$\mathbf{r}_1 = \left(\frac{1 + i}{2}, -\frac{1}{2}, \frac{1}{2}\right)$$

$$\mathbf{r}_2 = \left(\frac{i}{\sqrt{3}}, \frac{1}{\sqrt{3}}, \frac{-i}{\sqrt{3}}\right)$$

and

$$\mathbf{r}_3 = \left(\frac{3 + i}{2\sqrt{15}}, \frac{4 + 3i}{2\sqrt{15}}, \frac{5i}{2\sqrt{15}}\right)$$

We have $\|\mathbf{r}_1\| = \|\mathbf{r}_2\| = \|\mathbf{r}_3\| = 1$,

$$\mathbf{r}_1 \cdot \mathbf{r}_2 = \left[\frac{1 + i}{2}\right]\left[\frac{-i}{\sqrt{3}}\right] - \frac{1}{2} \cdot \frac{1}{\sqrt{3}} + \frac{1}{2} \cdot \frac{i}{\sqrt{3}} = 0$$

$$\mathbf{r}_2 \cdot \mathbf{r}_3 = \left[\frac{i}{\sqrt{3}}\right]\left[\frac{3 - i}{2\sqrt{15}}\right] + \frac{1}{\sqrt{3}} \cdot \frac{4 - 3i}{2\sqrt{15}} - \frac{i}{\sqrt{3}} \cdot \frac{-5i}{2\sqrt{15}} = 0$$

and

$$\mathbf{r}_1 \cdot \mathbf{r}_3 = \left[\frac{1 + i}{2}\right]\left[\frac{3 - i}{2\sqrt{15}}\right] - \frac{1}{2} \cdot \frac{4 - 3i}{2\sqrt{15}} + \frac{1}{2} \cdot \frac{-5i}{2\sqrt{15}} = 0$$

Hence, by Theorem 10.6.2, the matrix is unitary and thus

$$A^{-1} = A^{*} = \bar{A}^{t} = \begin{bmatrix} \frac{1 - i}{2} & \frac{-i}{\sqrt{3}} & \frac{3 - i}{2\sqrt{15}} \\ -\frac{1}{2} & \frac{1}{\sqrt{3}} & \frac{4 - 3i}{2\sqrt{15}} \\ \frac{1}{2} & \frac{i}{\sqrt{3}} & \frac{-5i}{2\sqrt{15}} \end{bmatrix}$$

6. Using Formula (11) of Section 10.3, we see that $\overline{e^{i\theta}} = e^{-i\theta}$ and hence that $\overline{e^{-i\theta}} = e^{i\theta}$. Let $\mathbf{r}_1 = \left(\frac{e^{i\theta}}{\sqrt{2}}, \frac{e^{-i\theta}}{\sqrt{2}}\right)$ and $\mathbf{r}_2 = \left(\frac{ie^{i\theta}}{\sqrt{2}}, \frac{-ie^{-i\theta}}{\sqrt{2}}\right)$. Then

$$\mathbf{r}_1 \cdot \mathbf{r}_1 = \mathbf{r}_2 \cdot \mathbf{r}_2 = \frac{1}{2} + \frac{1}{2} = 1$$

and

$$\begin{aligned}\mathbf{r}_1 \cdot \mathbf{r}_2 &= \frac{1}{2}\left[e^{i\theta}\left[-ie^{-i\theta}\right] + e^{-i\theta}\left[ie^{i\theta}\right]\right] \\ &= \frac{1}{2}\left[-i + i\right] = 0\end{aligned}$$

Hence, by Theorem 10.6.2, the row vectors from an orthonormal set and so the matrix is unitary.

7. The characteristic polynomial of A is

$$\det\begin{bmatrix} \lambda-4 & -1+i \\ -1-i & \lambda-5 \end{bmatrix} = (\lambda - 4)(\lambda - 5) - 2 = (\lambda - 6)(\lambda - 3)$$

Therefore, the eigenvalues are $\lambda = 3$ and $\lambda = 6$. To find the eigenvectors of A corresponding to $\lambda = 3$, we let

$$\begin{bmatrix} -1 \\ -1-i \end{bmatrix}\begin{bmatrix} x_1 \\ x_2 \end{bmatrix} = \begin{bmatrix} 0 \\ 0 \end{bmatrix}$$

This yields $x_1 = -(1 - i)s$ and $x_2 = s$ where s is arbitrary. If we put $s = 1$, we see that $\begin{bmatrix} -1+i \\ 1 \end{bmatrix}$ is a basis vector for the eigenspace corresponding to $\lambda = 3$. We normalize this vector to obtain

$$\mathbf{p}_1 = \begin{bmatrix} \frac{-1+i}{\sqrt{3}} \\ \frac{1}{\sqrt{3}} \end{bmatrix}$$

To find the eigenvectors corresponding to $\lambda = 6$, we let

$$\begin{bmatrix} 2 & -1+i \\ -1-i & 1 \end{bmatrix} \begin{bmatrix} x_1 \\ x_2 \end{bmatrix} = \begin{bmatrix} 0 \\ 0 \end{bmatrix}$$

This yields $x_1 = \frac{1 - i}{2} s$ and $x_2 = s$ where s is arbitrary. If we put $s = 1$, we have that $\begin{bmatrix} (1-i)/2 \\ 1 \end{bmatrix}$ is a basis vector for the eigenspace corresponding to $\lambda = 6$. We normalize this vector to obtain

$$\mathbf{p}_2 = \begin{bmatrix} \frac{1-i}{\sqrt{6}} \\ \frac{2}{\sqrt{6}} \end{bmatrix}$$

Thus

$$P = \begin{bmatrix} \frac{-1+i}{\sqrt{3}} & \frac{1-i}{\sqrt{6}} \\ \frac{1}{\sqrt{3}} & \frac{2}{\sqrt{6}} \end{bmatrix}$$

diagonalizes A and

$$P^{-1}AP = \begin{bmatrix} \frac{-1-i}{\sqrt{3}} & \frac{1}{\sqrt{3}} \\ \frac{1+i}{\sqrt{6}} & \frac{2}{\sqrt{6}} \end{bmatrix} \begin{bmatrix} 4 & 1-i \\ 1+i & 5 \end{bmatrix} \begin{bmatrix} \frac{-1+i}{\sqrt{3}} & \frac{1-i}{\sqrt{6}} \\ \frac{1}{\sqrt{3}} & \frac{2}{\sqrt{6}} \end{bmatrix}$$

$$= \begin{bmatrix} 3 & 0 \\ 0 & 6 \end{bmatrix}$$

9. The characteristic polynomial of A is

$$\det\begin{bmatrix} \lambda-6 & -2-2i \\ -2+2i & \lambda-4 \end{bmatrix} = (\lambda - 6)(\lambda - 4) - 8$$

$$= (\lambda - 8)(\lambda - 2)$$

Therefore the eigenvalues are $\lambda = 2$ and $\lambda = 8$. To find the eigenvectors of A corresponding to $\lambda = 2$, we let

$$\begin{bmatrix} -4 & -2-2i \\ -2+2i & -2 \end{bmatrix} \begin{bmatrix} x_1 \\ x_2 \end{bmatrix} = \begin{bmatrix} 0 \\ 0 \end{bmatrix}$$

This yields $x_1 = -\dfrac{1+i}{2}s$ and $x_2 = s$ where s is arbitrary. If we put $s = 1$, we have that $\begin{bmatrix} -(1+i)/2 \\ 1 \end{bmatrix}$ is a basis vector for the eigenspace corresponding to $\lambda = 2$. We normalize this vector to obtain

$$\mathbf{p}_1 = \begin{bmatrix} -\dfrac{1+i}{\sqrt{6}} \\ \dfrac{2}{\sqrt{6}} \end{bmatrix}$$

To find the eigenvectors corresponding to $\lambda = 8$, we let

$$\begin{bmatrix} 2 & -2-2i \\ -2+2i & 4 \end{bmatrix} \begin{bmatrix} x_1 \\ x_2 \end{bmatrix} = \begin{bmatrix} 0 \\ 0 \end{bmatrix}$$

This yields $x_1 = (1+i)s$ and $x_2 = s$ where s is arbitrary. If we set $s = 1$, we have that $\begin{bmatrix} 1+i \\ 1 \end{bmatrix}$ is a basis vector for the eigenspace corresponding to $\lambda = 8$. We normalize this vector to obtain

$$\mathbf{p}_2 = \begin{bmatrix} \dfrac{1+i}{\sqrt{3}} \\ \dfrac{1}{\sqrt{3}} \end{bmatrix}$$

Thus

$$P = \begin{bmatrix} -\dfrac{1+i}{\sqrt{6}} & \dfrac{1+i}{\sqrt{6}} \\ \dfrac{2}{\sqrt{6}} & \dfrac{1}{\sqrt{3}} \end{bmatrix}$$

diagonalizes A and

$$P^{-1}AP = \begin{bmatrix} \frac{-1+i}{\sqrt{6}} & \frac{2}{\sqrt{6}} \\ \frac{1-i}{\sqrt{3}} & \frac{1}{\sqrt{3}} \end{bmatrix} \begin{bmatrix} 6 & 2+2i \\ 2-2i & 4 \end{bmatrix} \begin{bmatrix} -\frac{1+i}{\sqrt{6}} & \frac{1+i}{\sqrt{3}} \\ \frac{2}{\sqrt{6}} & \frac{1}{\sqrt{3}} \end{bmatrix}$$

$$= \begin{bmatrix} 2 & 0 \\ 0 & 8 \end{bmatrix}$$

11. The characteristic polynomial of A is

$$\det \begin{bmatrix} \lambda-5 & 0 & 0 \\ 0 & \lambda+1 & 1-i \\ 0 & 1+i & \lambda \end{bmatrix} = (\lambda - 1)(\lambda - 5)(\lambda + 2)$$

Therefore, the eigenvalues are $\lambda = 1$, $\lambda = 5$, and $\lambda = -2$. To find the eigenvectors of A corresponding to $\lambda = 1$, we let

$$\begin{bmatrix} -4 & 0 & 0 \\ 0 & 2 & 1-i \\ 0 & 1+i & 1 \end{bmatrix} \begin{bmatrix} x_1 \\ x_2 \\ x_3 \end{bmatrix} = \begin{bmatrix} 0 \\ 0 \\ 0 \end{bmatrix}$$

This yields $x_1 = 0$, $x_2 = -\frac{1-i}{2}s$, and $x_3 = s$ where s is arbitrary. If we set $s = 1$, we have that $\begin{bmatrix} 0 \\ -(1-i)/2 \\ 1 \end{bmatrix}$ is a basis vector for the eigenspace corresponding to $\lambda = 1$. We normalize this vector to obtain

$$\mathbf{p}_1 = \begin{bmatrix} 0 \\ -\frac{1-i}{\sqrt{6}} \\ \frac{2}{\sqrt{6}} \end{bmatrix}$$

To find the eigenvectors corresponding to $\lambda = 5$, we let

$$\begin{bmatrix} 0 & 0 & 0 \\ 0 & 6 & 1-i \\ 0 & 1+i & 5 \end{bmatrix} \begin{bmatrix} x_1 \\ x_2 \\ x_3 \end{bmatrix} = \begin{bmatrix} 0 \\ 0 \\ 0 \end{bmatrix}$$

This yields $x_1 = s$ and $x_2 = x_3 = 0$ where s is arbitrary. If we let $s = 1$, we have that $\begin{bmatrix} 1 \\ 0 \\ 0 \end{bmatrix}$ is a basis vector for the eigenspace corresponding to $\lambda = 5$. Since this vector is already normal, we let

$$\mathbf{p}_2 = \begin{bmatrix} 1 \\ 0 \\ 0 \end{bmatrix}$$

To find the eigenvectors corresponding to $\lambda = -2$, we let

$$\begin{bmatrix} -7 & 0 & 0 \\ 0 & -1 & 1-i \\ 0 & 1+i & -2 \end{bmatrix} \begin{bmatrix} x_1 \\ x_2 \\ x_3 \end{bmatrix} = \begin{bmatrix} 0 \\ 0 \\ 0 \end{bmatrix}$$

This yields $x_1 = 0$, $x_2 = (1 - i)s$, and $x_3 = s$ where s is arbitrary. If we let $s = 1$, we have that $\begin{bmatrix} 0 \\ 1-i \\ 1 \end{bmatrix}$ is a basis vector for the eigenspace corresponding to $\lambda = -2$. We normalize this vector to obtain

$$\mathbf{p}_3 = \begin{bmatrix} 0 \\ \dfrac{1-i}{\sqrt{3}} \\ \dfrac{1}{\sqrt{3}} \end{bmatrix}$$

Thus

$$P = \begin{bmatrix} 0 & 1 & 0 \\ -\frac{1-i}{\sqrt{6}} & 0 & \frac{1-i}{\sqrt{3}} \\ \frac{2}{\sqrt{6}} & 0 & \frac{1}{\sqrt{3}} \end{bmatrix}$$

diagonalizes A and

$$P^{-1}AP = \begin{bmatrix} 0 & -\frac{1+i}{\sqrt{6}} & \frac{2}{\sqrt{6}} \\ 1 & 0 & 0 \\ 0 & \frac{1+i}{\sqrt{3}} & \frac{1}{\sqrt{3}} \end{bmatrix} \begin{bmatrix} 5 & 0 & 0 \\ 0 & -1 & -1+i \\ 0 & -1-i & 0 \end{bmatrix} \begin{bmatrix} 0 & 1 & 0 \\ -\frac{1-i}{\sqrt{6}} & 0 & \frac{1-i}{\sqrt{3}} \\ \frac{2}{\sqrt{6}} & 0 & \frac{1}{\sqrt{3}} \end{bmatrix}$$

$$= \begin{bmatrix} 1 & 0 & 0 \\ 0 & 5 & 0 \\ 0 & 0 & -2 \end{bmatrix}$$

13. The eigenvalues of A are the roots of the equation

$$\det\begin{bmatrix} \lambda-1 & -4i \\ -4i & \lambda-3 \end{bmatrix} = \lambda^2 - 4\lambda + 19 = 0$$

The roots of this equation, which are $\lambda = \dfrac{4 \pm \sqrt{16 - 4(19)}}{2}$, are not real. This shows that the eigenvalues of a symmetric matrix with nonreal entries need not be real. Theorem 10.6.6 applies only to matrices with real entries.

15. We know that $\det(A)$ is the sum of all the signed elementary products $\pm a_{1j_1}a_{2j_2}\cdots a_{nj_n}$, where a_{ij} is the entry from the ith row and j^{th} column of A. Since the ij^{th} element of $\overline{A}$ is $\overline{a}_{ij}$, then $\det(\overline{A})$ is the sum of the signed elementary products $\pm\,\overline{a}_{1j_1}\overline{a}_{2j_2}\cdots\overline{a}_{nj_n}$ or $\pm\,\overline{a_{1j_1}a_{2j_2}\cdots a_{nj_n}}$. That is, $\det(\overline{A})$ is the sum of the conjugates of the terms in $\det(A)$. But since the sum of the conjugates is the conjugate of the sum, we have $\det(\overline{A}) = \overline{\det(A)}$.

16. **(a)** We have

$$\begin{aligned} \det(A^*) &= \det(\overline{A}^{\,t}) \\ &= \det(\overline{A}) && \text{(since } \det(B) = \det(B^t)\text{)} \\ &= \overline{\det(A)} && \text{(by Exercise 15)} \end{aligned}$$

(b) If A is Hermitian, then $A = A^*$. Therefore,

$$\begin{aligned} \det(A) &= \det(A^*) \\ &= \overline{\det(A)} && \text{(by Part (b))} \end{aligned}$$

But $z = \overline{z}$ if and only if z is real.

(c) If A is unitary, then $A^{-1} = A^*$. Therefore,

$$\begin{aligned} \det(A) &= 1/\det(A^{-1}) \\ &= 1/\det(A^*) \\ &= 1/\overline{\det(A)} && \text{(by Part (a))} \end{aligned}$$

Thus $|\det(A)|^2 = 1$. But since $|z| \geq 0$ for all z, it follows that $|\det(A)| = 1$.

18. **(b)** We have

$$\begin{aligned}(A+B)^* &= (\overline{A+B})^t \\ &= (\overline{A}+\overline{B})^t \\ &= (\overline{A}^t+\overline{B}^t) \\ &= A^*+B^*\end{aligned}$$

(d) We have

$$\begin{aligned}(AB)^* &= (\overline{AB})^t \\ &= (\overline{A}\,\overline{B})^t \\ &= \overline{B}^t\overline{A}^t \\ &= B^*A^*\end{aligned}$$

19. If A is invertible, then

$$\begin{aligned}A^*(A^{-1})^* &= (A^{-1}A)^* && \text{(by Exercise 18(d))} \\ &= I^* = \overline{I}^t = I^t = I\end{aligned}$$

Thus we have $(A^{-1})^* = (A^*)^{-1}$.

20. Hint: use Exercise 18(a) and Exercise 19.

21. Let $\mathbf{r}_i$ denote the ith row of A and let $\mathbf{c}_j$ denote the jth column of A^*. Then, since $A^* = \overline{A}^t = \overline{(A^t)}$, we have $\mathbf{c}_j = \overline{\mathbf{r}}_j$ for $j = 1,\ldots,n$. Finally, let

$$\delta_{ij} = \begin{cases} 0 & \text{if} \quad i \neq j \\ 1 & \text{if} \quad i = j \end{cases}$$

Then A is unitary $\Leftrightarrow$ $A^{-1} = A^*$. But

$$
\begin{aligned}
A^{-1} = A^* \quad &\Leftrightarrow \quad AA^* = I \\
&\Leftrightarrow \quad \mathbf{r}_i \cdot \mathbf{c}_j = \delta_{ij} \text{ for all } i,j \\
&\Leftrightarrow \quad \mathbf{r}_i \cdot \overline{\mathbf{r}}_j = \delta_{ij} \text{ for all } i,j \\
&\Leftrightarrow \quad \{\mathbf{r}_1, \ldots, \mathbf{r}_n\} \text{ is an orthonormal set}
\end{aligned}
$$

22. We know that if A is unitary, then so is A^*. But since $(A^*)^* = A$, then if A^* is unitary, so is A. Moreover, the columns of A are the conjugates of the rows of A^*. Thus we have

A is unitary $\Leftrightarrow$ A^* is unitary

$\Leftrightarrow$ the rows of A^* form an orthonormal set

$\Leftrightarrow$ the conjugates of the columns of A form an orthonormal set

$\Leftrightarrow$ the columns of A form an orthonormal set

23. If $A = A^*$, then for $\mathbf{x}$ in C^n,

$$
\begin{aligned}
(\mathbf{x}^* A\mathbf{x})^* &= \mathbf{x}^* A^* (\mathbf{x}^*)^* \\
&= \mathbf{x}^* A\mathbf{x}
\end{aligned}
$$

Since $\mathbf{x}^* A\mathbf{x}$ is a 1×1 matrix, it is its own transpose. Thus we have

$$\overline{\mathbf{x}^* A\mathbf{x}} = \mathbf{x}^* A\mathbf{x}$$

so that the entry in $\mathbf{x}^* A\mathbf{x}$ must be real.

24. (a) We know that $A = A^*$, that $A\mathbf{x} = \lambda I\mathbf{x}$, and that $A\mathbf{y} = \mu I\mathbf{y}$. Therefore

$$\mathbf{x}^* A\mathbf{y} = \mathbf{x}^*(\mu I\mathbf{y}) = \mu(\mathbf{x}^* I\mathbf{y}) = \mu\mathbf{x}^*\mathbf{y}$$

and

$$\begin{aligned}\mathbf{x}^* A\mathbf{y} &= [(\mathbf{x}^* A\mathbf{y})^*]^* = [\mathbf{y}^* A^* \mathbf{x}]^* \\ &= [\mathbf{y}^* A\mathbf{x}]^* = [\mathbf{y}^*(\lambda I\mathbf{x})]^* \\ &= [\lambda\mathbf{y}^*\mathbf{x}]^* = \lambda\mathbf{x}^*\mathbf{y}\end{aligned}$$

The last step follows because λ, being the eigenvalue of an Hermitian matrix, is real.

(b) From the hint, we have

$$(\lambda - \mu)(\mathbf{x}^*\mathbf{y}) = 0$$

Since $\lambda \neq \mu$, the above equation implies that $\mathbf{x}^*\mathbf{y}$ is the 1×1 zero matrix. Let $\mathbf{x} = (x_1, \ldots, x_n)$ and $\mathbf{y} = (y_1, \ldots, y_n)$. Then we have just shown that

$$\overline{x}_1 y_1 + \cdots + \overline{x}_n y_n = 0$$

so that

$$x_1 \overline{y}_1 + \cdots + x_n \overline{y}_n = \overline{0} = 0$$

and hence $\mathbf{x}$ and $\mathbf{y}$ are orthogonal.

SUPPLEMENTARY EXERCISES 10

2. If a and b are not both zero, then $|a|^2 + |b|^2 \neq 0$ and the inverse is

$$\begin{bmatrix} \dfrac{\bar{a}}{|a|^2 + |b|^2} & \dfrac{b}{|a|^2 + |b|^2} \\ \dfrac{-\bar{b}}{|a|^2 + |b|^2} & \dfrac{a}{|a|^2 + |b|^2} \end{bmatrix}$$

This inverse is computed exactly as if the entries were real.

3. The system of equations has solution $x_1 = -is + t$, $x_2 = s$, $x_3 = t$. Thus

$$\begin{bmatrix} x_1 \\ x_2 \\ x_3 \end{bmatrix} = \begin{bmatrix} -i \\ 1 \\ 0 \end{bmatrix} s + \begin{bmatrix} 1 \\ 0 \\ 1 \end{bmatrix} t$$

where s and t are arbitrary. Hence

$$\begin{bmatrix} -i \\ 1 \\ 0 \end{bmatrix} \quad \text{and} \quad \begin{bmatrix} 1 \\ 0 \\ 1 \end{bmatrix}$$

form a basis for the solution space.

5. The eigenvalues are the solutions, λ, of the equation

$$\det\begin{bmatrix} \lambda & 0 & -1 \\ -1 & \lambda & -\omega - 1 - \dfrac{1}{\omega} \\ 0 & -1 & \lambda + \omega + 1 + \dfrac{1}{\omega} \end{bmatrix} = 0$$

or

$$\lambda^3 + \left[\omega + 1 + \frac{1}{\omega}\right]\lambda^2 - \left[\omega + 1 + \frac{1}{\omega}\right]\lambda - 1 = 0$$

But $\dfrac{1}{\omega} = \overline{\omega}$, so that $\omega + 1 + \dfrac{1}{\omega} = 2\,\mathrm{Re}(\omega) + 1 = 0$. Thus we have

$$\lambda^3 - 1 = 0$$

or

$$(\lambda - 1)(\lambda^2 + \lambda + 1) = 0$$

Hence $\lambda = 1$, ω, or $\overline{\omega}$. Note that $\overline{\omega} = \omega^2$.

6. **(c)** Following the hint, we let $z = \cos\theta + i\sin\theta = e^{i\theta}$ in Part (a). This yields

$$1 + e^{i\theta} + e^{2i\theta} + \cdots + e^{ni\theta} = \frac{1 - e^{(n+1)i\theta}}{1 - e^{i\theta}}$$

If we expand and equate real parts, we obtain

$$1 + \cos\theta + \cos 2\theta + \cdots + \cos n\theta = \mathrm{Re}\left[\frac{1 - e^{(n+1)i\theta}}{1 - e^{i\theta}}\right]$$

But

$$\text{Re}\left[\frac{1 - e^{(n+1)i\theta}}{1 - e^{i\theta}}\right] = \text{Re}\left[\frac{\left[1 - e^{(n+1)i\theta}\right]\left[1 - e^{-i\theta}\right]}{\left[1 - e^{i\theta}\right]\left[1 - e^{-i\theta}\right]}\right]$$

$$= \text{Re}\left[\frac{1 - e^{(n+1)i\theta} - e^{-i\theta} + e^{ni\theta}}{2 - 2\text{Re}(e^{i\theta})}\right]$$

$$= \frac{1}{2}\left[\frac{1 - \cos[(n+1)\theta] - \cos(-\theta) + \cos n\theta}{1 - \cos\theta}\right]$$

$$= \frac{1}{2}\left[\frac{1 - \cos\theta}{1 - \cos\theta} + \frac{\cos n\theta - \cos((n+1)\theta)}{1 - \cos\theta}\right]$$

$$= \frac{1}{2}\left[1 + \frac{\cos n\theta - [\cos n\theta\cos\theta - \sin n\theta\sin\theta]}{2\sin^2\frac{\theta}{2}}\right]$$

$$= \frac{1}{2}\left[1 + \frac{\cos n\theta(1-\cos\theta) + 2\sin n\theta\sin\frac{\theta}{2}\cos\frac{\theta}{2}}{2\sin^2\frac{\theta}{2}}\right]$$

$$= \frac{1}{2}\left[1 + \frac{2\cos n\theta\sin^2\frac{\theta}{2} + 2\sin n\theta\sin\frac{\theta}{2}\cos\frac{\theta}{2}}{2\sin^2\frac{\theta}{2}}\right]$$

$$= \frac{1}{2}\left[1 + \frac{\cos n\theta\sin\frac{\theta}{2} + \sin n\theta\cos\frac{\theta}{2}}{\sin\frac{\theta}{2}}\right]$$

$$= \frac{1}{2}\left[1 + \frac{\sin\left[\left(n + \frac{1}{2}\right)\theta\right]}{\sin\frac{\theta}{2}}\right]$$

Observe that because $0 < \theta < 2\pi$, we have not divided by zero.

7. Observe that $\omega^3 = 1$. Hence by Exercise 6(b), we have $1 + \omega + \omega^2 = 0$. Therefore

$$\mathbf{v}_1 \cdot \mathbf{v}_2 = \frac{1}{3}(1 + \overline{\omega} + \overline{\omega}^2) = \frac{1}{3}\overline{(1 + \omega + \omega^2)} = 0$$

$$\mathbf{v}_1 \cdot \mathbf{v}_3 = \frac{1}{3}\left[1 + \overline{\omega}^2 + \overline{\omega}^4\right] = \frac{1}{3}\overline{(1 + \omega^2 + \omega)} = 0$$

$$\begin{aligned}\mathbf{v}_2 \cdot \mathbf{v}_3 &= \frac{1}{3}\left[1 + \omega\overline{\omega}^2 + \omega^2\overline{\omega}^4\right] \\ &= \frac{1}{3}\left[1 + \omega^5 + \omega^{10}\right] \qquad \text{(since } \overline{\omega} = \omega^2\text{)} \\ &= \frac{1}{3}(1 + \omega^2 + \omega) \\ &= 0\end{aligned}$$

$$\mathbf{v}_1 \cdot \mathbf{v}_1 = \frac{1}{3}(1 + 1 + 1) = 1$$

$$\mathbf{v}_2 \cdot \mathbf{v}_2 = \frac{1}{3}\left[1 + \omega\overline{\omega} + \omega^2\overline{\omega}^2\right] = 1 \qquad \text{(since } \omega\overline{\omega} = 1\text{)}$$

$$\mathbf{v}_3 \cdot \mathbf{v}_3 = \frac{1}{3}\left[1 + \omega^2\overline{\omega}^2 + \omega^4\overline{\omega}^4\right] = 1 \qquad \text{(as above)}$$

Thus, the vectors form an orthonormal set in C^3.

8. Call the diagonal matrix D. Then

$$(UD)^* = (\overline{UD})^t = \overline{D}^t\,\overline{U}^t = \overline{D}U^* = \overline{D}U^{-1}$$

We need only check that $(UD)^* = (UD)^{-1}$. But

$$(UD)^*(UD) = (\overline{D}U^{-1})(UD) = \overline{D}D$$

and

$$\overline{D}D = \begin{bmatrix} |z_1|^2 & 0 & \cdots & 0 \\ 0 & |z_2|^2 & \cdots & 0 \\ \vdots & \vdots & & \vdots \\ 0 & 0 & \cdots & |z_n|^2 \end{bmatrix} = I$$

Hence $(UD)^* = (UD)^{-1}$ and so UD is unitary.

9. **(a)** Since $A^* = -A$, we have

$$(iA)^* = \overline{i}A^* = -i(-A) = iA$$

Hence iA is Hermitian.

(b) By Part (a), iA is Hermitian and hence unitarily diagonalizable with real eigenvalues. That is, there is a unitary matrix P and a diagonal matrix D with real entries such that

$$P^{-1}(iA)P = D$$

Thus

$$i(P^{-1}AP) = D$$

or

$$P^{-1}AP = -iD$$

That is, A is unitarily diagonalizable with eigenvalues which are $-i$ times the eigenvalues of iA, and hence pure imaginary.

10. **(b)** Since $a + bi = a(1) + b(i)$, the vectors 1 and i span the space. Since they are linearly independent, they form a basis, and therefore the space has dimension 2.